T0320380

FUNDAMENTALS OF COMPUTATIONAL INTELLIGENCE

FUNDAMENTALS OF COMPUTATIONAL INTELLIGENCE

NEURAL NETWORKS, FUZZY SYSTEMS, AND EVOLUTIONARY COMPUTATION

James M. Keller
Derong Liu
David B. Fogel

 IEEE Press Series on Computational Intelligence

WILEY

Copyright © 2016 by The Institute of Electrical and Electronics Engineers, Inc.

Published by John Wiley & Sons, Inc., Hoboken, New Jersey. All rights reserved
Published simultaneously in Canada

For general information on our other products and services or for technical support, please contact our
Customer Care Department within the United States at (800) 762–2974, outside the United States
at (317) 572–3993 or fax (317) 572–4002.

Wiley also publishes its books in a variety of electronic formats. Some content that appears in print may
not be available in electronic formats. For more information about Wiley products, visit our web site
at www.wiley.com.

Library of Congress Cataloging-in-Publication Data is available.

ISBN: 978-1-110-21434-2

Printed in the United States of America

10 9 8 7 6 5 4 3 2 1

Jim: To my grandsons, Mack, Shea, Jones, and Jameson who continuously remind me what is really important in life.

Derong: To my dear wife Mingshu for her love and support.

David: To Jonathan and Skyler, for helping me discover what it's all about.

◼◼ CONTENTS

ACKNOWLEDGMENTS

We are grateful to the many people who have helped us during the past years, who have contributed to work presented here, and who have offered critical reviews of prior publications. We also thank Wiley-IEEE Press for their assistance in publishing the manuscript, the IEEE Computational Intelligence Society for having it be part of the IEEE Press Series on Computational Intelligence, and Zhenbang Ju for his help in conducting simulations and program development in support of the chapters on evolutionary computation, as well as his assistance in formatting all the materials. We are also grateful to Andrew Buck and Alina Zare for using draft chapters from this book in teaching an Introduction to Computational Intelligence course at the University of Missouri, and for providing valuable feedback. Fernando Gomide also offered constructive criticisms that were helpful in improving the content. Jim Keller thanks Mihail Popescu and Derek Anderson for their assistance in implementing fuzzy clustering and fuzzy integral algorithms. Derong Liu thanks Ding Wang for his help with preparing the manuscript. David Fogel thanks the IEEE, Springer, MIT Press, and Morgan Kaufmann (Elsevier) for permissions to reprint materials (cited in the text). He also thanks SPIE for returning rights to materials published previously under its copyright. Finally, we acknowledge the assistance and friendship of our colleagues within the field of computational intelligence who hold the same passion for this material as we do. We hope the reader will enjoy as well.

■■■■■ **CHAPTER 1**

Introduction to Computational Intelligence

1.1 WELCOME TO COMPUTATIONAL INTELLIGENCE

Welcome to the world of computational intelligence (CI), which takes inspiration from nature to develop intelligent computer-based systems. Broadly, the field of CI encompasses three main branches of research and application: (1) neural networks, which model aspects of how brains function, (2) fuzzy systems, which model aspects of how people describe the world around them, and (3) evolutionary computation, which models aspects of variation and natural selection in the biosphere. These three approaches are often synergistic, working together to supplement each other and provide superior solutions to vexing problems.

1.2 WHAT MAKES THIS BOOK SPECIAL

A unique feature of this textbook is that each of us has been an editor-in-chief for an IEEE Transactions sponsored by the IEEE Computational Intelligence Society (CIS), the main technical society supporting research in CI around the world. This book offers the only systematic treatment of the entire field of CI from the perspectives of three experts who have guided peer-reviewed seminal research published in the top-tier journals in the area of CI.

The publications we've edited include the *IEEE Transactions on Neural Networks* (Derong Liu), the *IEEE Transactions on Fuzzy Systems* (James Keller), and the *IEEE Transactions on Evolutionary Computation* (David Fogel). These publications consistently present the most recent theoretical developments and practical implementations in the field of CI.

As you read through the book, you'll notice that each central area of CI is offered in its own style. That's because each of us has taken the primary lead on the material in our own area of expertise. We've made efforts to be consistent, but you'll certainly

Fundamentals of Computational Intelligence: Neural Networks, Fuzzy Systems, and Evolutionary Computation, First Edition. James M. Keller, Derong Liu, and David B. Fogel.

notice three distinct ways of conveying what we know. We believe that this is one of the advantages of our partnership—you get the whole story, but not from the standpoint of a single author. We made a deal to allow each of us to tell our story in our own way.

You may relate more to one of our styles over the others, but the content is solid and your efforts at studying this material will be rewarding. The theories and techniques described will allow you to create solutions to problems in pattern recognition, control, automated decision making, optimization, statistical modeling, and many other areas.

1.3 WHAT THIS BOOK COVERS

This introduction to CI covers basic and advanced material in neural networks, fuzzy systems, and evolutionary computation. Does it cover all of the possible topics within the field of computational intelligence? Certainly not!

Our goal is to provide fundamental material in the diverse and fast growing area of CI and give you a strong fundamental understanding of its basic concepts. We also provide some chapters with more advanced material. Each chapter offers exercises to test your knowledge and explore interesting research problems. When you master these chapters, you will be ready to dig deeper into the literature and create your own contributions to it.

1.4 HOW TO USE THIS BOOK

The best way for you to use this book is to study all of the chapters. (You knew we would say that, right?) We think that the development from neural networks to fuzzy systems to evolutionary computation provides a logical flow within the framework of a semester-long course. You'll find that each of the three main topics is described with basic chapters upfront, which cover theory, framework, and algorithms. These are followed by more advanced chapters covering more specific issues, fine points, and extensions of the basic constructions.

For instructors, presuming a typical 16-week U.S. university semester, you can easily construct three 4-week modules from the basic material with plenty of time remaining for in-class exercises, homework discussions, and computer projects. There's even some time available to pursue more advanced research in your own favorite area. (This is how the "Introduction to CI" class at the University of Missouri (MU) is organized.)

Alternatively, if you want to focus more on one area of CI, you can certainly use this book to do so. For example, if you wanted a course mainly on fuzzy systems, you could use all four of the chapters on fuzzy systems, and then sample from neural networks (to demonstrate the basis for neuro–fuzzy systems) and evolutionary computation (to develop optimization approaches in the design of fuzzy inference systems). By analogy, you could focus on neural networks or evolutionary computation, and then supplement those materials with the other chapters in the book.

1.5   FINAL THOUGHTS BEFORE YOU GET STARTED

An introductory course on computational intelligence has been taught at the University of Missouri (Jim's place) since 2005. Various texts have been used, including most recently draft chapters from this book. The class is colisted in the Electrical and Computer Engineering Department and the Computer Science Department and is available to both seniors and beginning graduate students.

In at least one semester, students were given a first-day assignment to provide a list of things that computers can't do as well as humans. The following are some of the items from the combined list:

Qualitative classification
Going from specific to general, or vice versa
Consciousness and emotion
Driving a car
Writing a poem
Chatting
Shopping
Handling inaccuracies in problems
Ethics
Natural language in conversation, with idioms
Face recognition
Aesthetics
Adaptivity
Learning (like humans do)

This was from a group of students with little or no background in intelligent systems. Depending on what you read and/or do, you might say that progress (significant in some cases) has been made on creating systems with attributes from that list, and you'd be right. Amazing things are happening. This book will provide you with the background and tools to join in the fun.

As editors-in-chief of the three main IEEE publications in the area of CI, we've had the good fortune to see novel advancements in our fields of interest even before they've been peer-reviewed and published. We've also had the joy of participating in making some of those advancements ourselves.

In fact, we've devoted our lives to advancing the theory and practice of the methods that you'll read about in this textbook. We've done that because we've often found these techniques to offer practical advantages as well as mental challenges. But in the end, we've pursued these lines of research primarily because they're a lot of fun.

We hope that you'll find not only a mathematically and practically challenging set of material in this book, but also that the material ultimately brings you as much enjoyment as it has brought for us, or even more!

Enjoy!

JAMES KELLER, Ph.D.
DERONG LIU, Ph.D.
DAVID FOGEL, Ph.D.

PART I

NEURAL NETWORKS

CHAPTER 2

Introduction and Single-Layer Neural Networks

All of us have a highly interconnected set of some 10^{11} neurons to facilitate our reading, breathing, motion, and thinking. Each of the biological neurons has the complexity of a microprocessor. Remarkably, the human brain is a highly complex, nonlinear, and parallel computer. It has the capability to organize its structural constituents, that is, neurons, so as to perform certain computations many times faster than the fastest digital computer in existence today.

Specifically, the human brain consists of a large number of highly connected elements (approximately 10^4 connections per element) called neurons [Hagan *et al.*, 1996]. For our purposes, these neurons have three principal components: the dendrites, the cell body, and the axon. The dendrites are tree-like receptive networks of nerve fibers that carry electrical signals into the cell body. The cell body effectively sums and thresholds these incoming signals. The axon is a single long fiber that carries the signal from the cell body out to other neurons. The point of contact between an axon of one cell and a dendrite of another cell is called a synapse. It is the arrangement of neurons and the strengths of the individual synapses, determined by a complex chemical process, that establishes the function of the neural network. Figure 2.1 shows a simplified schematic diagram of two biological neurons.

Actually, scientists have begun to study how biological neural networks operate. It is generally understood that all biological neural functions, including memory, are stored in the neurons and in the connections between them. Learning is viewed as the establishment of new connections between neurons or the modification of existing connections. Then, one may have the question: Although we have only a rudimentary understanding of biological neural networks, is it possible to construct a small set of simple artificial neurons and then train them to serve a useful function? The answer is "yes." This is accomplished using artificial neural networks, commonly referred to as neural networks, which have been motivated right from its inception by the recognition that the human brain computes in an entirely different way from the conventional digital computer. Figure 2.2 shows a simplified schematic diagram of

Fundamentals of Computational Intelligence: Neural Networks, Fuzzy Systems, and Evolutionary Computation, First Edition. James M. Keller, Derong Liu, and David B. Fogel.

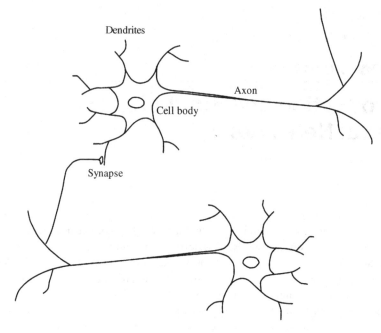

FIGURE 2.1   The schematic diagram of two biological neurons.

two artificial neurons. Here, the two artificial neurons are connected to be a simple artificial neural network and each artificial neuron contains some input and output signals.

The neurons that we consider here are not biological. They are extremely simple abstractions of biological neurons, realized as elements in a program or perhaps as circuits made of silicon. Networks of these artificial neurons do not have a fraction of the power of the human brain. However, they can be trained to perform useful functions. Note that even though biological neurons are very slow compared to electrical circuits, the brain is able to perform many tasks much faster than any conventional computer. One important reason is that the biological neural networks hold massively parallel structure and all of the neurons operate at the same time. Fortunately, the artificial neural networks share this parallel structure, which makes them useful in practice.

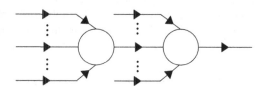

FIGURE 2.2   The schematic diagram of two artificial neurons.

Artificial neural networks do not approach the complexity of the brain. However, there are two main similarities between the biological neural networks and artificial neural networks. One is that the building blocks of both networks are simple computational devices (although artificial neurons are much simpler than biological neurons) that are highly interconnected. The other is that the connections between neurons determine the function of the network.

2.1 SHORT HISTORY OF NEURAL NETWORKS

Generally speaking, the history of neural networks has progressed through both conceptual innovations and implementation developments. However, these advancements seem to have occurred in fits and starts, rather than by steady evolution [Hagan et al., 1996].

Some of the background work for the field of neural networks occurred in the late nineteenth and early twentieth centuries. The primarily interdisciplinary work was conducted by many famous scientists from the fields of physics, psychology, and neurophysiology. In this stage, the research of neural networks emphasized general theories of learning, vision, conditioning, and so on, and did not include specific mathematical models of the neuron operation.

Then, the modern view of neural networks began in the 1940s with the work of Warren McCulloch and Walter Pitts, who showed that networks of artificial neurons could, in principle, compute any arithmetic or logical function. Notice that this important work is often regarded as the origin of the neural network community. Then, scientists proposed a mechanism for learning in biological neurons.

The first practical application of neural networks came in the late 1950s, with the invention of the perceptron network and the associated learning rule [Rosenblatt, 1958]. In this stage, the great success brought large interest to the research of neural networks. However, it was later shown that the basic perceptron network could only solve a limited class of problems. At the same time, scientists introduced a new learning algorithm and used it to train an adaptive linear neural network, which is still in use today. In fact, it was similar in structure and capability to Rosenblatt's perceptron.

Unfortunately, Rosenblatt's networks suffered from the same inherent limitation of what class of problems could be learned. Though Rosenblatt was aware of the limitation and proposed new networks to overcome it, he was not able to modify the learning algorithm to accommodate training more complex networks. Therefore, many people believed that further research on neural networks was a dead end. What's more, considering the fact that there were no powerful digital computers to conduct experiment, the research on neural networks was largely suspended.

Interest in neural networks faltered during the late 1960s because of the lack of new ideas and powerful computers with which to experiment. However, during the 1980s, these impediments were gradually overcome. Hence, research on neural networks increased dramatically. In this stage, new personal computers and workstations, which rapidly grew in capability, became widely available. More importantly, some new concepts were introduced. Among those, two novel concepts were most responsible for the rebirth of the neural network field. One was the use of statistical mechanics to explain the operation of a certain class of recurrent networks, which could be used as an associative memory. The other was the development of the backpropagation algorithm, which was introduced for helping to train multilayer perceptron networks [Rumelhart et al., 1986a, 1986b; Werbos, 1974, 1994].

These new developments reinvigorated the neural network community. In the last several decades, thousands of excellent papers have been written. The neural network technique has found many applications. Now, the field is buzzing with new theoretical and practical work. It is important to notice that many of the advances in neural networks have been related to new concepts, such as innovative architectures and training rules. In addition, the availability of powerful new computers, which test the new concepts, is also of great significance [Hagan et al., 1996].

It is apparent that a neural network derives its computing power through its massively parallel distributed structure and also its ability to learn and therefore generalize. The characteristic of generalization refers to the neural network producing reasonable outputs for inputs that were not encountered during training. These two information processing capabilities make it possible for neural networks to solve complex and large-scale problems that are currently intractable. However, in practice, neural networks cannot provide the solution by working individually. Instead, they need to be integrated into a consistent system engineering approach. Specifically, a complex problem of interest is decomposed into a number of relatively simple tasks, and some neural networks are assigned a subset of the tasks that match their inherent capabilities. However, it is important to recognize that we still have a long way to go before we can build a computer architecture that mimics a human brain. Consequently, how to achieve the true brain intelligence via artificial neural network is one of the main research objectives of scientists.

2.2 ROSENBLATT'S NEURON

Artificial neural networks, commonly referred to as "neural networks," represent a technology rooted in many disciplines: neurosciences, mathematics, statistics, physics, computer science, and engineering. Neural networks are potentially massively parallel distributed structures and have the ability to learn and generalize. Generalization denotes the neural network's production of reasonable outputs for inputs not encountered during learning process. Therefore, neural networks can be applied to diverse fields, such as modeling, time series analysis, pattern recognition, signal processing, and system control.

The neuron is the information processing unit of a neural network and the basis for designing numerous neural networks. A fundamental neural model consists of the following basic elements [Haykin, 2009]:

> ➢ A set of synapses, or connecting links, each of which is characterized by a weight or strength of its own.

➤ An adder for summing the input signals, weighted by the respective synaptic strengths of the neuron.

➤ An activation function for limiting the amplitude of the output of a neuron.

➤ An externally applied bias, which has the effect of increasing or lowering the net input of the activation function.

The most fundamental network architecture is a single-layer neural network, where the "single-layer" refers to the output layer of computation neurons. Note that we do not count the input layer of source nodes because no computation is performed there.

In neural network community, a signal flow graph is often used to provide a complete description of signal flow in a network. A signal flow graph is a network of directed links that are interconnected at certain points called nodes. The flow of signals in the various parts of the graph complies with the following three rules [Haykin, 1999]:

1. A signal flows along a link only in the direction indicated by the arrow on the link.
2. A node signal equals the algebraic sum of all signals entering the pertinent node via the incoming links.
3. The signal at a node is transmitted to each outgoing link originating from that node, with the transmission being entirely independent of the transfer functions of the outgoing links.

It should be pointed out that there are two different types of links, namely, synaptic links and activation links.

➤ The behavior of synaptic links is governed by a linear input–output relation. See Figure 2.3, the node signal x_j is multiplied by the synaptic weight w_{kj} to produce the node signal y_k, that is, $y_k = w_{kj}x_j$.

➤ The behavior of activation links is governed by a nonlinear input–output relation. This is illustrated in Figure 2.4, where $\phi(\cdot)$ is called the nonlinear activation function, that is, $y_k = \phi(x_j)$.

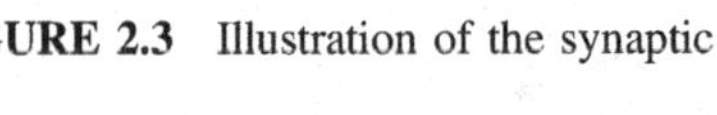

$$x_j \circ \xrightarrow{\quad w_{kj} \quad} \circ \; y_k = w_{kj}x_j$$

FIGURE 2.3   Illustration of the synaptic link.

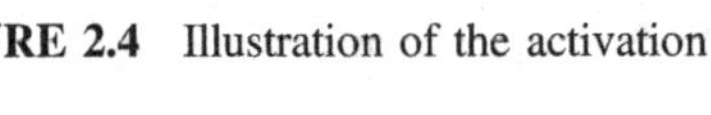

$$x_j \circ \xrightarrow{\quad \phi(\cdot) \quad} \circ \; y_k = \phi(x_j)$$

FIGURE 2.4   Illustration of the activation link.

Another expression method that can also be utilized to depict a network is called the architectural graph. Unlike the signal flow graph, the architectural graph possesses the following characteristics [Haykin, 1999]:

1. Source nodes supply input signals to the graph.
2. Each neuron is represented by a signal node called a computation node.
3. The communication links interconnecting the source and computation nodes of the graph carry no weight. They merely provide directions of signal flow in the graph.

Now, we introduce Rosenblatt's neuron [Haykin, 1999, 2009; Rosenblatt, 1958]. Rosenblatt's perceptron occupies a special place in the historical development of neural networks. It was the first algorithmically described neural network, which was built around a nonlinear neuron, namely, the McCulloch–Pitts model. Incidentally, the McCulloch–Pitts model is a neuron stated in recognition of the pioneering work done by McCulloch–Pitts. Rosenblatt's perceptron is an algorithm for learning a binary classifier: a function that maps its input x (a real-valued vector) to an output value f(x) (a single binary value):

$$f(x) = \begin{cases} 1, & \text{if } w \cdot x + b > 0 \\ 0, & \text{otherwise} \end{cases} \tag{2.1}$$

where w is a vector of real-valued weights, $w \cdot x$ is the dot product (which here computes a weighted sum), and b (a real scalar) is the "bias," a constant term that does not depend on any input value. Figure 2.5 shows the signal flow graph of the Rosenblatt's perceptron.

Here, the harder limiter input (i.e., the induced local field) of the neuron is $w \cdot x + b$.

The value of f(x) (0 or 1) is used to classify x as either a positive or a negative instance, in the case of a binary classification problem. The decision rule for the classification is to assign the point represented by the inputs $x_1, x_2, \ldots, x_n$ to class $\aleph_1$ if the perceptron output y is +1 and to class $\aleph_2$ if it is 0. Note that for the case of two input variables, the decision boundary takes the form of a straight line in a two-dimensional plane (see Figure 2.6). If b is negative, then the weighted combination of inputs must produce a positive value greater than |b| in order to push the classifier

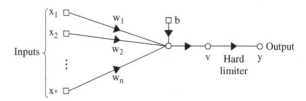

FIGURE 2.5 Signal flow graph of the perceptron.

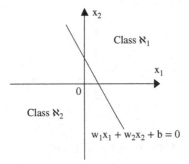

FIGURE 2.6   Illustration of the two-dimensional, two-class pattern classification problem.

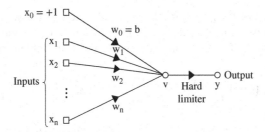

FIGURE 2.7   Equivalent signal flow graph of the perceptron.

neuron over the 0 threshold. Spatially, the bias alters the position (although not the orientation) of the decision boundary.

We now give an equivalent model of the perceptron described in Figure 2.5. Here, the bias b is viewed as a synaptic weight driven by a fixed input equal to +1. Then, the signal flow graph is shown in Figure 2.7.

2.3   PERCEPTRON TRAINING ALGORITHM

Now we consider the performance of the perceptron network and are in a position to introduce the perceptron learning rule. This learning rule is an example of supervised training, in which the learning rule is provided with a set of examples of proper network behavior:

$$\{p_1, t_1\}, \{p_2, t_2\}, \ldots, \{p_q, t_q\} \tag{2.2}$$

Here p_i is an input to the network and t_i is the corresponding target output. As each input is applied to the network, the network output is compared with the target. The learning rule then adjusts the weights and biases of the network in order to move the network output closer to the target.

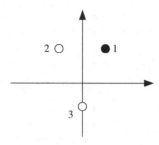

FIGURE 2.8   The test problem.

2.3.1   Test Problem

In our presentation of the perceptron learning rule, we will begin with a simple test problem and will experiment with possible rules to develop some intuition about how the rule should work. The input–target pairs for our test problem are

$$\left\{ p_1 = \begin{bmatrix} 1 \\ 2 \end{bmatrix}, \; t_1 = 1 \right\}, \quad \left\{ p_2 = \begin{bmatrix} -1 \\ 2 \end{bmatrix}, \; t_2 = 0 \right\}, \quad \left\{ p_3 = \begin{bmatrix} 0 \\ -1 \end{bmatrix}, \; t_3 = 0 \right\}$$

where p_i represents the input and t_i represents the corresponding target output. The problem is displayed graphically in Figure 2.8, where the two input vectors whose target is 0 are represented with a light circle ○, and the vector whose target is 1 is represented with a dark circle ●. This is a very simple problem, and we could almost obtain a solution by inspection. This simplicity will help us gain some intuitive understanding of the basic concepts of the perceptron learning rule.

The network for this problem should have two inputs and one output. To simplify our development of the learning rule, we will begin with a network without a bias. The network will then have just two parameters $w_{1,1}$ and $w_{1,2}$, as shown in Figure 2.9.

By removing the bias, we are left with a network whose decision boundary must pass through the origin. We need to ensure that this network is still able to solve the test problem. There must be an allowable decision boundary that can separate the

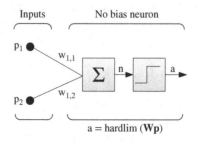

FIGURE 2.9   Test problem network.

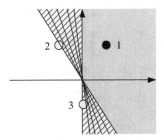

FIGURE 2.10 The boundaries.

vectors p_2 and p_3 from the vector p_1. Figure 2.10 illustrates that there are indeed an infinite number of such boundaries.

Figure 2.11 shows the weight vectors that correspond to the allowable decision boundaries. (Recall that the weight vector is orthogonal to the decision boundary.) We would like a learning rule that will find a weight vector that points to one of these directions. Remember that the length of the weight vector does not matter; only its direction is important.

2.3.2 Constructing Learning Rules

Training begins by assigning some initial values to the network parameters. In this case, we are training a two-input/single-output network without a bias, so we can only initialize its two weights. Here we set the elements of the weight vector $_1w$ to the following randomly generated values:

$$_1w^T = \begin{bmatrix} 1.0, & -0.8 \end{bmatrix} \tag{2.3}$$

We will now begin presenting the input vectors to the network and find the corresponding outputs, which we will call ϑ. We begin with p_1:

$$\vartheta = \text{hardlim}(_1w^T p_1) = \text{hardlim}\left([1.0 - 0.8] \begin{bmatrix} 1 \\ 2 \end{bmatrix} \right) \tag{2.4}$$
$$\vartheta = \text{hardlim}(-0.6) = 0$$

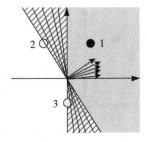

FIGURE 2.11 The weight vectors and boundaries.

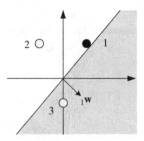

FIGURE 2.12   The classification result for the first input.

The network has not returned the correct value. The network output is 0, while the target response t_1 is 1.

We can see what happened in Figure 2.12. The initial weight vector results in a decision boundary that incorrectly classifies the vector p_1. We need to alter the weight vectors so that it points more toward p_1, so that in the future it has a better chance of classifying it correctly.

One approach would be to set $_1w$ equal to p_1. This is simple and would ensure that p_1 was classified properly in the future. Unfortunately, it is easy to construct a problem for which this rule cannot find a solution. Figure 2.13 shows a problem that cannot be solved with the weight vectors pointing directly at either of the two class 1 vectors. If we apply the rule $_1w = p$ every time one of these vectors is misclassified, the network's weights will simply oscillate back and forth and will never find a solution.

Another possibility would be to add p_1 to $_1w$. Adding p_1 to $_1w$ would make $_1w$ point more in the direction of p_1. Repeated presentations of p_1 would cause the direction of $_1w$ to asymptotically approach the direction of p_1. This rule can be stated:

$$\text{If } t = 1 \text{ and } \vartheta = 0, \text{ then } _1w^{\text{new}} = _1w^{\text{old}} + p \tag{2.5}$$

Applying this rule to our test problem results in new values for $_1w$:

$$_1w^{\text{new}} = _1w^{\text{old}} + p_1 = \begin{bmatrix} 1.0 \\ -0.8 \end{bmatrix} + \begin{bmatrix} 1 \\ 2 \end{bmatrix} = \begin{bmatrix} 2.0 \\ 1.2 \end{bmatrix} \tag{2.6}$$

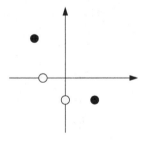

FIGURE 2.13   Another test problem that poses a challenge for setting $_1w = p_1$.

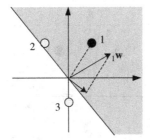

FIGURE 2.14 New values of the weight after adjusting by Eq. 2.6.

This operation is illustrated in Figure 2.14.

We now move on to the next input vector and continue making changes to the weights and cycling through the inputs until they are all classified correctly.

The next input vector is p_2. When it is presented to the network, we find

$$\vartheta = \text{hardlim}(_1w^T p_2) = \text{hardlim}\left([2.0\;1.2]\begin{bmatrix} -1 \\ 2 \end{bmatrix}\right) \tag{2.7}$$
$$= \text{hardlim}(0.4) = 1$$

The target t_2 associated with p_2 is 0 and the output ϑ is 1. A class 0 vector was misclassified as a 1.

Since we now find that we'd like to move the weight vector $_1w$ away from the input, we can simply change the addition in Eq. 2.5 to subtraction:

$$\text{If } t = 0 \text{ and } \vartheta = 1, \text{ then } _1w^{new} = {_1w^{old}} - p \tag{2.8}$$

If we apply this to the test problem, we find

$$_1w^{new} = {_1w^{old}} - p_2 = \begin{bmatrix} 2.0 \\ 1.2 \end{bmatrix} - \begin{bmatrix} -1 \\ 2 \end{bmatrix} = \begin{bmatrix} 3.0 \\ -0.8 \end{bmatrix} \tag{2.9}$$

which is illustrated in Figure 2.15.

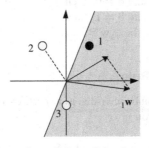

FIGURE 2.15 New values of the weight after adjusting by Eq. 2.9.

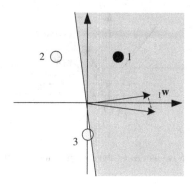

FIGURE 2.16   The classification result for the updated perceptron on all three vectors.

Now we present the third vector p_3:

$$\vartheta = \text{hardlim}(_1w^T p_3) = \text{hardlim}\left([3.0 - 0.8]\begin{bmatrix} 0 \\ -1 \end{bmatrix}\right)$$
$$= \text{hardlim}(0.8) = 1 \tag{2.10}$$

The current $_1w$ results in a decision boundary that misclassifies p_3. This is a situation for which we already have a rule, so $_1w$ will be updated again, according to Eq. 2.8:

$$_1w^{new} = _1w^{old} - p_3 = \begin{bmatrix} 3.0 \\ -0.8 \end{bmatrix} - \begin{bmatrix} 0 \\ -1 \end{bmatrix} = \begin{bmatrix} 3.0 \\ 0.2 \end{bmatrix} \tag{2.11}$$

Figure 2.16 shows that the perceptron has finally learned to classify the three vectors properly. If we present any of the input vectors to the neuron, it will output the correct class for that input vector.

This brings us to our third and final rule: If it works, don't fix it.

$$\text{If } t = \vartheta, \text{ then } _1w^{new} = _1w^{old} \tag{2.12}$$

Here are the three rules, which cover all possible combinations of output and target values:

$$\begin{aligned} &\text{If } t = 1 \text{ and } \vartheta = 0, \text{ then } _1w^{new} = _1w^{old} + p \\ &\text{If } t = 0 \text{ and } \vartheta = 1, \text{ then } _1w^{new} = _1w^{old} - p \\ &\text{If } t = \vartheta, \text{ then } _1w^{new} = _1w^{old} \end{aligned} \tag{2.13}$$

2.3.3   Unified Learning Rule

The three rules in Eq. 2.13 can be rewritten as a single expression. First, we will define a new variable, the perceptron error e:

$$e = t - \vartheta \tag{2.14}$$

We can now rewrite the three rules of Eq. 2.13 as follows:

$$\text{If } e = 1, \text{ then } {}_1w^{new} = {}_1w^{old} + p$$
$$\text{If } e = -1, \text{ then } {}_1w^{new} = {}_1w^{old} - p \qquad (2.15)$$
$$\text{If } e = 0, \text{ then } {}_1w^{new} = {}_1w^{old}$$

Looking carefully at the first two rules in Eq. 2.15, we can see that the sign of p is the same as the sign on the error e. Furthermore, the absence of p in the third rule corresponds to an e of 0. Thus, we can unify the three rules into a single expression:

$$ {}_1w^{new} = {}_1w^{old} + ep = {}_1w^{old} + (t - \vartheta)p \qquad (2.16)$$

This rule can be extended to train the bias by noting that a bias is simply a weight whose input is always 1. We can thus replace the input p in Eq. 2.16 with the input to the bias, which is 1. The result is the perceptron rule for a bias:

$$b^{new} = b^{old} + e \qquad (2.17)$$

2.3.4 Training Multiple-Neuron Perceptrons

The perceptron rule, as given by Eqs. 2.16 and 2.17, updates the weight vector of a single-neuron perceptron. We can generalize this rule for the multiple-neuron perceptron of Figure 2.17 as follows:

To update the ith row of the weight matrix, use

$$ {}_iw^{new} = {}_iw^{old} + e_ip \qquad (2.18)$$

To update the ith element of the bias vector, use

$$b_i^{new} = b_i^{old} + e_i \qquad (2.19)$$

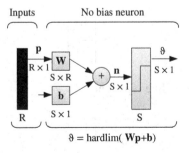

$$\vartheta = \text{hardlim}(\ \mathbf{Wp+b})$$

FIGURE 2.17   Test problem multiple-neuron network.

Perceptron rule: This rule can be written conveniently in matrix notation:

$$W^{new} = W^{old} + ep^T \tag{2.20}$$

and

$$b^{new} = b^{old} + e \tag{2.21}$$

To test the perceptron learning rule, consider the apple/orange recognition problem.

2.3.4.1 Problem Statement A producer dealer has a warehouse that stores a variety of fruits and vegetables. When fruit is brought to the warehouse, various types of fruit may be mixed together. The dealer wants a machine that will sort the fruit according to the type. There is a conveyer belt on which the fruit is loaded. This conveyer passes through a set of sensors that measure three properties of the fruit: *shape, texture,* and *weight*. These sensors are somewhat primitive. The shape sensor will output a 1 if the fruit is approximately round and a −1 if it is more elliptical. The texture sensor will output a 1 if the surface of the fruit is smooth and a −1 if it is rough. The weight sensor will output a 1 if the fruit is more than 1 lb and a −1 if it is less than 1 lb.

The three sensor outputs will then be input to a neural network. The purpose of the network is to decide which kind of fruit is on the conveyer, so that the fruit can be directed to the correct storage bin. To make the problem even simpler, let's assume that there are only two kinds of fruit on the conveyer: apples and oranges.

As each fruit passes through the sensors, it can be represented by a three-dimensional vector. The first element of the vector represents shape, the second element represents texture, and the third element represents weight:

$$p = \begin{bmatrix} shape \\ texture \\ weight \end{bmatrix} \tag{2.22}$$

Therefore, a prototype orange would be represented by

$$p_1 = \begin{bmatrix} 1 \\ -1 \\ -1 \end{bmatrix} \tag{2.23}$$

and a prototype apple would be represented by

$$p_2 = \begin{bmatrix} 1 \\ 1 \\ -1 \end{bmatrix} \tag{2.24}$$

The neural network will receive one three-dimensional input vector for each fruit on the conveyer and must make a decision as to whether the fruit is an orange (p_1) or an apple (p_2).

For the apple and orange problem, the input/output prototype vectors will be

$$\left\{ p_1 = \begin{bmatrix} 1 \\ -1 \\ -1 \end{bmatrix}, \; t_1 = [0] \right\}, \quad \left\{ p_2 = \begin{bmatrix} 1 \\ 1 \\ -1 \end{bmatrix}, \; t_2 = [1] \right\} \tag{2.25}$$

(Note that we are using 0 as the target output for the orange pattern p_1 instead of -1, as was used in previous statements. This is because we are using the *hardlim* transfer function.)

Typically the weights and biases are initialized to small random numbers. Suppose that here we start with the initial weight matrix and bias:

$$W = \begin{bmatrix} 0.5 & -1 & -0.5 \end{bmatrix}, \quad b = 0.5 \tag{2.26}$$

The first step is to apply the first input vector p_1 to the network:

$$\vartheta = \text{hardlim}(Wp_1 + b) = \text{hardlim} \left(\begin{bmatrix} 0.5 & -1 & -0.5 \end{bmatrix} \begin{bmatrix} 1 \\ -1 \\ -1 \end{bmatrix} + 0.5 \right) \tag{2.27}$$

$$\vartheta = \text{hardlim}(2.5) = 1$$

Then we calculate the error:

$$e = t_1 - \vartheta = 0 - 1 = -1 \tag{2.28}$$

The weight update is

$$W^{\text{new}} = W^{\text{old}} + ep^T = \begin{bmatrix} 0.5 & -1 & -0.5 \end{bmatrix} + (-1) \begin{bmatrix} 1 & -1 & -1 \end{bmatrix} \tag{2.29}$$

$$= \begin{bmatrix} -0.5 & 0 & 0.5 \end{bmatrix}$$

The bias update is

$$b^{\text{new}} = b^{\text{old}} + e = 0.5 + (-1) = -0.5 \tag{2.30}$$

This completes the first iteration.

The second iteration of the perceptron rule is

$$\vartheta = \text{hardlim}(Wp_2 + b) = \text{hardlim} \left(\begin{bmatrix} -0.5 & 0 & 0.5 \end{bmatrix} \begin{bmatrix} 1 \\ 1 \\ -1 \end{bmatrix} + (-0.5) \right) \tag{2.31}$$

$$= \text{hardlim}(-1.5) = 0$$

$$e = t_2 - \vartheta = 1 - 0 = 1 \tag{2.32}$$

$$W^{new} = W^{old} + ep^T = \begin{bmatrix} -0.5 & 0 & 0.5 \end{bmatrix} + 1\begin{bmatrix} 1 & 1 & -1 \end{bmatrix} = \begin{bmatrix} 0.5 & 1 & -0.5 \end{bmatrix} \tag{2.33}$$

$$b^{new} = b^{old} + e = -0.5 + 1 = 0.5 \tag{2.34}$$

The third iteration begins again with the first input vector:

$$\vartheta = \text{hardlim}(Wp_1 + b) = \text{hardlim}\left(\begin{bmatrix} 0.5 & 1 & -0.5 \end{bmatrix} \begin{bmatrix} 1 \\ -1 \\ -1 \end{bmatrix} + 0.5 \right)$$
$$= \text{hardlim}(0.5) = 1 \tag{2.35}$$

$$e = t_1 - \vartheta = 0 - 1 = -1 \tag{2.36}$$

$$W^{new} = W^{old} + ep^T = \begin{bmatrix} 0.5 & 1 & -0.5 \end{bmatrix} + (-1)\begin{bmatrix} 1 & -1 & -1 \end{bmatrix}$$
$$= \begin{bmatrix} -0.5 & 2 & 0.5 \end{bmatrix} \tag{2.37}$$

$$b^{new} = b^{old} + e = 0.5 - 1 = -0.5 \tag{2.38}$$

If you continue with more iterations, you will find that both input vectors will be classified correctly. The algorithm has converged to a solution, whose boundary correctly classifies the two input vectors.

We can now summarize the perceptron training algorithm. First, we define some variables:

- $y = f(x)$ denotes the *output* from the perceptron for an input vector x.
- b is the *bias* term.
- $D = \{(x(1), d(1)), \ldots, (x(s), d(s))\}$ is the *training set* of s samples, where
- $x(k)$ is the n-dimensional input vector.
- $d(k)$ is the desired output value of the perceptron for that input.

We show the values of the features as follows:

- $x_i(k)$ is the value of the ith feature of the kth training *input vector*.
- $x_0(k) = 1$.

To represent the weights:

- w_i is the ith value in the *weight vector*, to be multiplied by the value of the ith input feature.
- Because $x_0(k) = 1$, w_0 is effectively a learned bias that we use instead of the bias constant b.

To show the time dependence of w, we use the following:

- $w_i(k)$ is the weight i at time k.
- α is the *learning rate*, where $0 < \alpha \leq 1$.

Too high a learning rate makes the perceptron periodically oscillate around the solution unless additional steps are taken.

Then, the training algorithm goes iteratively in accordance with the following procedure [Haykin, 1999, 2009]:

1. Initialize the weights and the threshold. Weights may be initialized to 0 or to a small random value. In the example below, we use 0.
2. For each example k in our training set D, perform the following steps over the input x(k) and desired output d(k):
3. Calculate the actual output:

$$y(k) = f[w(t) \cdot x(k)] = f[w_0(k) + w_1(k)x_1(k) + w_2(k)x_2(k) + \cdots + w_n(k)x_n(k)] \tag{2.39}$$

4. Update the weights:

$$w_i(k + 1) = w_i(k) + \alpha(d_i(k) - y_i(k))x_i(k) \tag{2.40}$$

for all features $0 \leq i \leq n$.

In the following section, the proof of the perceptron convergence algorithm in the case of $\alpha = 1$ and the fixed increment convergence theorem will be discussed.

2.4 THE PERCEPTRON CONVERGENCE THEOREM

Given the initial condition $w(0) = 0$, we now present the proof of the perceptron convergence algorithm [Haykin, 1999, 2009].

Let $\mathfrak{A}_1$ be the subspace of training vectors that belong to class $\aleph_1$, while $\mathfrak{A}_2$ be the subspace of training vectors that belong to class $\aleph_2$.

Suppose the perceptron incorrectly classifies the training vectors x(1), x(2), ... For example, $w^T(k)x(k) < 0$ for $k = 1, 2, \ldots$, but x(1), x(2), ... belongs to the subset $\mathfrak{A}_1$. Considering $\alpha(k) = 1$, the second equation in (2.40) can be written as

$$w(k + 1) = w(k) + x(k), \quad \text{for } x(k) \text{ belongs to class } \aleph_1 \qquad (2.41)$$

Expanding (2.41), we can further obtain

$$w(k + 1) = \sum_{j=1}^{k} x(j) \qquad (2.42)$$

Because $\aleph_1$ and $\aleph_2$ are assumed to be linearly separable, there exists a solution w^0 such that $w^{0T}x(k) > 0$ for the training vectors $x(1), x(2), \ldots, x(k)$ belonging to the subset $\mathfrak{A}_1$. Define

$$\theta = \min_{x(k) \in \mathfrak{A}_1} w^{0T}x(k) \qquad (2.43)$$

Then we can derive that

$$w^{0T}w(k + 1) = \sum_{j=1}^{k} w^{0T}x(j) \geq k\theta \qquad (2.44)$$

Using the Cauchy–Schwarz inequality, we can obtain

$$\left\| w^0 \right\|^2 \left\| w(k + 1) \right\|^2 \geq \left(w^{0T}w(k + 1) \right)^2 \geq k^2\theta^2 \qquad (2.45)$$

where $\|\cdot\|$ denotes the Euclidean norm of the enclosed argument vector. Therefore, we have

$$\left\| w(k + 1) \right\|^2 \geq \frac{k^2\theta^2}{\left\| w^0 \right\|^2} \qquad (2.46)$$

On the other hand, we can rewrite (2.41) as

$$w(l + 1) = w(l) + x(l), \quad \text{for } l = 1, 2, \ldots, k \text{ and } x(l) \in \mathfrak{A}_1 \qquad (2.47)$$

Then, we can obtain

$$\left\| w(l + 1) \right\|^2 = \left\| w(l) \right\|^2 + \left\| x(l) \right\|^2 + 2w^T(l)x(l) \qquad (2.48)$$

According to the supposition $w^T(k)x(k) < 0$, we can further derive that

$$\left\| w(l + 1) \right\|^2 \leq \left\| w(l) \right\|^2 + \left\| x(l) \right\|^2 \qquad (2.49)$$

which is equivalent to

$$\|w(1+1)\|^2 - \|w(1)\|^2 \leq \|x(1)\|^2 \qquad (2.50)$$

Define

$$\delta = \min_{x(1) \in \mathfrak{A}_1} \|x(1)\|^2 \qquad (2.51)$$

Then, in accordance with Eq. 2.50, we can obtain

$$\|w(1+1)\|^2 \leq \sum_{1=1}^{k} \|x(1)\|^2 \leq k\delta \qquad (2.52)$$

By making a comparison between Eqs. 2.46 and 2.52, we can find that the two equations hold simultaneously only if

$$k \leq k_{max} = \delta \frac{\|w^0\|^2}{\theta^2} \qquad (2.53)$$

Otherwise, the two equations conflict with each other. As a consequence, we have proved that for $\alpha(k) = 1$ for all k and $w(0) = 0$, and given that a solution vector w^0 exists, the rule for updating the synaptic weights of the perceptron must terminate after at most k_{max} iterations. Incidentally, the value of w^0 or k_{max} is not unique, which can be seen from Eqs. 2.43, 2.51, and 2.53.

Now, we expound the fixed increment convergence theorem for the perceptron as follows [Rosenblatt, 1962].

Theorem 2.1 Let the subsets of training vectors $\mathfrak{A}_1$ and $\mathfrak{A}_2$ be linearly separable. Let the inputs presented to the perceptron originate from these two subsets. The perceptron converges after some k_0 iterations, in the sense that

$$w(k_0) = w(k_0 + 1) = w(k_0 + 2) = \cdots$$

is a solution vector for $k_0 \leq k_{max}$.

2.5 COMPUTER EXPERIMENT USING PERCEPTRONS

Table 2.1 describes two classes of patterns in the two-dimensional plane.

Consider the single-layer perceptron illustrated in Figure 2.18. The corresponding input–output mapping is defined as

$$y = \phi(v) = \begin{cases} 1, & \text{if } w_1x_1 + w_2x_2 + b \geq 0 \\ 0, & \text{if } w_1x_1 + w_2x_2 + b < 0 \end{cases} \qquad (2.54)$$

TABLE 2.1   **Pattern Classification**

x_1	x_2	d	Class
2	2	0	$\aleph_1$
1	-2	1	$\aleph_2$
-2	2	0	$\aleph_1$
-1	0	1	$\aleph_2$

Now, we describe the iterative process of the training algorithm as follows, with the purpose of classifying the patterns:

1. Set w(0) = 0.
2. Compute

$$y(0) = \phi\Big(\begin{bmatrix} 0 & 0 & 0 \end{bmatrix} \begin{bmatrix} 1 & 2 & 2 \end{bmatrix}^T \Big) = \phi(0) = 1$$

Since the actual response is not equal to the desired response, we update the weight and bias as

$$w(1) = \begin{bmatrix} 0 & 0 & 0 \end{bmatrix}^T + (0-1) \begin{bmatrix} 1 & 2 & 2 \end{bmatrix}^T = \begin{bmatrix} -1 & -2 & -2 \end{bmatrix}^T$$

3. Compute

$$y(1) = \phi\Big(\begin{bmatrix} -1 & -2 & -2 \end{bmatrix} \begin{bmatrix} 1 & 1 & -2 \end{bmatrix}^T \Big) = \phi(1) = 1$$

The actual response is equal to the desired response, so we do not need to update the weight and bias.

4. Compute

$$y(2) = \phi\Big(\begin{bmatrix} -1 & -2 & -2 \end{bmatrix} \begin{bmatrix} 1 & -2 & 2 \end{bmatrix}^T \Big) = \phi(-1) = 0$$

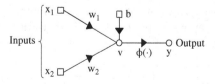

FIGURE 2.18   Structure of the single-layer perceptron.

Here, the actual response is equal to the desired response. Therefore, we do not need to update the weight and bias.

5. Compute

$$y(3) = \phi\left(\begin{bmatrix} -1 & -2 & -2 \end{bmatrix} \begin{bmatrix} 1 & -1 & 0 \end{bmatrix}^T\right) = \phi(1) = 1$$

Since the actual response is also equal to the desired response, we keep the weight and bias in their present values.

6. Compute

$$y(4) = \phi\left(\begin{bmatrix} -1 & -2 & -2 \end{bmatrix} \begin{bmatrix} 1 & 2 & 2 \end{bmatrix}^T\right) = \phi(-9) = 0$$

We see that the actual response is equal to the desired response. Thus, we do not need to update the weight and bias. Besides, we can further observe that the weight and bias here can make the actual response equal to the desired response for all input patterns. Accordingly, no update is required.

By considering

$$-2x_1 - 2x_2 - 1 = 0 \tag{2.55}$$

we can obtain a line

$$x_2 = -x_1 - \frac{1}{2} \tag{2.56}$$

to classify the inputs patterns in Table 2.1, as shown in Figure 2.19.

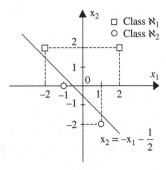

FIGURE 2.19   Illustration of the computer experiment that creates a line to separate the two classes.

2.6 ACTIVATION FUNCTIONS

In biologically inspired neural networks, the activation function is usually an abstraction representing the rate of action potential firing in the cell. In its simplest form, this function is binary, that is, the neuron is either firing or not.

The activation function defines the output of a neuron in terms of the induced local field [Haykin, 2009]. In this part, we identify two basic types of activation functions: a threshold function and a sigmoid function.

2.6.1 Threshold Function

The threshold function is defined as

$$\phi(v) = \begin{cases} 1, & \text{if } v \geq 0; \\ 0, & \text{if } v < 0 \end{cases} \tag{2.57}$$

In engineering, this form of a threshold function is commonly referred to as a Heaviside function. See Figure 2.20.

The output of neuron m employing such a threshold function is expressed as

$$y_m = \begin{cases} 1, & \text{if } v_m \geq 0 \\ 0, & \text{if } v_m < 0 \end{cases} \tag{2.58}$$

where

$$v_m = \sum_{i=1}^{n} w_{mi} x_i + b_m \tag{2.59}$$

Notice in this model, the output of a neuron takes on the value of 1 if the induced local field of that neuron is nonnegative, and 0 otherwise. Such a neuron is referred to as the McCulloch–Pitts model.

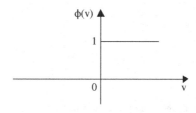

FIGURE 2.20 The threshold function from Eq. 2.57.

2.6.2 Sigmoid Function

The sigmoid function is the most common form of activation function used in the construction of neural networks. It is a strictly increasing function, which holds an excellent balance between linear and nonlinear behavior. The logistic function, a typical example of such function, is defined by

$$\phi(v) = \frac{1}{1 + e^{-av}} \tag{2.60}$$

where a is the slope parameter of the sigmoid function. It is depicted in Figure 2.21. Note that a threshold function assumes the value of 0 or −1, while a sigmoid function assumes a continuous range of values from 0 to 1. Another important property of the sigmoid function is that it is differentiable, whereas the threshold function is not.

In (2.57) and (2.60), the activation functions range from 0 to +1. However, it is sometimes desirable to have the activation function range from −1 to 1. In this case, the threshold function can be defined as the signum function, which is formulated as

$$\phi(v) = \begin{cases} 1, & \text{if } v > 0 \\ 0, & \text{if } v = 0 \\ -1, & \text{if } v < 0 \end{cases} \tag{2.61}$$

and described in Figure 2.22.

The corresponding form of a sigmoid function is hyperbolic tangent function defined as

$$\phi(v) = \tanh(v) = \frac{e^v - e^{-v}}{e^v + e^{-v}} \tag{2.62}$$

and plotted in Figure 2.23. It allows the activation function to assume negative values and can bring practical benefits.

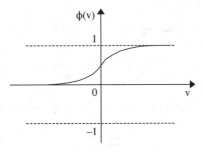

FIGURE 2.21 The logistic function from Eq. 2.60.

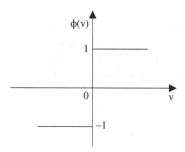

FIGURE 2.22 The signum function.

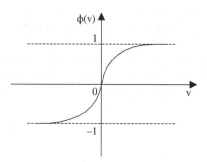

FIGURE 2.23 The hyperbolic tangent function.

EXERCISES

2.1. Suppose the inputs applied to neuron j, which is built around the McCulloch–Pitts model, are 10, −20, 4, −2. The synaptic weights of connecting to the neuron are 0.8, 0.2, −1, −0.9. Given that the externally applied bias is 0, compute the induced local field and the output of neuron j, respectively.

2.2. Study how the graph of the logistic function is affected by varying the value of slope parameter a in Eq. 2.60. Calculate the slope of the logistic function at origin. Then point out what the logistic function will be when the slope parameter approaches infinity.

2.3. The limiting values of the algebraic sigmoid function

$$\phi(v) = \frac{v}{\sqrt{1 + v^2}}$$

are −1 and +1. Show that the derivative of $\phi(v)$ with respect to v is given by

$$\frac{d\phi}{dv} = \frac{\phi^3(v)}{v^3}$$

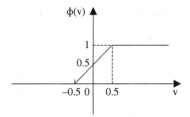

FIGURE 2.24   The piecewise-linear function of Exercise 2.4.

Calculate the value of the derivative at the origin.

2.4. Another activation function

$$\phi(v) = \begin{cases} 0, & \text{if } v \leq -\dfrac{1}{2} \\[2mm] v, & \text{if } -\dfrac{1}{2} < v < \dfrac{1}{2} \\[2mm] 1, & \text{if } v \geq \dfrac{1}{2} \end{cases}$$

is depicted in Figure 2.24 and is called the piecewise-linear function. Here, the amplification factor inside the linear region of operation is assumed to be unity. Based on this expression, consider the activation functions presented in Figures 2.25 and 2.26 and, and study the following problems, respectively.

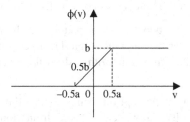

FIGURE 2.25   The activation function of Exercise 2.4 (case 1).

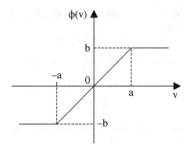

FIGURE 2.26   The activation function of Exercise 2.4 (case 2).

TABLE 2.2   Pattern Classification of Exercise 2.6

x_1	x_2	d	Class
0	2	1	$\aleph_1$
1	0	1	$\aleph_1$
0	-2	0	$\aleph_2$
2	0	0	$\aleph_2$

1. Formulate $\phi(v)$ as a function of v.

2. Show what will happen to $\phi(v)$ if a is allowed to approach zero.

2.5. Suppose that in the signal flow graph of the perceptron shown in Figure 2.7, the hard limiter is prescribed as the sigmoidal nonlinearity:

$$\phi(v) = \tanh\left(\frac{v}{2}\right)$$

where v is the induced local field. The classification decisions made by the perceptron are defined as

➢ observation vector x that belongs to class $\aleph_1$ if the output y > ξ, where ξ is a threshold; otherwise, x belongs to class $\aleph_2$.

Show that the decision boundary so constructed is a hyperplane.

2.6. Table 2.2 describes the two classes of patterns in the two-dimensional plane. Classify the two classes by using the single-layer perceptron. Given the detailed iterative process of the training algorithm, draw the separating line on the x_1, x_2 plane.

2.7. Show the single-layer perceptron can classify the patterns described in Table 2.3 successfully. However, a basic limitation of the perceptron is

TABLE 2.3   Pattern Classification of Exercise 2.7

x_1	x_2	d
0	0	0
0	1	0
1	0	0
1	1	1

that it cannot implement the EXCLUSIVE OR function. Explain the reason for this limitation.

2.8. Consider two one-dimensional, Gaussian-distributed classes $\aleph_1$ and $\aleph_2$ that have a common variance equal to 1. Their mean values are $\rho_1 = -10$, $\rho_2 = 10$, respectively. These two classes are essentially linearly separable. Design a classifier that separates these two classes.

CHAPTER 3

Multilayer Neural Networks and Backpropagation

Rosenblatt's perceptron that we studied in Chapter 2 is basically a single-layer neural network and it is limited to the classification of linearly separable patterns. In this chapter, in order to overcome the practical limitations of the perceptron, we look to a new neural network structure called a multilayer perceptron.

The following are the basic features of multilayer perceptrons [Haykin, 2009]:

- Each neuron model in the network includes a nonlinear activation function that is differentiable.
- The network contains one or more layers that are hidden from both the input and output modes.
- The network holds a high degree of connectivity, the extent of which is determined by synaptic weights of the network.

The architectural graph of a multilayer perceptron with two hidden layers and an output layer is depicted in Figure 3.1.

A computationally effective method for training the multilayer perceptrons is the backpropagation algorithm [Rumelhart et al., 1986a, 1986b; Werbos, 1974, 1994], which is regarded as a landmark in the development of neural network.

3.1 UNIVERSAL APPROXIMATION THEORY

A multilayer perceptron trained with the backpropagation algorithm can be considered as a practical means for performing a nonlinear input–output mapping of a general nature. Let n_0 denote the number of input (source) nodes of a multilayer perceptron, and let $N = n_l$ denote the number of neurons in the output layer of the network. Then, the input–output relationship of the network defines a mapping from an n_0-dimensional Euclidean input space to an N-dimensional Euclidean output

Fundamentals of Computational Intelligence: Neural Networks, Fuzzy Systems, and Evolutionary Computation, First Edition. James M. Keller, Derong Liu, and David B. Fogel.
© 2016 by The Institute of Electrical and Electronics Engineers, Inc. Published 2016 by John Wiley & Sons, Inc.

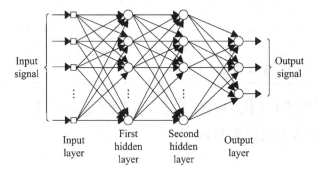

FIGURE 3.1   Architectural graph of a multilayer perceptron with two hidden layers.

space, which is infinitely continuously differentiable when the activation function is likewise continuously differentiable. The universal approximation theorem for a nonlinear input–output mapping is directly applicable to multilayer perceptrons.

Theorem 3.1 [Haykin, 1999]   Let $\phi(\cdot)$ be a nonconstant, bounded, and monotone-increasing continuous function. Let I_{n_0} denote the n_0-dimensional unit hypercube $[0, 1]^{n_0}$. The space of continuous functions on I_{n_0} is denoted by $C(I_{n_0})$. Then, given any function $f \in C(I_{n_0})$ and $\varepsilon > 0$, there exist an integer n_1 and sets of real constants μ_i and b_i, and w_{ij}, where $i = 1, 2, \ldots, n_1$ and $j = 1, 2, \ldots, n_0$ such that we may define

$$F(x_1, x_2, \ldots, x_{n_0}) = \sum_{i=1}^{n_1} \mu_i \phi \left(\sum_{j=1}^{n_0} w_{ij} x_j + b_i \right) \tag{3.1}$$

as an approximate realization of the function $f(\cdot)$; that is,

$$\left| F(x_1, x_2, \ldots, x_{n_0}) - f(x_1, x_2, \ldots, x_{n_0}) \right| < \varepsilon \tag{3.2}$$

for all $x_1, x_2, \ldots, x_{n_0}$ that lie in the input space.

Here, we notice that the hyperbolic tangent function used as the nonlinearity in a neural model for the construction of a multilayer perceptron is indeed a nonconstant, bounded, and monotone-increasing function. Therefore, it satisfies the conditions imposed on the function $\phi(\cdot)$. In addition, Eq. 3.1 represents the output of a multilayer perceptron described as follows:

- The network has n_0 input nodes and a single hidden layer consisting of n_1 neurons, while the inputs are $x_1, x_2, \ldots, x_{n_0}$.
- Hidden neuron i has synaptic weights $w_{i1}, w_{i2}, \ldots, w_{in_0}$ and bias b_i.
- The network output is a linear combination of the outputs of the hidden neurons, with $\mu_1, \mu_2, \ldots, \mu_{n_1}$ being the synaptic weights of the output layer.

Note that the universal approximation theorem is an existence theorem in the sense that it provides the mathematical justification for the approximation of an arbitrary continuous function as opposed to an exact representation.

3.2   THE BACKPROPAGATION TRAINING ALGORITHM

3.2.1   The Description of the Algorithm

Now, we present the backpropagation algorithm.

Consider Figure 3.2, the neuron j is fed by a set of function signals produced by a layer of neurons to its left. The induced local field $v_j(k)$ produced at the input of the activation function associated with neuron j is

$$v_j(k) = \sum_{i=0}^{n} w_{ji}(k)y_i(k) \tag{3.3}$$

Here, n is the total number of inputs (excluding the bias) applied to neuron j. Note the synaptic weight w_{j0} related to the fixed input $y_0 = +1$ equals the bias b_j applied to neuron j. The function signal $y_j(k)$ appearing at the output of neuron j at iteration k is

$$y_j(k) = \phi_j(v_j(k)) \tag{3.4}$$

Then, the error signal produced at the output of neuron j is defined by

$$e_j(k) = d_j(k) - y_j(k) \tag{3.5}$$

where $d_j(k)$ is the corresponding desired signal. We define the instantaneous error energy of neuron j as

$$E_j(k) = \frac{1}{2}e_j^2(k) \tag{3.6}$$

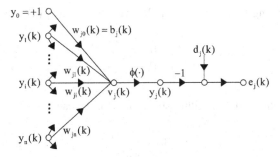

FIGURE 3.2   Signal flow graph highlighting the details of output neuron j.

By summing the error energy contributions of all the neurons in the output layer, we express the total instantaneous error energy of the whole network as the following form:

$$E(k) = \sum_j E_j(k) = \frac{1}{2} \sum_j e_j^2(k) \tag{3.7}$$

Let

$$\ell = \{x(k), d(k)\}_{k=1}^K \tag{3.8}$$

be the training sample used to train the network. Basing on the total instantaneous error energy given in Eq. 3.7, we define the error energy averaged over the training sample, or the empirical risk as

$$\bar{E}(k) = \frac{1}{K} \sum_{k=1}^K E(k) = \frac{1}{2K} \sum_{k=1}^K \sum_j e_j^2(k) \tag{3.9}$$

where K denotes the number of examples that the training sample consists.

The backpropagation algorithm applied a correction $\Delta w_{ji}(k)$ to the synaptic weight $w_{ji}(k)$. Therefore, it is important to compute the partial derivative $\partial E(k)/\partial w_{ji}(k)$. According to the chain rule of calculus, we can express the gradient as follows:

$$\frac{\partial E(k)}{\partial w_{ji}(k)} = \frac{\partial E(k)}{\partial e_j(k)} \frac{\partial e_j(k)}{\partial y_j(k)} \frac{\partial y_j(k)}{\partial v_j(k)} \frac{\partial v_j(k)}{\partial w_{ji}(k)} \tag{3.10}$$

First, considering Eq. 3.6, we can find that

$$\frac{\partial E(k)}{\partial e_j(k)} = e_j(k) \tag{3.11}$$

Then, differentiating both sides of Eq. 3.5 with respect to $y_j(k)$, we obtain

$$\frac{\partial e_j(k)}{\partial y_j(k)} = -1 \tag{3.12}$$

Next, differentiating Eq. 3.4 with respect to $v_j(k)$, we can get

$$\frac{\partial y_j(k)}{\partial v_j(k)} = \phi_j'(v_j(k)) \tag{3.13}$$

At last, from Eq. 3.3, we derive that

$$\frac{\partial v_j(k)}{\partial w_{ji}(k)} = y_i(k) \tag{3.14}$$

Substituting Eqs. 3.11–3.14 to Eq. 3.10 yields

$$\frac{\partial E(k)}{\partial w_{ji}(k)} = -e_j(k)\phi_j'(v_j(k))y_i(k) \tag{3.15}$$

The correction $\Delta w_{ji}(k)$ applied to $w_{ji}(k)$ is defined by the delta rule, that is,

$$\Delta w_{ji}(k) = -\alpha \frac{\partial E(k)}{\partial w_{ji}(k)} \tag{3.16}$$

where α is the learning rate parameter of the backpropagation algorithm. Combining Eqs. 3.15 and 3.16, we have

$$\Delta w_{ji}(k) = \alpha \delta_j(k) y_i(k) \tag{3.17}$$

where the local gradient $\delta_j(k)$ is defined by

$$\begin{aligned}
\delta_j(k) &= -\frac{\partial E(k)}{\partial v_j(k)} \\
&= -\frac{\partial E(k)}{\partial e_j(k)} \frac{\partial e_j(k)}{\partial y_j(k)} \frac{\partial y_j(k)}{\partial v_j(k)} \\
&= e_j(k)\phi_j'(v_j(k))
\end{aligned} \tag{3.18}$$

From Eq. 3.18, we observe that the local gradient $\delta_j(k)$ for output neuron j is equal to the product of the corresponding error signal $e_j(k)$ for the neuron and the derivative $\phi_j'(v_j(k))$ of the associated activation function. We have two cases to be considered to compute it according to the location of neuron j, as shown in the following [Haykin, 2009]:

Case 1 Neuron j is an output node. When neuron j is located in the output layer of the network, it is supplied with a desired response of its own. We can use Eq. 3.5 to compute the error signal $e_j(k)$ associated with this neuron. After that, we may compute the local gradient $\delta_j(k)$ straightforwardly by using Eq. 3.18.

Case 2 Neuron j is a hidden node. When neuron j is located in a hidden layer of the network, there is no specified desired response for that neuron. Then, the backpropagation algorithm becomes complicated. The error signal for a hidden neuron would have to be determined recursively and working backward in terms of the error signals of all the neurons to which that hidden neuron is directly connected.

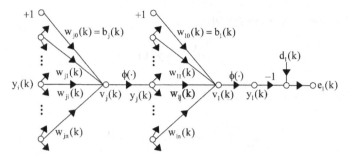

FIGURE 3.3   Signal flow graph highlighting the details of output neuron l connected to the hidden neuron j.

See Figure 3.3, where neuron j is a hidden node of the network. The error signal at the output of neuron l at iteration k is defined by

$$
\begin{aligned}
e_l(k) &= d_l(k) - y_l(k) \\
&= d_l(k) - \phi_l(v_l(k))
\end{aligned}
\tag{3.19}
$$

where

$$
v_l(k) = \sum_{j=0}^{n} w_{lj}(k) y_j(k)
\tag{3.20}
$$

Notice in Eq. 3.20, the synaptic weight $w_{l0}(k)$ is equal to the bias $b_l(k)$ applied to neuron l, while the corresponding input is $+1$. Here, the instantaneous sum of squared errors of the network is

$$
E(k) = \frac{1}{2} \sum_l e_l^2(k)
\tag{3.21}
$$

Then, we can obtain

$$
\begin{aligned}
\frac{\partial E(k)}{\partial y_j(k)} &= \sum_l e_l \frac{\partial e_l(k)}{\partial y_j(k)} \\
&= \sum_l e_l \frac{\partial e_l(k)}{\partial v_l(k)} \frac{\partial v_l(k)}{\partial y_j(k)}
\end{aligned}
\tag{3.22}
$$

From Eq. 3.19, we have

$$
\frac{\partial e_l(k)}{\partial v_l(k)} = -\phi_l'(v_l(k))
\tag{3.23}
$$

Besides, according to Eq. 3.20, we can find that

$$\frac{\partial v_l(k)}{\partial y_j(k)} = w_{lj}(k) \tag{3.24}$$

Substituting Eqs. 3.23 and 3.24 to Eq. 3.22, we derive the partial derivative:

$$\begin{aligned}
\frac{\partial E(k)}{\partial y_j(k)} &= -\sum_l e_l \phi_l'(v_l(k)) w_{lj}(k) \\
&= -\sum_l \delta_l(k) w_{lj}(k)
\end{aligned} \tag{3.25}$$

where $\delta_l(k) = e_l \phi_l'(v_l(k))$. Therefore, considering Eqs. 3.13 and 3.18, we can further obtain

$$\begin{aligned}
\delta_j(k) &= -\frac{\partial E(k)}{\partial v_j(k)} \\
&= -\frac{\partial E(k)}{\partial y_j(k)} \frac{\partial y_j(k)}{\partial v_j(k)} \\
&= \phi_j'(v_j(k)) \sum_l \delta_l(k) w_{lj}(k)
\end{aligned} \tag{3.26}$$

Now, we summarize the relations that we have derived for the backpropagation algorithm. According to Eq. 3.17, the correction $\Delta w_{lj}(k)$ applied to the synaptic weight connecting neuron i to neuron j is given by the following delta rule:

$$\Delta w_{ji}(k) = \begin{cases} \alpha e_j(k) \phi_j'(v_j(k)) y_i(k), & \text{if neuron j is an output node} \\ \alpha \phi_j'(v_j(k)) \sum_l \big(e_l \phi_j'(v_l(k)) w_{lj}(k) \big) y_i(k), & \text{if neuron j is a hidden node} \end{cases} \tag{3.27}$$

From Eq. 3.27, we can clearly find that the computation of the correction $\Delta w_{ji}(k)$ requires knowledge of the derivative of the activation function $\phi(\cdot)$. Evidently, we require the function $f(\cdot)$ to be continuous to ensure its derivative exists. Actually, differentiability is the only requirement that an activation function must satisfy. The logistic function and hyperbolic tangent function are two representative instances.

1. *Logistic function.* According to the definition of logistic function given in (2.60), we can express the induced local field of neuron j as

$$\phi_j(v_j(k)) = \frac{1}{1 + \exp(-a v_j(k))} \tag{3.28}$$

where $a > 0$ is an adjustable parameter. By using Eq. 3.4, we can write the derivative $\phi_j'(v_j(k))$ as follows:

$$
\begin{aligned}
\phi_j'(v_j(k)) &= \frac{a \exp(-av_j(k))}{(1 + \exp(-av_j(k)))^2} \\
&= ay_j(k)(1 - y_j(k))
\end{aligned}
\tag{3.29}
$$

Let $o_j(k)$ be the jth element of the output vector of the multilayer perceptron. When the neuron j is located in the output layer, we have

$$
y_j(k) = o_j(k)
\tag{3.30}
$$

Then, according to Eqs. 3.18 and 3.29, we obtain the local gradient for neuron j as

$$
\begin{aligned}
\delta_j(k) &= e_j(k)\phi_j'(v_j(k)) \\
&= a(d_j(k) - o_j(k))o_j(k)(1 - o_j(k))
\end{aligned}
\tag{3.31}
$$

When the neuron j is located in the hidden layer, from Eqs. 3.26 and 3.29, the local gradient can be expressed as

$$
\begin{aligned}
\delta_j(k) &= \phi_j'(v_j(k)) \sum_l \delta_l(k)w_{lj}(k) \\
&= ay_j(k)(1 - y_j(k)) \sum_l \delta_l(k)w_{lj}(k)
\end{aligned}
\tag{3.32}
$$

2. *Hyperbolic tangent function.* Another commonly used sigmoidal nonlinearity is the hyperbolic tangent function, which in general, can be defined as

$$
\phi_j(v_j(k)) = a \tan h(bv_j(k))
\tag{3.33}
$$

where a and b are positive constants. Considering Eq. 3.4, we obtain the derivative of the hyperbolic tangent function as

$$
\begin{aligned}
\phi_j'(v_j(k)) &= ab \operatorname{sech}^2(bv_j(k)) \\
&= ab(1 - \tan h^2(bv_j(k))) \\
&= \frac{b}{a}(a - y_j(k))(a + y_j(k))
\end{aligned}
\tag{3.34}
$$

Then, using Eqs. 3.18 and 3.34, we derive the local gradient as

$$
\begin{aligned}
\delta_j(k) &= e_j(k)\phi_j'(v_j(k)) \\
&= \frac{b}{a}(d_j(k) - o_j(k))(a - o_j(k))(a + o_j(k))
\end{aligned}
\tag{3.35}
$$

when the neuron j is an output node. Similarly, from Eqs. 3.26 and 3.34, the local gradient is

$$\delta_j(k) = \phi_j'(v_j(k)) \sum_l \delta_l(k)w_{lj}(k)$$
$$= \frac{b}{a}(a - y_j(k))(a + y_j(k)) \sum_l \delta_l(k)w_{lj}(k) \tag{3.36}$$

when the neuron j is a hidden node.

Using Eqs. 3.31 and 3.32 for the logistic function and Eqs. 3.35 and 3.36 for the hyperbolic tangent function, we can compute the local gradient $\delta_j(k)$, and then get the correction $\Delta w_{ji}(k)$ applied to $w_{ji}(k)$.

3.2.2   The Strategy for Improving the Algorithm

In the backpropagation algorithm, the smaller we set the learning rate parameter, the smaller the changes to the synaptic weights in the network will be from one iteration to the next, and the smoother will be the trajectory in the weight space. However, this result is attained at the cost of a slower rate of learning. On the other hand, if we set the learning rate parameter too large in order to speed up the rate of learning, the corresponding large changes in the synaptic weights may assume such a form that the network becomes unstable.

For the purpose of increasing the rate of network learning while avoiding the appearance of instability, a momentum term may be added to the delta rule (Eq. 3.17). This results in a generalized delta rule formulated as

$$\Delta w_{ji}(k) = \beta \Delta w_{ji}(k-1) + \alpha \delta_j(k)y_i(k) \tag{3.37}$$

where the parameter β is usually positive, called the momentum constant [Haykin, 2009]. This represents a minor modification to the weight update in the back-propagation algorithm.

Next, in order to observe the effect of the sequence of pattern presentations on the synaptic weights due to the momentum constant, we rewrite Eq. 3.37 as a time series with index t. Then, $\Delta w_{ji}(k)$ can be further denoted as a sum of $\delta_j(t)y_i(t)$ with t growing from 0 to k, that is,

$$\Delta w_{ji}(k) = \alpha \sum_{t=0}^{k} \beta^{k-t} \delta_j(t)y_i(t) \tag{3.38}$$

Obviously, the momentum constant must satisfy $0 \leq |\beta| < 1$ to ensure the time series to be convergent. The case $\beta = 0$ reveals that the backpropagation algorithm operates without momentum, namely, Eq. 3.37 becomes Eq. 3.17. Moreover, by making a comparison between Eq. 3.16 and Eq. 3.17, we can obtain

$$\Delta w_{ji}(k) = -\alpha \sum_{t=0}^{k} \beta^{k-t} \frac{\partial E(t)}{\partial w_{ji}(t)} \qquad (3.39)$$

which shows that the current adjustment $\Delta w_{ji}(k)$ represents the sum of an exponentially weighted time series.

The inclusion of a momentum term has a great impact on the backpropagation algorithm in terms of finding a proper equilibrium between the learning speed and stability during the training process of network. When the partial derivative $\partial E(t)/\partial w_{ji}(t)$ has the same algebraic sign on consecutive iterations, the exponentially weighted sum $\Delta w_{ji}(k)$ grows in magnitude, and therefore the weight $w_{ji}(k)$ is adjusted by a large amount. In this case, the inclusion of momentum in backpropagation algorithm tends to accelerate adjustment in steady directions. Conversely, when the partial derivative $\partial E(t)/\partial w_{ji}(t)$ has opposite sign on consecutive iterations, the exponentially weighted sum $\Delta w_{ji}(k)$ shrinks in magnitude, and therefore the weight $w_{ji}(k)$ is adjusted by a small amount. Here, the inclusion of momentum in backpropagation algorithm has a stabilizing effect.

3.2.3 The Design Procedure of the Algorithm

For implementing the backpropagation algorithm online, the sequential updating approach of network weights is performed here [Haykin, 1999]. The design procedure of backpropagation algorithm via the training sample $\{x(k), d(k)\}_{k=1}^{K}$ is described as follows. Notice $x(k)$ is the input vector applied to the input layer and $d(k)$ is the desired response vector presented to the output layer.

1. *Initialization.* Start with a reasonable network configuration. Set the synaptic weights and threshold levels of the network to small random numbers that are uniformly distributed.

2. *Presentation of training samples.* Present the network with an epoch of training examples. For each example in the sample, perform the forward and backward computations, as described in steps 3 and 4.

3. *Forward computation.* For a training example denoted by $(x(k), d(k))$, compute the induced local fields and function signals of the network by proceeding forward through the network, layer-by-layer. The induced local field $v_j^{(h)}$ for neuron j in layer h is

$$v_j^{(h)}(k) = \sum_{i=0}^{n} w_{ji}^{(h)}(k) y_i^{(h-1)}(k) \qquad (3.40)$$

where $y_i^{(h-1)}(k)$ is the output signal of neuron i at iteration k, and $w_{ji}^{(h)}(k)$ is the synaptic weight of neuron j in layer h that is fed from neuron i in layer $h - 1$.

For $i = 0$, we have $y_0^{(h-1)}(k) = +1$ and $w_{j0}^{(h)}(k) = b_j^{(h)}(k)$, where $b_j^{(h)}(k)$ is the bias applied to neuron j in layer h. Then, the output signal of neuron j in layer h is

$$y_j^{(h)}(k) = \phi_j(v_j^{(h)}(k)) \tag{3.41}$$

If neuron j is in the first hidden layer (i.e., $h = 1$), set

$$y_j^{(0)}(k) = x_j(k) \tag{3.42}$$

If neuron j is in the output layer (i.e., $h = H$, where H is referred to as the depth of the network), set

$$y_j^{(H)}(k) = o_j(k) \tag{3.43}$$

Then, the error signal can be obtained by

$$e_j^{(H)}(k) = d_j(k) - o_j(k) \tag{3.44}$$

4. *Backward computation.* Compute the local gradients of the network according to

$$\delta_j^{(h)}(k) = \begin{cases} e_j^{(H)}(k)\phi_j'(v_j^{(H)}(k)), & \text{if neuron j is located in output layer H} \\ \phi_j'(v_j^{(h)}(k))\sum_l \delta_l^{(h+1)}(k)w_{lj}^{(h+1)}(k), & \text{if neuron j is located in hidden layer h} \end{cases} \tag{3.45}$$

Here, $\phi_j'(v_j^{(h)}(k))$ denotes the differentiation with respect to the argument. Update the synaptic weights of the network in layer h in accordance with the generalized delta rule:

$$w_{ji}^{(h)}(k+1) = w_{ji}^{(h)}(k) + \beta\left(w_{ji}^{(h)}(k) - w_{ji}^{(h)}(k-1)\right) + \alpha\delta_j^{(h)}(k)y_i^{(h-1)}(k) \tag{3.45}$$

where α is the learning rate parameter and β is the momentum constant.

5. *Iteration.* Let $k = k + 1$. Iterate the forward and backward computations in steps 3 and 4 by presenting new epochs of training examples to the network until $\overline{E}(k)$ satisfies the prespecified requirement. The order of presentation of training examples should be randomized from epoch-to-epoch.

3.3 BATCH LEARNING AND ONLINE LEARNING

Now, we present two different learning methods, batch learning and online learning, on the basis of how the supervised learning of the multilayer perceptron is actually performed.

3.3.1   Batch Learning

In batch learning, adjustments to the synaptic weights of the multilayer perceptron are performed after presenting all the K examples in the training sample ℓ that constitute one epoch of training. That is to say, the cost function for batch learning is defined by the average error energy $\overline{E}$. Adjustments to the synaptic weights of the multilayer perceptron are made on an epoch-by-epoch basis. Then, one realization of the learning curve is obtained by plotting $\overline{E}$ versus the number of epochs. Note that for each epoch of training, the examples in the training sample ℓ are randomly shuffled. Therefore, the learning curve is computed by ensemble averaging a large enough number of such realizations, where each realization is performed for a different set of initial conditions chosen at random.

The advantages of batch learning are as follows when the gradient descent method is used to implement the training process:

- It can give an accurate estimation of the gradient vector, that is, the derivative of the cost function $\overline{E}$ with respect to the weight vector w, thereby guaranteeing, under simple conditions, convergence of the steepest descent method to a local minimum.
- It ensures the parallelization of the learning process.

Nevertheless, from a practical perspective, batch learning is rather demanding in terms of storage requirements.

Besides, in a statistical context, batch learning may be viewed as a form of statistical inference. Therefore, it is well suited for solving nonlinear regression problems.

3.3.2   Online Learning

In online learning, adjustments to the synaptic weights of the multilayer perceptron are performed on the example-by-example basis. Thus, the cost function to be minimized is the total instantaneous error energy E(k).

Consider an epoch of K training examples arranged in the order $\{x(1), d(1)\}$, $\{x(2), d(2)\}, \ldots, \{x(K), d(K)\}$. The first example pair $\{x(1), d(1)\}$ in the epoch is presented to the network, and the weight adjustments are performed using the gradient descent method. Then, the second example $\{x(2), d(2)\}$ in the epoch is presented to the network, which leads to further adjustments to weights in the network. This procedure is continued until the last example $\{x(K), d(K)\}$ is considered. Unfortunately, such a procedure works against the parallelization of online learning.

For a given set of initial conditions, a single realization of the learning curve is obtained by plotting the final value E(k) versus the number of epochs used in the training session. The training examples are randomly shuffled after each epoch. As with batch learning, the learning curve for online learning is computed by ensemble averaging such realizations over a large enough number of initial conditions chosen at random. Naturally, for a given network structure, the learning curve obtained under online learning will be quite different from that under batch learning.

Given that the training examples are presented to the network in a random manner, the use of online learning makes the search in the multidimensional weight space stochastic in nature. Note that it is for this reason that the method of online learning is sometimes referred to as a stochastic method. This stochasticity has the desirable effect of making it less likely for the learning process to be trapped in a local minimum, which is a definite advantage of online learning over batch learning. Another advantage of online learning is the fact that it requires much less storage than batch learning.

Moreover, when the training data are redundant (i.e., the training sample ℓ contains several copies of the same example), we find that, unlike batch learning, online learning is able to take advantage of this redundancy because the examples are presented one at a time.

Another useful property of online learning is its ability to track small changes in the training data, particularly when the environment responsible for generating the data is nonstationary.

To summarize, despite the disadvantage of online learning, it is highly popular for solving pattern classification problems due to the following two important practical reasons [Haykin, 2009]:

- Online learning is simple to implement.
- It provides effective solutions to large-scale and difficult pattern classification problems.

Accordingly, we can find that much of the material presented in the chapter is devoted to online learning.

3.4 CROSS-VALIDATION AND GENERALIZATION

3.4.1 Cross-Validation

From the above sections, we know that the essence of backpropagation learning is to encode an input–output mapping (represented by a set of labeled examples) into the synaptic weights and thresholds of a multilayer perceptron. It is hoped that the network becomes well trained so that it learns enough about the past to generalize to the future. From such a perspective, the learning process amounts to a choice of network parameterization for a given set of data. In other words, we may view the network selection problem as choosing, within a set of candidate model structures (parameterizations), the "best" one according to a certain criterion. Here, cross-validation, which is a standard tool in statistics, provides an appealing guiding principle.

First, the available data set is randomly partitioned into a training sample and a test set. The training sample is further partitioned into two disjoint subsets:

- an estimation subset, used to select the model;
- a validation subset, used to test or validate the model.

The motivation is to validate the model on a data set different from the one used for parameter estimation. In this way, we may use the training sample to assess the performance of various candidate models and then choose the "best" one. However, there is a distinct possibility that the model with the best-performing parameter values may end up overfitting the validation subset. For the purpose of guarding against this possibility, the generalization performance of the selected model is measured on the test set, which is different from the validation subset [Haykin, 2009].

Cross-validation is appealing particularly when we have to design a large neural network with good generalization as the goal in different ways.

1. *Network complexity.* The problem of choosing network complexity measured in terms of the number of hidden neurons used in a multilayer perceptron can be interpreted as that of choosing the size of the parameter set used to model the data set. Measured in terms of the ability of the network to generalize, there is obviously a limit on the size of the network. This follows from the basic observation that it may not be an optimal strategy to train the network to perfection on a given data set, because of the ill-posedness of any finite set of data representing a target function, a condition that is true for both "noisy" and "clean" data. Rather, it would be better to train the network in order to produce the "best" generalization. To do so, we may use cross-validation. Specifically, the training data set is partitioned into training and test subsets, in which case "overtraining" will show up as poorer performance on the cross-validation set.

2. *Size of training set.* Another direction in which cross-validation can be used is to decide when the training of a network on the training set should be actually stopped. In this case, the error performance of the network on generalization is exploited to determine the size of the data set used in training. The idea of cross-validation used here is illustrated in Figure 3.4, where two curves are shown for the mean-squared error in generalization, plotted versus the number of epochs used in training. In Figure 3.4, one curve relates to the use of few adjustable parameters (i.e., underfitting), while the other relates to the use of many parameters (i.e., overfitting). In addition, we can find that the error performance on generalization exhibits a minimum point, and the minimum mean-squared error for overfitting is smaller and better defined than that for underfitting. Therefore, we may obtain good generalization even if the neural network designed has too many parameters, provided that training of the network on the training set is stopped at a number of epochs corresponding to the minimum point of the error performance curve on cross-validation.

3. *Size of learning rate parameter.* Cross-validation may also be employed to adjust the size of the learning rate parameter of a multilayer perceptron, with backpropagation learning used as a pattern classifier. In particular, the network is first trained on the subtraining set, and then the cross-validation set is used to

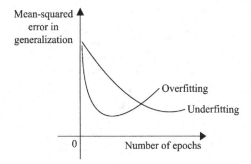

FIGURE 3.4   Illustrating the idea of cross-validation.

validate the training after each epoch. When the classification performance of the network on the cross-validation set fails to improve by a certain amount, the size of the learning rate parameter is reduced. After each succeeding epoch, the learning rate parameter is further reduced, until once again there is no further improvement in classification performance on the cross-validation set. The training of the network is halted as soon as that point is reached.

3.4.2 Generalization

During the process of backpropagation learning, we typically start with a training sample and use the backpropagation algorithm to compute the synaptic weights of a multilayer perceptron by loading (encoding) as many of the training examples as possible into the network. It is hoped that the designed neural network will generalize well. A network is said to generalize well when the input–output mapping computed by the network is correct (or nearly so) for test data never used in creating or training the network. Note that the term "generalization" is borrowed from psychology. Here, it is assumed that the test data are drawn from the same population used to generate the training data [Haykin, 2009].

Training a neural network may be considered as a "curve fitting" problem. Besides, the network itself may be viewed simply as a nonlinear input–output mapping. In this sense, we can regard generalization as the effect of a good nonlinear interpolation of the input data. The network performs useful interpolation primarily because multilayer perceptrons with continuous activation functions lead to output functions that are also continuous.

Figure 3.5 describes how generalization may occur in a hypothetical network. The nonlinear input–output mapping represented by the curve depicted in the figure is computed by the network as a result of learning the points labeled as "training data." The generalization point is seen as the result of interpolation performed by the network.

A neural network that is designed to generalize well will produce a correct input–output mapping even when the input is slightly different from the examples used to

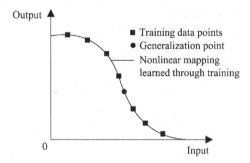

FIGURE 3.5   Properly fitted nonlinear mapping with good generalization.

train the network. However, if a neural network learns too many input–output examples, the network may end up memorizing the training data. It may do so by finding a feature that is present in the training data, but not true of the underlying function that is to be modeled. Such a phenomenon is referred to as overfitting or overtraining. When the network is overtrained, it loses the ability to generalize between similar input–output patterns.

Ordinarily, loading data into a multilayer perceptron in this way requires the use of more hidden neurons than are actually necessary, with the result that undesired contributions to the input space due to noise are stored in synaptic weights of the network. For the same data as depicted in Figure 3.5, an example of how poor generalization due to memorization in a neural network may occur is illustrated in Figure 3.6. "Memorization" is essentially a "lookup table," which implies that input–output mapping computed by the neural network is not smooth. In input–output mapping, on the contrary, it is important to select the "simplest" function in the absence of any prior knowledge [Poggio and Girosi, 1990a, 1990b]. The simplest function signifies the smoothest function that approximates the mapping for a given error criterion, because such a choice generally demands the fewest computational resources. In addition, smoothness is also natural in many applications, depending on

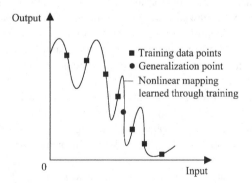

FIGURE 3.6   Overfitted nonlinear mapping with poor generalization.

the scale of the phenomenon being studied. Therefore, it is meaningful to seek a smooth nonlinear mapping for ill-posed input–output relationships, so that the network is able to classify novel patterns correctly with respect to the training patterns.

Overall, generalization is influenced by the following three factors:

- the size of the training sample and how representative the training sample is of the environment of interest;
- the architecture of the neural network; and
- the physical complexity of the problem at hand.

Clearly, we have no control over the third factor. Thus, we can view the issue of generalization from the following two perspectives [Haykin, 2009]:

- When the architecture of the network is fixed, the issue to be considered is that of determining the size of the training sample needed for a good generalization to occur.
- When the size of the training sample is fixed, the issue of interest is that of determining the best architecture of network for achieving good generalization.

In practice, it seems that all we really need for a good generalization is to have the size of the training sample K satisfy the condition

$$K = O\left(\frac{W}{\varepsilon}\right) \tag{3.47}$$

where W is the total number of free parameters in the network, including synaptic weights and biases, ε denotes the fraction of classification errors permitted on test data, and $O(\cdot)$ denotes the order of quantity enclosed within.

3.4.3 Convolutional Neural Networks

In this chapter, we know that the basic idea of backpropagation is that gradients can be computed efficiently by propagation from the output to the input. Needless to say, backpropagation is by far the most widely used neural network learning algorithm, and probably the most widely used learning algorithm of any form. In this part, we provide an extended introduction of the convolutional neural networks [LeCun et al., 1998].

The ability of multilayer networks trained with gradient descent to learn complex, high-dimensional, nonlinear mappings from large collections of examples makes them obvious candidates for image recognition tasks. In the traditional model of pattern recognition, a hand-designed feature extractor gathers relevant information from the input and eliminates irrelevant variabilities. A trainable classifier then categorizes the resulting feature vectors into classes. In this scheme, standard, fully connected multilayer networks can be used as classifiers. A potentially more interesting scheme is to rely as much as possible on learning in the feature extractor itself. In the case of character recognition, a network could be fed with almost raw inputs (e.g., size-normalized images). While this can be done with an ordinary fully connected feedforward network with some success for tasks such as character recognition, there are problems shown as follows.

For one thing, typical images are large, often with several hundred variables (pixels). A fully connected first layer with, for example, 100 hidden units in the first layer would already contain several tens of thousands of weights. Such a large number of parameters increase the capacity of the system and therefore it requires a larger training set. In addition, the memory requirement to store so many weights may rule out certain hardware implementations. But the main deficiency of unstructured nets for image or speech applications is that they have no built-in invariance with respect to translations or local distortions of the inputs. Before being sent to the fixed-size input layer of a neural network, character images, or other two- or one-dimensional signals, must be approximately size normalized and centered in the input field. Unfortunately, no such preprocessing can be perfect: Handwriting is often normalized at the word level, which can cause size, slant, and position variations for individual characters. This, combined with variability in writing style, will cause variations in the position of distinctive features in input objects. In principle, a fully connected network of sufficient size could learn to produce outputs that are invariant with respect to such variations. However, learning such a task would probably result in multiple units with similar weight patterns positioned at various locations in the input so as to detect distinctive features wherever they appear on the input. Learning these weight configurations requires a very large number of training instances to cover the space of possible variations. In convolutional neural networks, as described below, shift invariance is automatically obtained by forcing the replication of weight configurations across space.

For another thing, a deficiency of fully connected architectures is that the topology of the input is entirely ignored. The input variables can be presented in any (fixed) order without affecting the outcome of the training. On the contrary, images (or time–frequency representations of speech) have a strong two-dimensional local structure: Variables (or pixels) that are spatially or temporally nearby are highly correlated. Local correlations are the reasons for the well-known advantages of extracting and combining local features before recognizing spatial or temporal objects, because configurations of neighboring variables can be classified into a small number of categories (e.g., edges, corners). Convolutional neural networks force the extraction of local features by restricting the receptive fields of hidden units to be local.

Convolutional neural networks provide an efficient method to constrain the complexity of feedforward neural networks by weight sharing and restriction to local connections. Convolutional networks combine three architectural ideas to ensure some degree of shift, scale, and distortion invariance: (i) local receptive fields; (ii) shared weights (or weight replication); and (iii) spatial or temporal subsampling. The kernel of the convolution is the set of connection weights used by the units in the feature map. An interesting property of convolutional layers is that if the input image is shifted, the feature map output will be shifted by the same amount, but it will be left unchanged otherwise. This property is at the basis of the robustness of convolutional networks to shifts and distortions of the input. Besides, since all the weights are learned with backpropagation, convolutional networks can be seen as synthesizing their own feature extractor. The weight sharing technique has the interesting side effect of reducing the number of free parameters, thereby reducing the "capacity" of the machine and reducing the gap between the test error and training error.

The field of deep machine learning focuses on computational models for information representation that exhibit similar characteristics to that of the neo-cortex. Actually, the structure of convolutional neural networks is well established in the current deep learning field and shows great promise for future work. Convolutional neural networks are the first truly successful deep learning approach where many layers of a hierarchy are successfully trained in a robust manner. A convolutional neural network is a choice of topology or architecture that leverages spatial relationships to reduce the number of parameters that must be learned and thus improves upon general feedforward backpropagation training. Convolutional neural networks were proposed as a deep learning framework that was motivated by minimal data preprocessing requirements. In convolutional neural networks, small portions of the image (dubbed a local receptive field) are treated as inputs to the lowest layer of the hierarchical structure. Information generally propagates through the different layers of the network whereby at each layer digital filtering is applied in order to obtain salient features of the data observed. The method provides a level of invariance to shift, scale, and rotation as the local receptive field allows the neuron or processing unit access to elementary features such as oriented edges or corners. The advancements made with respect to developing deep machine learning systems will undoubtedly shape the future of machine learning and artificial intelligence systems in general.

3.5 COMPUTER EXPERIMENT USING BACKPROPAGATION

In this section, we reconsider the exclusive-OR (XOR) problem [Haykin, 2009]. From Chapter 2, we know that the Rosenblatt's single-layer perceptron has no hidden neuron. As a result, it cannot classify input patterns that are not linearly separable. A typical example is the XOR problem.

The XOR problem can be viewed as a special case of the problem of classifying points in the unit hypercube. Each point in the hypercube is in either class 0 or class 1. In a more special case of the XOR problem, we need to consider only the four corners of a unit square that correspond to the input patterns $(0, 0)$, $(0, 1)$, $(1, 0)$, and $(1, 1)$.

Denote $\oplus$ as the exclusive-OR Boolean function operator. Since

$$0 \oplus 0 = 0 \tag{3.48}$$

$$1 \oplus 1 = 0 \tag{3.49}$$

TABLE 3.1   **Pattern Classification**

x_1	x_2	y
0	0	0
0	1	1
1	0	1
1	1	0

and

$$0 \oplus 1 = 1 \tag{3.50}$$

$$1 \oplus 0 = 1 \tag{3.51}$$

the input patterns $(0, 0)$ and $(1, 1)$ are in class 0, while $(0, 1)$ and $(1, 0)$ are in class 1. See Table 3.1. Obviously, the input patterns $(0, 0)$ and $(1, 1)$ are at opposite corners of the unit square, but they produce the identical output. This is the same for the input patterns $(0, 1)$ and $(1, 0)$.

We know that the use of a single neuron with two inputs results in a straight line for a decision boundary in the input space. For all points on one side of this line, the neuron outputs 1, while for all points on the other side of the line, it outputs 0. The position and orientation of the line in the input space are determined by the synaptic weights of the neuron connected to the input nodes and the bias applied to the neuron. As the input patterns $(0, 0)$ and $(1, 1)$ are located on opposite corners of the unit square, and likewise for the other two input patterns $(0, 1)$ and $(1, 0)$, we cannot construct a straight line for a decision boundary so that $(0, 0)$ and $(1, 1)$ lie in one decision region and $(0, 1)$ and $(1, 0)$ lie in the other decision region. This implies that the single-layer perceptron cannot solve the XOR problem.

However, we can solve the XOR problem by using a network with a single hidden layer with two neurons, as plotted in Figure 3.7. The corresponding signal flow graph of the network is shown in Figure 3.8. Notice that the following two assumptions are required.

- Each neuron is represented by a McCulloch–Pitts model, which utilizes the threshold function as activation function.

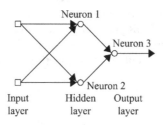

FIGURE 3.7   Architectural graph of network for solving the XOR problem.

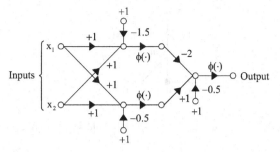

FIGURE 3.8 Signal flow graph of network for solving the XOR problem.

- Bits 0 and 1 are represented by the levels 0 and +1, respectively.

In Figure 3.7, the top neuron, labeled as "Neuron 1" in the hidden layer, is characterized as

$$w_{11} = w_{12} = +1$$

$$b_1 = -\frac{3}{2}$$

The slope of the decision boundary constructed by this hidden neuron is equal to -1 and positioned as in Figure 3.9. The bottom neuron, labeled as "Neuron 2" in the hidden layer, is characterized as

$$w_{21} = w_{22} = +1$$

$$b_2 = -\frac{1}{2}$$

The orientation and position of the decision boundary constructed by this hidden neuron are shown in Figure 3.10. The output neuron, labeled as "Neuron 3," is characterized as

$$w_{31} = -2$$
$$w_{32} = +1$$
$$b_3 = -\frac{1}{2}$$

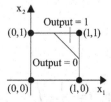

FIGURE 3.9 Decision boundary constructed by hidden neuron 1 of the network in Figure 3.7.

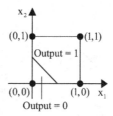

FIGURE 3.10 Decision boundary constructed by hidden neuron 2 of the network in Figure 3.7.

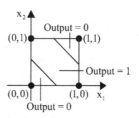

FIGURE 3.11 Decision boundaries constructed by the complete network in Figure 3.7.

The function of the output neuron is to construct a linear combination of the decision boundaries formed by the two hidden neurons. The computation result is shown in Figure 3.11. See Figure 3.8, we find that the bottom hidden neuron has a positive connection to the output neuron, whereas the top hidden neuron has a negative connection to the output neuron. When both hidden neurons are off, which occurs when the input pattern is $(0, 0)$, the output neuron remains off. When both hidden neurons are on, which occurs when the input pattern is $(1, 1)$, the output neuron is switched off again because the inhibitory effect of the larger negative weight connected to the top hidden neuron overpowers the excitatory effect of the positive weight connected to the bottom hidden neuron. When the top hidden neuron is off and the bottom neuron is on, which occurs when the input patterns is $(0, 1)$ or $(1, 0)$, the output neuron is switched on because of the excitatory effect of the positive weight connected to the bottom hidden neuron. Accordingly, the network described in Figure 3.7 does indeed solve the XOR problem.

EXERCISES

3.1. Consider the functions

$$\phi(x) = \frac{1}{\sqrt{2\pi}} \int_{-\infty}^{x} e^{-(t^2/2)} dt$$

and

$$\phi(x) = \frac{2}{\pi}\tan^{-1}(x)$$

Explain why both the functions fit the requirement of a sigmoid function. What is the difference between the two functions?

3.2. Consider a two-layer network containing no hidden neurons. Assume that the network has q inputs and a single output neuron. Let x_i denote the ith input signal and define the corresponding output as

$$y = \phi\left(\sum_{i=0}^{q} w_i x_i\right)$$

where w_i is a threshold and

$$\phi(v) = \frac{1}{1 + \exp(-v)}$$

Show that this network implements a linear decision boundary that consists of a hyperplane in the input space $\mathfrak{R}^q$. Illustrate your conclusion when $q = 2$.

3.3. The network shown in Figure 3.12 has been trained to classify correctly a set of two-dimensional, two-class patterns. Identify the function performed by the classifier, assuming initially that the neurons have function

$$\phi(v) = \frac{1}{1 + \exp(-av)}$$

Draw the resulting separating lines between the two classes on the x_1, x_2 plane.

3.4. Figure 3.13 shows a neural network involving a single hidden neuron for solving the XOR problem. It can be viewed as an alternative to the network considered in Section 3.5. Show that the network of Figure 3.13 solves the XOR problem by constructing decision regions and a truth table for the network.

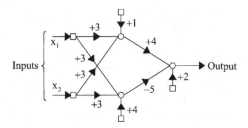

FIGURE 3.12 The network of Exercise 3.3.

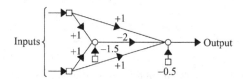

FIGURE 3.13   The network of Exercise 3.4.

3.5. Use the backpropagation algorithm for computing a set of synaptic weights and bias levels for a neural network structured as in Figure 3.7 to solve the XOR problem. Assume the use of a logistic function for the nonlinearity.

3.6. Consider the network shown in Figure 3.14, with the initial weight and basis are chosen as $w_1 = 1$, $b_1 = 1$, $w_2 = -2$, $b_2 = 1$.

The activation function $\phi(\cdot)$ is set the same as Exercise 3.2. Assume the input and the desired responses of the network are $x = 1$, $d = 1$, respectively.

1. Calculate the total instantaneous error energy E.

2. Compute $\partial E/\partial w_1$ based on the result of 1.

3. Recompute $\partial E/\partial w_1$ by using the backpropagation algorithm and compare the result with (2).

3.7. Linearly nonseparable patterns as shown in Figure 3.15 have to be classified into two categories by using a layered network. Construct the separating planes in the pattern space and draw patterns in the image space. Calculate all weights and

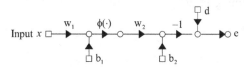

FIGURE 3.14   The network of Exercise 3.6.

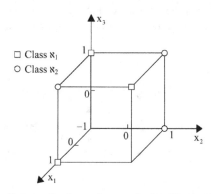

FIGURE 3.15   Patterns for layered network classification for Exercise 3.7.

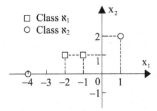

FIGURE 3.16 Pattern classification of Exercise 3.9.

threshold values of related units. Use the minimum number of threshold units to perform the classification.

3.8. Investigate the use of backpropagation learning algorithm employing a sigmoidal nonlinearity to achieve one-to-one mappings, as described below:

1. $f(x) = \dfrac{1}{x}$, $1 \le x \le 100$

2. $f(x) = \lg x$, $1 \le x \le 10$

3. $f(x) = \exp(-x)$, $1 \le x \le 10$,

4. $f(x) = \sin x$, $0 \le x \le \dfrac{\pi}{2}$

For each mapping, do the following things.

1. Set up two sets of data, one for network training and the other for testing.

2. Use the training data set to compute the synaptic weights of the network, assuming it has a single hidden layer.

3. Evaluate the computation accuracy of the network by using the test data.

Use a single hidden layer, but with a variable number of hidden neurons. Investigate how the network performance is affected by varying the size of the hidden layer.

3.9. Classify the two classes of input patterns depicted in Figure 3.16 by using backpropagation training algorithm.

CHAPTER 4

Radial-Basis Function Networks

As described in Chapter 3, the backpropagation algorithm for the design of a multilayer perceptron can be viewed as an application of an optimization method known in statistics as stochastic approximation. In this chapter, we adopt a different approach by viewing the design of a neural network as a curve-fitting (approximation) problem in a high-dimensional space. In this sense, learning is equivalent to finding a surface in a multidimensional space that provides a best fit to the training data. Correspondingly, generalization is equivalent to the use of this multidimensional surface to interpolate the test data. Such a viewpoint is indeed the motivation behind the method of radial-basis function(RBF) in the sense that it draws upon research work on traditional strict interpolation in a multidimensional space. In RBFs, the hidden units provide a set of "functions" that constitute an arbitrary "basis" for the input pattern (vectors) when they are expanded into the hidden unit space. With first introduced in solving the real multivariate interpolation problem, the radial-basis functions are now one of the main fields of research in numerical analysis. In this chapter, we focus on the radial-basis function network as an alternative to multilayer perceptrons. It will be interesting to find that in a multilayer perceptron, the function approximation is defined by a nested set of weighted summations, while in a RBF network, the approximation is defined by a single weighted sum. This can be regarded as the basic structural difference between the two networks.

4.1 RADIAL-BASIS FUNCTIONS

A radial-basis function is a real-valued function whose value depends only on the distance from the origin, so that $\phi(x) = \phi(\|x\|)$, or alternatively on the distance from some other point c, called a center, so that $\phi(x, c) = \phi(\|x - c\|)$. Some commonly used radial-basis functions are as follows:

Fundamentals of Computational Intelligence: Neural Networks, Fuzzy Systems, and Evolutionary Computation, First Edition. James M. Keller, Derong Liu, and David B. Fogel.
© 2016 by The Institute of Electrical and Electronics Engineers, Inc. Published 2016 by John Wiley & Sons, Inc.

1. Multiquadrics

$$\phi(r) = \left(r^2 + c^2\right)^{1/2} \tag{4.1}$$

where $c > 0$ and $r \in \Re$.

2. Inverse multiquadrics

$$\phi(r) = \frac{1}{\left(r^2 + c^2\right)^{1/2}} \tag{4.2}$$

where $c > 0$ and $r \in \Re$.

3. Gaussian functions

$$\phi(r) = \exp\left(-\frac{r^2}{2\sigma^2}\right) \tag{4.3}$$

where $\sigma > 0$ and $r \in \Re$.

4.2 THE INTERPOLATION PROBLEM

Consider a feedforward network with an input layer, a single hidden layer, and an output layer consisting of a single unit. We choose a single output unit to simplify the exposition without loss of generality. The network is designed to perform a nonlinear mapping from the input space to the hidden space, followed by a linear mapping from the hidden space to the output space. Let n_0 denote the dimension of the input space. Then, the network represents a map from the n_0-dimensional input space to the single-dimensional output space, which can be written as

$$\mathfrak{F} : \Re^{n_0} \to \Re^1 \tag{4.4}$$

Here, we can think of the map $\mathfrak{F}$ as a hypersurface (graph) $\Gamma \subset \Re^{n_0+1}$. For example, we think of the elementary map $\mathfrak{F} : \Re^1 \to \Re^1$, where $\mathfrak{F}(x) = x^2$, as a parabola drawn in $\Re^2$ space. The surface Γ is a multidimensional plot of the output as a function of the input. In a practical situation, the surface Γ is unknown and the training data are usually contaminated with noise. The training phase and generalization phase of the learning process, respectively, may be viewed as follows:

- The training phase constitutes the optimization of a fitting procedure for the surface Γ, based on known data points presented to the network in the form of input–output examples (patterns).
- The generalization phase is synonymous with interpolation between the data points, with the interpolation being performed along the constrained surface generated by the fitting procedure as the optimum approximation to the true surface Γ.

Now, we introduce the theory of the well-known multivariable interpolation in high-dimensional space. The interpolation problem can be stated in the following form:

Given a set of K different points $\{x^i \in \Re^{n_0} | i = 1, 2, \ldots, K\}$ and a corresponding set of K real numbers $\{d_i \in \Re^1 | i = 1, 2, \ldots, K\}$, find a function $F : \Re^{n_0} \to \Re^1$ that satisfies the interpolation condition:

$$F(x^i) = d_i, \quad i = 1, 2, \ldots, K \tag{4.5}$$

Note that the interpolating surface, that is, the function F, is constrained to pass through all the training data points. The $d_i (i = 1, 2, \ldots, K)$ is called the desired response scalar.

The technique of radial-basis functions consists of choosing a function F that has the form

$$F(x) = \sum_{i=1}^{K} w_i \phi(\|x - x^i\|) \tag{4.6}$$

where $\{\phi(\|x - x^i\|) | i = 1, 2, \ldots, K\}$ is a set of K arbitrary (generally nonlinear) functions, known as radial-basis functions, and $\|\cdot\|$ denotes a norm that is usually Euclidean.

By combining Eqs. 4.5 and 4.6, and denoting

$$\phi_{ji} = \phi(\|x^j - x^i\|), \quad i, j = 1, 2, \ldots, K \tag{4.7}$$

we obtain

$$
\begin{bmatrix}
\phi_{11} & \phi_{12} & \cdots & \phi_{1K} \\
\phi_{21} & \phi_{22} & \cdots & \phi_{2K} \\
\vdots & \vdots & \vdots & \vdots \\
\phi_{K1} & \phi_{K2} & \cdots & \phi_{KK}
\end{bmatrix}
\begin{bmatrix}
w_1 \\
w_2 \\
\vdots \\
w_K
\end{bmatrix}
=
\begin{bmatrix}
d_1 \\
d_2 \\
\vdots \\
d_K
\end{bmatrix}
\tag{4.8}
$$

Denote

$$
\phi =
\begin{bmatrix}
\phi_{11} & \phi_{12} & \cdots & \phi_{1K} \\
\phi_{21} & \phi_{22} & \cdots & \phi_{2K} \\
\vdots & \vdots & \vdots & \vdots \\
\phi_{K1} & \phi_{K2} & \cdots & \phi_{KK}
\end{bmatrix}
\tag{4.9}
$$

$$w = \begin{bmatrix} w_1 & w_2 & \cdots & w_K \end{bmatrix}^T$$

$$d = \begin{bmatrix} d_1 & d_2 & \cdots & d_K \end{bmatrix}^T$$

where ϕ, w, and d represent the $K \times K$ interpolation matrix, $K \times 1$ linear weight vector, and $K \times 1$ desired response vector, respectively. Then, (4.8) can be written in compact form as

$$\phi w = d \tag{4.10}$$

Assuming that ϕ is nonsingular, the inverse matrix ϕ^{-1} exists. Then, we can solve the weight vector w as

$$w = \phi^{-1}d \tag{4.11}$$

The vital question is how can we be sure that the interpolation matrix ϕ is nonsingular? It turns out that for a large class of radial-basis functions and under certain conditions, the answer to this basic question is given in the following Micchelli's theorem.

Theorem 4.1 [Micchelli, 1986]   Let $\{x^i\}_{i=1}^{K}$ be a set of distinct points in $\Re^{n_0}$. Then, the $K \times K$ interpolation matrix ϕ, whose ji-th element is $\phi_{ji} = \phi(\|x^j - x^i\|)$, is nonsingular.

In order to ensure that the radial-basis functions listed in Eqs. 4.1–4.3 are nonsingular, the input points $\{x^i\}_{i=1}^{K}$ must all be different. This is all that is required for nonsingularity of the interpolation matrix ϕ, whatever the values of size K of the data points or dimensionality n_0 of the vectors (points) x^i happen to be.

Actually, we can solve the linear weight vector in the light of the generalized inverse theory if the interpolation matrix is singular, that is,

$$w = \phi^{+1}d \tag{4.12}$$

Note that the inverse multiquadrics (4.2) and the Gaussian functions (4.3) share a common property. That is, they are both localized functions, in the sense that $\phi(r) \to 0$ as $r \to \infty$. For the two cases, the interpolation matrix ϕ is positive definite. However, the multiquadrics (4.1) are nonlocal, because $\phi(r)$ becomes unbounded as $r \to \infty$. In addition, the corresponding interpolation matrix ϕ has $K - 1$ negative eigenvalues and only one positive eigenvalue, with the result that it is not positive definite [Micchelli, 1986]. Anyway, an interpolation matrix ϕ based on multiquadrics is nonsingular and therefore suitable for use in the design of RBF networks (see Exercise 4.1).

In the next section, we introduce a new network, namely, the RBF network, according to (4.5)–(4.11), and derive its training algorithm.

4.3 TRAINING ALGORITHMS FOR RADIAL-BASIS FUNCTION NETWORKS

4.3.1 Layered Structure of a Radial-Basis Function Network

We now envision an RBF network in the form of a layered structure, as illustrated in Figure 4.1. Specifically speaking, the three layers are described as follows [Haykin, 2009]:

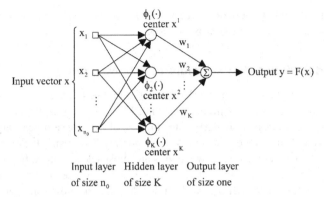

Input layer   Hidden layer   Output layer
of size n_0 of size K   of size one

FIGURE 4.1 Structure of an RBF network, based on the interpolation theory.

1. *Input layer:* The input layer consists of n_0 source nodes, where n_0 is the dimensionality of the input vector x.
2. *Hidden layer:* The hidden layer consists of the same number of computation units as the size of the training sample, namely, K. Each unit is mathematically described by a radial-basis function:

$$\phi_j(x) = \phi\big(\|x - x^j\|\big), \quad j = 1, 2, \dots, K \tag{4.13}$$

The jth input data point x^j defines the center of the radial-basis function, and the vector x is the signal (pattern) applied to the input layer. Besides, we find that unlike the multilayer perceptron, the source nodes are directly connected with the hidden units with no weights.

3. *Output layer:* The output layer consists of a single computational unit. Clearly, there is no restriction on the size of the output layer, except to say that typically the size of the output layer is much smaller than that of the hidden layer.

Henceforth, we focus on the use of a Gaussian function as the radial-basis function, in which case each computational unit in the hidden layer of the network is defined by

$$\phi_j(x) = \phi\big(\|x - x^j\|\big)$$
$$= \exp\left(-\frac{1}{2\sigma_j^2}\|x - x^j\|^2\right), \quad j = 1, 2, \dots, K \tag{4.14}$$

where σ_j is a measure of the width of the jth Gaussian function with center x^j. Typically, but not always, all the Gaussian hidden units are assigned a common width σ. In such a case, the parameter that distinguishes one hidden unit from another is the center x^j. The reason behind the choice of the Gaussian function as the radial-basis function in building RBF networks is that it has many desirable properties, which will become evident as the discussion progresses.

4.3.2 Modification of the Structure of RBF Network

In the formulation of the RBF network depicted in Figure 4.1, the size of hidden layer is set the same as the size of training sample. However, this could be wasteful of computational resources, particularly when the size of training sample is large. In fact, the training sample may inherently contain redundancy. As a result, the redundancy of neurons in hidden layer may occur when they are chosen according to Eq. 4.14. Under such circumstances, it is necessary to make the size of the hidden layer a part of that of the training sample. The RBF network under this idea is shown in Figure 4.2. By comparing it with Figure 4.1, we find that the size of hidden layer of the two networks is different, namely, S < K.

Unlike the multilayer neural networks, the training process of both of these two networks does not involve the backpropagation of error signals. Besides, the approximating function realized by them has the same mathematical form

$$F(x) = \sum_{j=1}^{S} w_j \phi(x, x^j)$$

$$= \sum_{j=1}^{S} w_j \exp\left(-\frac{1}{2\sigma^2}\left\|x - x^j\right\|^2\right)$$

(4.15)

where the dimensionality of the input vector x is n_0 and each hidden unit is characterized by the radial-basis function $\phi(x, x^j)$ with $j = 1, 2, \ldots, S$. The output layer is assumed to consist of a single unit and characterized by the weight vector w, whose dimensionality is also S.

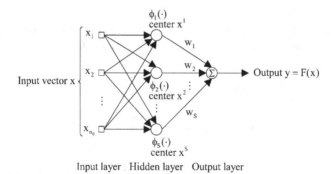

FIGURE 4.2   Structure of the modified RBF network. Note that the size of the hidden layer in Figure 4.2 is smaller than that in Figure 4.1, that is, S < K.

Note that when the training sample $\{x^i, d_i\}_{i=1}^K$ is noiseless, the design of the hidden layer in Figure 4.1 is solved simply by using the input vector x^j to define the center of the radial-basis function $\phi(x, x^j)$ for $j = 1, 2, \ldots, K$. Otherwise, we have to come up with a new procedure in order to design the hidden layer in Figure 4.2, as stated in the following section.

4.3.3 Hybrid Learning Process

The hybrid learning process for RBF network consists of two different stages: estimating appropriate locations for the centers of the radial-basis functions in the hidden layer and estimating the linear weights of the output layer.

In the learning process, we need a clustering algorithm to partition the given set of data points into subgroups, each of which should be as homogeneous as possible. One such algorithm is the k-means clustering algorithm, which places the centers of the radial-basis functions in only those regions of the input space where significant data are present.

First, we partition the training samples into S clusters, where S is smaller than K and may be determined through experimentation. Then, let $\{c_j(k)\}_{j=1}^S$ denote the centers of the radial-basis functions at iteration k of the algorithm. Next, we describe the design procedure of k-means clustering algorithm as follows:

1. *Initialization*: Choose random values for the initial center $c_j(0)$ for $j = 1, 2, \ldots, S$. These initial values should be different.
2. *Sampling:* Draw a sample vector x from the input space with a certain probability, and then use it as the input of the algorithm at iteration k.
3. *Similarity matching:* Let $j(x)$ denote the index of the best matching center for input vector x. Find $j(x)$ at iteration k by using

$$j(x) = \arg\min_j \|x(k) - c_j(k)\|, \quad j = 1, 2, \ldots, S \qquad (4.16)$$

where $c_j(k)$ is the center of the jth radial-basis function at iteration k.
4. *Updating:* Adjust the centers of the radial-basis functions according to the update rule:

$$c_j(k+1) = \begin{cases} c_j(k) + \alpha_c(x(k) - c_j(k)), & \text{if } j = j(x) \\ c_j(k), & \text{if } j \neq j(x) \end{cases} \qquad (4.17)$$

where α_c is the learning rate parameter and $0 < \alpha_c < 1$.
5. *Continuation:* Set $k = k + 1$ and go back to step 2. Continue the procedure until no noticeable changes are observed in the centers $c_j(k)$ for $j = 1, 2, \ldots, S$.

Subsequently, let d_{max} be the maximum distance between the obtained S centers of k-means clustering algorithm. To simplify the design, the standard width of all the

Gaussian radial-basis functions is fixed at

$$\sigma = \frac{d_{max}}{\sqrt{2S}} \tag{4.18}$$

This formula ensures that the individual radial-basis functions are not too peaked or too flat.

After identifying the individual centers of the Gaussian radial-basis functions and their common width, the next stage of the hybrid leaning process is to estimate the weights of the output layer.

Define

$$\phi_{ij} = \phi(\|x^i - x^j\|), \quad i = 1, 2, \dots, K, \quad j = 1, 2, \dots, S \tag{4.19}$$

Let the error signal produced at the output layer at iteration K is

$$e_i(k) = d_i(k) - \sum_{j=1}^{S} w_j(k)\phi_{ij} \tag{4.20}$$

Then, the instantaneous error energy at the output layer is

$$E(k) = \frac{1}{2} \sum_{i=1}^{K} e_i^2(k) \tag{4.21}$$

Hence, we can obtain the gradient

$$\frac{\partial E(k)}{\partial w_j(k)} = -\sum_{i=1}^{K} e_i(k)\phi_{ij} \tag{4.22}$$

Like Eq. 3.16, the correction $\Delta w_j(k)$ applied to $w_j(k)$ is defined by the delta rule:

$$\Delta w_j(k) = -\alpha \frac{\partial E(k)}{\partial w_j(k)} \tag{4.23}$$

where α is the learning rate parameter. Accordingly, the weights of output layer can be updated as

$$w_j(k + 1) = w_j(k) + \alpha \sum_{i=1}^{K} e_i(k)\phi_{ij} \tag{4.24}$$

for $j = 1, 2, \dots, S$.

4.4 UNIVERSAL APPROXIMATION

We discussed the universal approximation properties of multilayer perceptron in the last chapter. Similarly, radial-basis functions also have good approximation properties. Actually, the family of RBF network is broad enough to uniformly approximate any continuous function on a compact set.

Let $G : \Re^{n_0} \rightarrow \Re^1$ be an integrable bounded function such that G is continuous and

$$\int_{\Re^{n_0}} G(x) \, dx \neq 0 \tag{4.25}$$

Let $\mathfrak{A}_G$ denote the family of RBF networks consisting of functions $F : \Re^{n_0} \rightarrow \Re^1$ represented by

$$F(x) = \sum_{i=1}^{n_1} w_i G \left(\frac{x - c_i}{\sigma} \right) \tag{4.26}$$

where $\sigma > 0$, $w_i \in \Re^1$, and $c_i \in \Re^{n_0}$ for $i = 1, 2, \ldots, n_1$. Then, the universal approximation theorem of RBF networks can be stated as follows.

Theorem 4.2 [Park and Sandberg, 1991] For any continuous input–output mapping function $f(x)$, there is an RBF network with a set of centers $\{c_i\}_{i=1}^{n_1}$ and a common width $\sigma > 0$, such that the input–output mapping function $F(x)$ realized by the RBF networks is close to $f(x)$ in the L_p norm, $p \in [1, \infty]$.

Note in the stated theorem, the $G : \Re^{n_0} \rightarrow \Re^1$ is not required to satisfy the property of radial symmetry. Hence, the theorem is stronger than necessary for RBF networks. Most importantly, it provides the theoretical basis for design of neural networks in practice based on radial-basis functions.

Here, it is essential to clarify the comparison between RBF networks and multilayer perceptrons [Haykin, 1999]. On the one hand, they are both universal approximators. Accordingly, there always exists an RBF network that is capable of accurately mimicking a specified multilayer perceptron, or vice versa. On the other hand, there are also some remarkable differences between the two networks, as stated in the following respects:

- An RBF network has a single hidden layer, while a multilayer perceptron may have one or more hidden layers.
- Typically, the computation nodes of a multilayer perceptron located in a hidden or an output layer share a common neuronal model. However, the computation nodes in the hidden layer of an RBF network are quite different and serve a different purpose from those in the output layer of the network.
- The hidden layer of an RBF network is nonlinear, while the output layer is linear. However, the hidden and output layers of a multilayer perceptron used as a pattern classifier are usually all nonlinear.

- The argument of the activation function of each hidden unit in an RBF network computes the distance between the input vector and the center of the unit. However, the activation function of each hidden unit in a multilayer perceptron computes the inner product of the input vector and the synaptic weight vector of that unit.
- The multilayer perceptrons construct global approximations to nonlinear input–output mapping. Nevertheless, the RBF networks construct local approximations to nonlinear input–output mappings, using exponentially decaying localized nonlinearities.

The differences also reveal that for the approximation of a nonlinear input–output mapping, multilayer perceptrons require a smaller number of parameters than an RBF network for the same degree of accuracy. In addition, the RBF differs from the perceptron in that it is capable of implementing arbitrary nonlinear transformations of the input space. This significant property can be illustrated by revisiting the XOR problem, which cannot be solved by a single perceptron. See Exercise 4.4.

At last, it is important to point out that RBF networks and multilayer perceptrons can be trained in alternative ways besides those presented. For multilayer perceptrons, the backpropagation algorithm is simple to compute locally and it performs stochastic gradient descent in weight space when the algorithm is implemented in an online learning mode. Hence, the backpropagation algorithm has become a computationally efficient and useful algorithm for the training of multilayer perceptrons. The algorithm derives its name from the fact that the partial derivatives of the cost function with respect to the free parameters of the network are determined by backpropagating the error signals through the network layer by layer. However, it relies on the gradient vector as the only source of local first-order information about the error surface, which always brings in a slow rate of convergence, particularly in the case of large-scale problems. Therefore, in order to produce a significant improvement in the convergence performance of a multilayer perceptron, it is better to use the higher order information in the training process. This can be done by invoking a quadratic approximation of the error surface around the current point, which is the essence of Newton's method. In fact, there are some efficient algorithms for training multilayer perceptrons, like quasi-Newton method, conjugate-gradient method, and Levenberg–Marquardt method, as alternative ways of the classical backpropagation algorithm. For RBF networks, there are also some alternative methods for the training purpose in the literature, using different algorithms for the two stages of the procedure.

4.5   KERNEL REGRESSION

The description of RBF network presented in the foregoing sections is built on the notion of interpolation. Now, we take another viewpoint to study the network. It is kernel regression, which is built on the notion of density estimation.

First, we point out that the Gaussian function $\phi(x, x^j)$ may be interpreted as a kernel, a term that is widely used in the statistics literature.

Consider a function dependent on an input vector x, with its center located at the origin of the Euclidean space. Basic to the formulation of this function is a kernel, denoted by $k(x)$. It has the properties similar to those associated with the probability density function of a random variable.

- The kernel $k(x)$ is a continuous, bounded, and real function of x and symmetric about the origin, where it attains its maximum value.
- The total volume under the surface of the kernel $k(x)$ is unity, that is,

$$\int_{\mathfrak{R}^n} k(x)dx = 1 \tag{4.27}$$

      for an n $-$ dimensional vector x.

Except for a scaling factor, the Gaussian function $\phi(x, x^j)$ satisfies both of these properties for the center x^j located at the origin. For a nonzero value of x^j, the above two properties still hold except for the fact that x^j replaces the origin.

Then, consider a nonlinear regression model defined by

$$y_i = f(x^i) + \varepsilon(i), \quad i = 1, 2, \ldots, K \tag{4.28}$$

where $\varepsilon(i)$ is an additive white-noise term of zero mean and variance σ_ε^2. As a reasonable estimate of the unknown regression function $f(x)$, we may take the mean of observable near a point x. For this approach to be successful, however, the local average should be confined to observations in a small neighborhood around the point x, because the observations related to points away from x will have different mean values. More precisely, we find that the unknown function $f(x)$ is equal to the conditional mean of the observable y given the regressor x, that is,

$$\begin{aligned} f(x) &= \mathbb{E}\{y|x\} \\ &= \int_{-\infty}^{\infty} y p_{Y|X}(y|x)dy \end{aligned} \tag{4.29}$$

where $p_{Y|X}(y|x)$ is the conditional probability density function of the random variable Y given that the random vector X is assigned the value x. In accordance with the probability theory, we have

$$p_{Y|X}(y|x) = \frac{p_{X,Y}(x, y)}{p_X(x)} \tag{4.30}$$

where $p_X(x)$ is the probability density function of $p_X(x)$ and $p_{X,Y}(x, y)$ is the joint probability density function of X and Y. Combining (4.29) and (4.30), the regression function can be written as

$$f(x) = \frac{\int_{-\infty}^{\infty} y p_{X,Y}(x, y) dy}{p_X(x)} \qquad (4.31)$$

In the situation that the joint probability density function $p_{X,Y}(x, y)$ is unknown while all that we have available is the training sample $\{(x^i, y_i)\}_{i=1}^{K}$, we use a nonparametric estimator, known as the Parzen–Rosenblatt density estimator, to estimate $p_{X,Y}(x, y)$ and $p_X(x)$. Basic to the formulation of this estimator is the availability of a kernel $k(x)$. Assuming that the observations $x^1, x^2, \ldots, x^K$ are statistically independent and identically distributed, we may formally define the Parzen–Rosenblatt density estimate of $p_X(x)$ as

$$\hat{p}_X(x) = \frac{1}{Kh^{n_0}} \sum_{i=1}^{K} k\left(\frac{x - x^i}{h}\right), \quad \text{for } x \in \mathbb{R}^{n_0} \qquad (4.32)$$

where the smoothing parameter h is a positive number called bandwidth, or simply width. Note that h controls the size of the kernel. An important property of the Parzen–Rosenblatt density estimator is that it is a consistent estimator, that is, asymptotically unbiased, in the sense that if $h = h(K)$ is chosen as a function of K such that

$$\lim_{K \to \infty} h(K) = 0 \qquad (4.33)$$

then

$$\lim_{K \to \infty} \mathbb{E}\{\hat{p}_X(x)\} = p_X(x) \qquad (4.34)$$

In Eq. 4.34, x should be a point of continuity for $\hat{p}_X(x)$.

In a manner similar to that described in Eq. 4.32, we may formulate the Parzen–Rosenblatt density estimate of the joint probability density function $p_{X,Y}(x, y)$ as

$$\hat{p}_{X,Y}(x, y) = \frac{1}{Kh^{n_0+1}} \sum_{i=1}^{K} k\left(\frac{x - x^i}{h}\right) k\left(\frac{y - y_i}{h}\right), \quad \text{for } x \in \mathfrak{R}^{n_0} \quad \text{and} \quad y \in \mathfrak{R} \qquad (4.35)$$

Integrating $\hat{p}_{X,Y}(x, y)$ with respect to y, we get

$$\int_{-\infty}^{\infty} y \hat{p}_{X,Y}(x, y) dy = \frac{1}{Kh^{n_0+1}} \sum_{i=1}^{K} k\left(\frac{x - x^i}{h}\right) \int_{-\infty}^{\infty} y k\left(\frac{y - y_i}{h}\right) dy \qquad (4.36)$$

Set $\tau = (y - y_i)/h$, make use of Eq. 4.27, and then we can derive that

$$\int_{-\infty}^{\infty} y\hat{p}_{X,Y}(x,y)dy = \frac{1}{Kh^{n_0}} \sum_{i=1}^{K} y_i k\left(\frac{x - x^i}{h}\right) \tag{4.37}$$

Hence, by using Eqs. 4.32 and 4.37, we can obtain an estimate of the regression function f(x) in Eq. 4.31 as

$$
\begin{aligned}
F(x) &= \hat{f}(x) \\
&= \frac{\sum_{i=1}^{K} y_i k\left((x - x^i)/h\right)}{\sum_{j=1}^{K} k\left((x - x^j)/h\right)}
\end{aligned}
\tag{4.38}$$

Next, we assume spherical symmetry of the kernel k(x), in which case we may set

$$k\left(\frac{x - x^i}{h}\right) = k\left(\frac{\|x - x^i\|}{h}\right), \quad \text{for all } i \tag{4.39}$$

where $\|\cdot\|$ denotes the Euclidean norm of the enclosed vector. Define the normalized radial-basis function as

$$\Psi_K(x, x^i) = \frac{k\left((\|x - x^i\|)/h\right)}{\sum_{j=1}^{K} k\left((\|x - x^j\|)/h\right)}, \quad i = 1, 2, \ldots, K \tag{4.40}$$

with

$$\sum_{i=1}^{K} \Psi_K(x, x^i) = 1, \text{for all } x \tag{4.41}$$

The subscript K in $\Psi_K(x, x^i)$ signifies the use of normalization.

The linear weights w_i applied to the basic functions $\Psi_K(x, x^i)$ are simply the observable y_i of the regression model for the input data x^i. Accordingly, letting

$$y_i = w_i, \quad i = 1, 2, \ldots, K \tag{4.42}$$

we may rewrite the approximating function of Eq. 4.38 in the general form:

$$F(x) = \sum_{i=1}^{K} w_i \Psi_K(x, x^i) \tag{4.43}$$

Equation 4.43 represents the input–output mapping of a normalized RBF network [Xu et al., 1994]. Note that

$$0 \leq \Psi_K(x, x^i) \leq 1, \quad \text{for all } x \text{ and } x^i \tag{4.44}$$

Therefore, $\Psi_K(x, x^i)$ may be interpreted as the probability of an event described by the input vector x, conditional on x^i.

The basic difference between the normalized radial-basis function $\Psi_K(x, x^i)$ of Eq. 4.40 and an ordinary radial-basis function, such as Eq. 4.14, is a denominator term that constitutes the normalization factor. This normalization factor is an estimate of the underlying probability density function of the input vector x. Thus, the basis function $\Psi_K(x, x^i)$ for $i = 1, 2, \ldots, K$ sum to unity for all x, as described in Eq. 4.41.

In general, a variety of kernel functions can be utilized. However, the theoretical and practical considerations may limit the choice. A widely used kernel is the multivariate Gaussian distribution

$$k(x) = \frac{1}{(2\pi)^{n_0/2}} \exp\left(-\frac{\|x\|^2}{2}\right) \tag{4.45}$$

where n_0 is the dimension of the input vector x. The spherical symmetry of the kernel $k(x)$ can be observed in Eq. 4.45. Assuming the use of a common width σ that plays the role of smoothing parameter h for a Gaussian distribution, and centering the kernel on a data point x^i, we may write

$$k\left(\frac{x - x^i}{h}\right) = \frac{1}{(2\pi\sigma^2)^{n_0/2}} \exp\left(-\frac{\|x - x^i\|^2}{2\sigma^2}\right), \quad i = 1, 2, \ldots, K \tag{4.46}$$

Then, according to Eqs. 4.40, 4.43, and 4.46, we derive that the input–output mapping function of the normalized RBF network can be written as

$$F(x) = \frac{\sum_{i=1}^{K} w_i \exp\left(-(\|x - x^i\|^2)/2\sigma^2\right)}{\sum_{j=1}^{K} \exp\left(-(\|x - x^j\|^2)/2\sigma^2\right)} \tag{4.47}$$

where the denominator term, representing the Parzen–Rosenblatt density estimator, consists of the sum of K multivariate Gaussian distributions centered on the data points x^1, x^2, ..., x^K [Specht, 1991].

We can see that in Eq. 4.47, the centers of the normalized radial-basis functions coincide with the data points $\{x^i\}_{i=1}^{K}$. As with ordinary radial-basis functions, a smaller number of normalized radial-basis functions can be used, with their centers treated as free parameters to be chosen according to some heuristic manners, for example, the K-means clustering algorithm.

EXERCISES

4.1. For the inverse multiquadrics (4.2) and the Gaussian functions (4.3), the corresponding interpolation matrix ϕ is positive definite. However, the ϕ related to the multiquadrics (4.1) is not positive definite. Prove these propositions.

4.2. There is a remarkable fact that radial-basis functions that grow at infinity, such as multiquadrics, can be used to approximate a smooth input–output mapping with greater accuracy than those that yield a positive definite interpolation matrix. Try to explain the statement.

4.3. The multiquadrics (4.1) and the inverse multiquadrics (4.2) provide two possible choices for radial-basis functions. An RBF network using the inverse multiquadrics constructs local approximation to nonlinear input–output mapping. However, the use of a multiquadrics represents a counterexample to this property of RBF networks. Justify the validity of these two statements.

4.4. Revisit the XOR problem described in Table 3.1. Here, we solve it using an RBF network, whose structure is given in Figure 4.3.

Define a pair of Gaussian hidden functions as

$$\phi_j(x) = \exp\left(-\|x - c_j\|^2\right), \quad j = 1, 2$$

where the centers c_1 and c_2 are

$$c_1 = \begin{bmatrix} 1 \\ 1 \end{bmatrix}, \quad c_2 = \begin{bmatrix} 0 \\ 0 \end{bmatrix}$$

1. Calculate the weight vector of the RBF network and explain why this kind of network can be utilized to deal with the XOR problem successfully.
2. Study the necessity of adding the bias b. Take off the bias and the fixed input, and then point out if the corresponding network can still solve the problem or not.

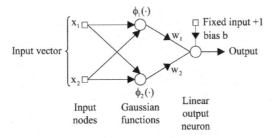

FIGURE 4.3 Structure of the RBF network in Exercise 4.4.

4.5. Reconsider the XOR problem by using an RBF network with four hidden units, with each radial-basis function center being determined by each piece of input data.

1. Construct the interpolation matrix ϕ for the resulting RBF network and then compute the inverse matrix ϕ^{-1}.

2. Calculate the linear weights of the output layer of the network.

4.6. The input–output relationship of a Gaussian-based RBF network is defined by

$$y(i) = \sum_{j=1}^{K} w_j(k) \exp\left(-\frac{\|x(i) - \mu^j(k)\|^2}{2\sigma^2(k)}\right), \quad i = 1, 2, \ldots, k$$

where $\mu^j(k)$ is the center point of the jth Gaussian unit, the width $\sigma(k)$ is common to all the K units, and $w_j(k)$ is the linear weight assigned to the output of the jth unit. Note that all these parameters are measured at time k. The cost function used to train the network is defined by

$$E = \frac{1}{2}\sum_{i=1}^{k} e^2(i)$$

where

$$e(i) = d(i) - y(i)$$

The cost function E is a convex function of the linear weights in the output layer, but nonconvex with respect to the centers and the width of the Gaussian units.

1. Evaluate the partial derivatives of the cost function with respect to each of the network parameters $w_j(k)$, $\mu^j(k)$, and $\sigma(k)$, for all i.

2. Express the update formulas for all the network parameters $w_j(k)$, $\mu^j(k)$, and $\sigma(k)$, provided the learning rates of them are α_w, α_μ, and α_σ, respectively.

3. The gradient vector $\partial E / \partial \mu^j(k)$ has an effect on the input data that is similar to clustering. Justify this statement.

CHAPTER 5

Recurrent Neural Networks

The multilayer perceptron and the RBF network considered in Chapters 3 and 4 represent two important examples of a class of neural networks known as nonlinear layered feedforward networks. In this chapter, we consider another class of neural networks that have a recurrent structure.

As we have seen from the preceding chapters, time plays a critical role in learning. When time is built into the operation of a neural network through the use of global feedback, which encompasses one or more layers of hidden neurons, or the whole network, it results in the recurrent neural network.

The recurrent neural networks incorporate a static multilayer perceptron or parts thereof, and exploit the nonlinear mapping capability of the multilayer perceptron as well.

5.1 THE HOPFIELD NETWORK

The Hopfield network is a form of recurrent artificial neural network invented by John Hopfield in 1982. It consists of a set of neurons and a corresponding set of unit time delays, formatting a multiple-loop feedback system. A simple example of the architectural graph can be seen in Figure 5.1. The number of feedback loops is equal to the number of neurons. Basically, there is no self-feedback in the model. The output of each neuron is fed back, via a unit time delay element, to each of the other neurons in the network.

The equations that describe the network operation are

$$x(k) = p \tag{5.1}$$

and

$$x(k + 1) = satlins(Wp + b) \tag{5.2}$$

where satlins is the transfer function that is linear in the range $[-1, 1]$, and saturates at 1 for inputs greater than 1 and at -1 for inputs less than -1.

Fundamentals of Computational Intelligence: Neural Networks, Fuzzy Systems, and Evolutionary Computation, First Edition. James M. Keller, Derong Liu, and David B. Fogel.

77

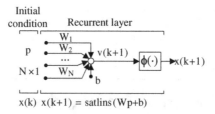

FIGURE 5.1   Architectural graph of the Hopfield network consisting of N neurons.

To illustrate the operation of the network, we have determined a weight matrix and a bias vector that can solve our orange and apple pattern recognition problem. They are given in

$$W = \begin{bmatrix} 0.2 & 0 & 0 \\ 0 & 1.2 & 0 \\ 0 & 0 & 0.2 \end{bmatrix}, \quad b = \begin{bmatrix} 0.9 \\ 0 \\ -0.9 \end{bmatrix} \tag{5.3}$$

Although the procedure for computing the weights and biases for the Hopfield network is beyond the scope of this chapter, we can say a few things about why the parameters in (5.3) work for the apple and orange example.

We want the network output to converge to either the orange pattern, p_1, or the apple pattern, p_2. In both patterns, the first element is 1 and the third element is -1. The difference between the patterns occurs in the second element. Therefore, no matter what pattern is input to the network, we want the first element to converge to -1, and the second element to go to either 1 or -1, whichever is closer to the second element of the input vector.

The equations of operation of the Hopfield network, using the parameters given in (5.3), are

$$\begin{aligned} x_1(k+1) &= satlins(0.2x_1(k) + 0.9) \\ x_2(k+1) &= satlins(1.2x_2(k)) \\ x_3(k+1) &= satlins(0.2x_3(k) - 0.9) \end{aligned} \tag{5.4}$$

Regardless of the initial values, $x_1(k)$, the first element will be increased until it saturates at 1, and the third element will be decreased until it saturates at -1. The second element is multiplied by a number larger than 1. Therefore, if it is initially negative, it will eventually saturate at -1; if it is initially positive, it will saturate at 1.

Let us take our oblong orange to test the Hopfield network. The outputs of the Hopfield network for the first three iterations would be

$$x(k) = \begin{bmatrix} -1 \\ -1 \\ -1 \end{bmatrix}, \quad x(k+1) = \begin{bmatrix} 0.7 \\ -1 \\ -1 \end{bmatrix}, \quad x(k+2) = \begin{bmatrix} 1 \\ -1 \\ -1 \end{bmatrix}, \quad x(k+3) = \begin{bmatrix} 1 \\ -1 \\ -1 \end{bmatrix} \tag{5.5}$$

The network has converged to the orange pattern.

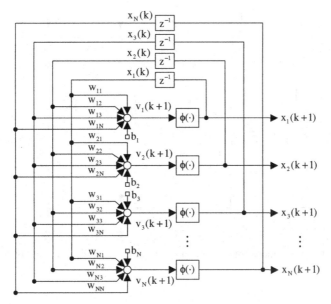

FIGURE 5.2   Discrete model of the Hopfield network consisting of N neurons.

Now, we will introduce a normalized discrete Hopfield network based on the McCulloch–Pitts model, which is shown in Figure 5.2.

The induced local field of neuron j at time step k + 1 is denoted by

$$v_j(k + 1) = \sum_{i=1}^{N} w_{ji}x_i(k) + b_j \tag{5.6}$$

where w_{ji} and b_j are the synaptic weight and the bias of neuron j, respectively. Additionally, $w_{ji} = 0$ when i = j, which in fact means that the self-feedback does not exist in the model. Then, the output of neuron j by applying the signum function is

$$x_j(k + 1) = \phi(v_j(k + 1)) = \begin{cases} 1, & \text{if } v_j(k+1) > 0; \\ -1, & \text{if } v_j(k+1) < 0 \end{cases} \tag{5.7}$$

Note that neuron j remains in its previous state if $v_j(k + 1)$ is zero.

Next, we consider the circuit model of the continuous Hopfield network depicted in Figure 5.3, where N denotes the number of neurons and $\phi_j(\cdot), j = 1, 2, \ldots, N$, represent the activation functions. The corresponding physical terms in Figure 5.3 are defined as follows:

I_j: external current

w_{ji}: conductance

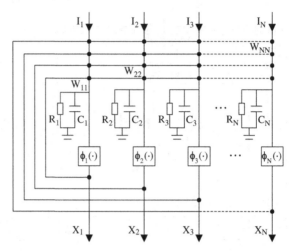

FIGURE 5.3 Circuit model of the continuous Hopfield network consisting of N neurons.

R_j: leakage resistance
C_j: leakage capacitance
x_j: potential

Let the induced local field of neuron j be denoted by

$$v_j(t) = \sum_{i=1}^{N} w_{ji} x_i(t) + I_j \tag{5.8}$$

where $x_i(t)$ is the potential at time t. Also, we can determine the output of neuron j by using the nonlinear relation

$$x_j(t) = \phi_j(v_j(t)) \tag{5.9}$$

Then, according to the Kirchoff's current law, we have

$$C_j(t)\frac{dv_j(t)}{dt} + \frac{v_j(t)}{R_j} = \sum_{i=1}^{N} w_{ji} x_i(t) + I_j, \quad j = 1, 2, \ldots, N \tag{5.10}$$

Hence, the model of the Hopfield network can be given in the following form:

$$C_j(t)\frac{dv_j(t)}{dt} = -\frac{v_j(t)}{R_j} + \sum_{i=1}^{N} w_{ji} \phi_i(v_i(t)) + I_j, \quad j = 1, 2, \ldots, N \tag{5.11}$$

5.2 THE GROSSBERG NETWORK

In this section, we will present the Grossberg network [Grossberg, 1998]. This network was inspired by the operation of the mammalian visual system. Grossberg networks are so heavily influenced by biology that it is difficult to discuss his networks without putting them in their biological context. In this section we want to provide a brief introduction to vision, so that the function of the network will be more understandable.

First, we see Figure 5.4 [Hagan, Demuth, and Beale, 1996]. In part (a), we see an edge as it is originally perceived by the rods and cones, with missing sections. But usually we do not see edges as displayed in part (a). The neural systems in our visual pathway must be performing some operation that compensates for the distortions and completes the image.

Grossberg suggests that there are two primary types of compensatory processing involved. The first, which he calls *emergent segmentation*, completes missing boundaries. The second, which he calls *featural filling-in*, fills in the color and brightness inside the resulting boundaries. Consider, for example, the two figures in Figure 5.5. In part (a) you should be able to see a bright white triangle lying on the top of several other black objects. In fact, no such triangle exists in the figure. It is purely a creation of the emergent segmentation and featural filling-in process of your visual system. The same is true of the bright white circle that appears to lie on the top of the lines in part (b) of the figure.

In addition to emergent segmentation and featural filling-in, there are two other phenomena that give us an indication of what operations are being performed in the

FIGURE 5.4 Compensatory processing.

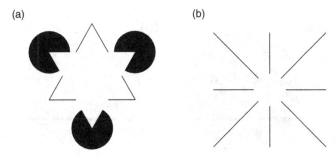

FIGURE 5.5 Emergent segmentation and featural filling-in.

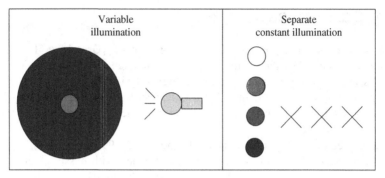

FIGURE 5.6   Test of brightness constancy.

early vision system: *brightness constancy* and *brightness contrast*. The brightness constancy effect is shown as small gray disk inside a darker gray annulus, which is illuminated by white light of a certain intensity. The subject is asked to indicate the brightness of the central disk by looking at a series of gray disks, separately illuminated, and selecting the disk with the same brightness. Next, the brightness of the light illuminating the gray disk and dark annulus is increased, and the subject is again asked to select the disk as matching the original central disk. Even though the total light entering the subject's eye is 10–100 times brighter, it is only the relative brightness that registers (Figure 5.6).

Another phenomenon of the vision system, which is closely related to brightness constancy, is brightness contrast. This effect is illustrated by the two figures in Figure 5.7. At the centers of the two figures, we have two small disks with equivalent gray scale. The small disk in part (a) of the figure is surrounded by a darker annulus, while the small disk in part (b) is surrounded by a lighter annulus. Even though both disks have the same gray scale, the one inside the darker annulus appears brighter. This is because our vision system is sensitive to relative intensities. It would seem that the total activity across the image is held constant.

The properties of the brightness constancy and brightness contrast are very important to our vision system. Since we see things in so many different lighting conditions, if we were not able to compensate for the absolute intensity of a scene, we

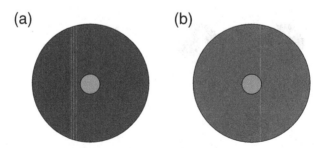

FIGURE 5.7   Test of brightness contrast.

would never learn to recognize things. Grossberg calls this process of normalization "discounting the illuminant."

5.2.1 Basic Nonlinear Model

Before we introduce the Grossberg network, we will begin by looking at some of the building blocks that make up the network [Hagan, Demuth, and Beale, 1996]. The first building block is the "leaky" integrator, which is shown in Figure 5.8.

The response of the leaky integrator to an arbitrary input p(t) is

$$n(t) = e^{-(t/\varepsilon)}n(0) + \frac{1}{\varepsilon}\int_0^t e^{-(t-\tau)/\varepsilon}p(t-\tau)d\tau \tag{5.12}$$

The leaky integrator forms the nucleus of one of Grossberg's fundamental neural models: *the shunting model*, which is shown in Figure 5.9. The equation of operation of this network is

$$\varepsilon\frac{dn}{dt} = -n + (b^+ - n)p^+ - (n + b^-)p^- \tag{5.13}$$

where p^+ is a nonnegative value representing the *excitatory* input to the network, and p^- is a nonnegative value representing the *inhibitory* input. The biases b^+ and b^- are

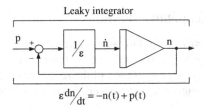

$$\varepsilon\,dn/dt = -n(t) + p(t)$$

FIGURE 5.8   Leaky integrator.

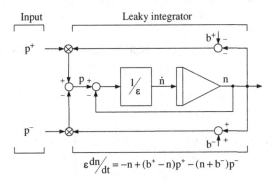

$$\varepsilon\,dn/dt = -n + (b^+ - n)p^+ - (n + b^-)p^-$$

FIGURE 5.9   Shunting model.

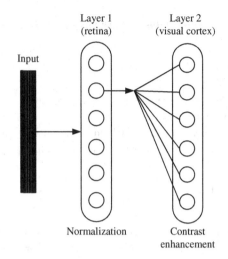

Layer 1 (retina)   Layer 2 (visual cortex)

Input

Normalization   Contrast enhancement

FIGURE 5.10 Grossberg competitive network.

nonnegative constants that determine the upper and lower limits on the neuron response, respectively. From the operation of the shunting model (5.7), we can see that if $n(0)$ falls between b^+ and b^-, then $n(t)$ will remain between these limits.

5.2.2 Two-Layer Competitive Network

We are now ready to present the Grossberg competitive network. There are three components to the Grossberg network: Layer 1, Layer 2, and the adaptive weights. Layer 1 is a rough model of the operation of the retina, while Layer 2 represents the visual cortex. A block diagram of the network is shown in Figure 5.10.

5.2.2.1 Layer 1 Layer 1 of the Grossberg network receives external inputs and normalizes the intensity of the input pattern (Figure 5.11).

The equation of operation of layer 1 is

$$\varepsilon(dn^1/dt) = -n^1 + (^+b^+ - n^1)[^+W^1]p - (n^1 + {}^-b^1)[^-W^1]p \qquad (5.14)$$

where

$$^+W^1 = \begin{bmatrix} 1 & 0 & \cdots & 0 \\ 0 & 1 & \cdots & 0 \\ \vdots & \vdots & \ddots & \vdots \\ 0 & 0 & \cdots & 1 \end{bmatrix}$$

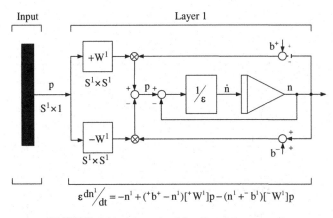

FIGURE 5.11   Layer 1 of the Grossberg network.

Therefore, the excitatory input to neuron i is the ith element of the input vector. The inhibitory input to layer 1 is $[^-W^1]p$ where

$$^-W^1 = \begin{bmatrix} 0 & 1 & \cdots & 1 \\ 1 & 0 & \cdots & 1 \\ \vdots & \vdots & \ddots & \vdots \\ 1 & 1 & \cdots & 0 \end{bmatrix}$$

Thus, the inhibitory input to neuron i is the sum of all elements of the input vector, expect the ith element.

To illustrate the performance of Layer 1, consider the case of two neurons, with the inhibitory bias $^-b^1 = 0$, $^+b^1 = 1$, $\varepsilon = 0.1$:

$$(0.1)\frac{dn_1^1(t)}{dt} = -n_1^1(t) + (1 - n_1^1(t))p_1 - n_1^1(t)p_2 \tag{5.15}$$

$$(0.1)\frac{dn_2^1(t)}{dt} = -n_2^1(t) + (1 - n_2^1(t))p_1 - n_2^1(t)p_2s \tag{5.16}$$

For two inputs $P_1 = \begin{bmatrix} 2 \\ 8 \end{bmatrix}$ and $P_2 = \begin{bmatrix} 10 \\ 40 \end{bmatrix}$, we can compute that the response of the network maintains the relative intensities of the inputs, while limiting the total response. The total response $(n_1^1(t) + n_2^1(t))$ will always be less than 1.

5.2.2.2 *Layer 2*   Layer 2 of the Grossberg network, which is a layer of continuous-time instars, performs several functions. Figure 5.12 is a diagram of layer 2. As with layer 1, the shunting model forms the basis for layer 2. The main difference between layer 2 and layer 1 is that layer 2 uses feedback connections. The feedback enables the

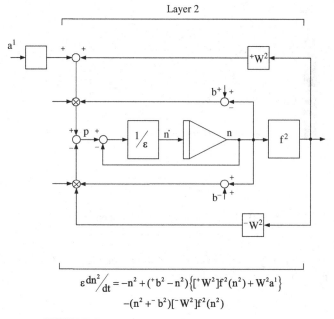

$$\varepsilon \frac{dn^2}{dt} = -n^2 + (^+b^2 - n^2)\{[^+W^2]f^2(n^2) + W^2a^1\}$$
$$-(n^2 + {}^-b^2)[^-W^2]f^2(n^2)$$

FIGURE 5.12 Layer 2 of the Grossberg network.

network to store a pattern, even after the input has been removed. The feedback also performs the competition that causes the contrast enhancement of pattern.

The equation of operation of layer 2 is

$$\varepsilon \frac{dn^2}{dt} = -n^2 + (^+b^2 - n^2)\{[^+W^2]f^2(n^2) + W^2a^1\} \qquad (5.17)$$
$$-(n^2 + {}^-b^2)[^-W^2]f^2(n^2)$$

To illustrate the performance of layer 2, consider a two-neuron layer with

$$\varepsilon = 0.1, {}^+b^2 = \begin{bmatrix} 1 \\ 1 \end{bmatrix}, {}^-b^2 = \begin{bmatrix} 0 \\ 0 \end{bmatrix}, W^2 = \begin{bmatrix} (_1w^2)^T \\ (_2w^2)^T \end{bmatrix} = \begin{bmatrix} 0.9 & 0.45 \\ 0.45 & 0.9 \end{bmatrix} \qquad (5.18)$$

and

$$f(n^2) = \frac{10(n)^2}{1 + (n)^2} \qquad (5.19)$$

The equations of operation of the layer will be

$$(0.1)\frac{dn_1^2(t)}{dt} = -n_1^2(t) + (1 - n_1^2(t))\{f^2(n_1^2(t)) + (_1w^2)^Ta^1\} - n_1^2(t)f^2(n_2^2(t)) \qquad (5.20)$$

$$(0.1)\frac{dn_2^2(t)}{dt} = -n_2^2(t) + (1 - n_2^2(t))\{f^2(n_2^2(t)) + (_2w^2)^T a^1\} - n_2^2(t)f^2(n_1^2(t)) \quad (5.21)$$

Define the input vector $a^1 = [0.2, 0.8]^T$. Then the inputs of layer 2 are

$$(_1w^2)^T a^1 = \begin{bmatrix} 0.9 & 0.45 \end{bmatrix} \begin{bmatrix} 0.2 \\ 0.8 \end{bmatrix} = 0.54$$

$$(_2w^2)^T a^1 = \begin{bmatrix} 0.45 & 0.9 \end{bmatrix} \begin{bmatrix} 0.2 \\ 0.8 \end{bmatrix} = 0.81 \qquad (5.22)$$

Therefore, the second neuron has 1.5 times as much input as the first neuron. However, after 0.25 s the output of the second neuron is 6.34 times the output of the first neuron. The second characteristic of response is that after the input has been set to zero, the network further enhances the contrast and stores the pattern. After the input is removed, the output of the first neuron decays to zero, while the output of the second reaches a steady state value of 0.79. This output is maintained, even after the input is removed.

5.2.2.3 *Learning Law* The third component of the Grossberg network is the learning law for the adaptive weights W^2. Grossberg calls these adaptive weights the long-term memory (LTM). This is because the rows of W^2 will represent patterns that have been stored and that the network will be able to recognize. The learning for W^2 is given by

$$\frac{dw_{i,j}^2(t)}{dt} = \alpha\{-w_{i,j}^2(t) + n_i^2(t)n_j^1(t)\} \qquad (5.23)$$

Now we summarize the Grossberg network.

Basic Nonlinear Model: Leaky Integrator

$$\varepsilon\frac{dn}{dt} = -n(t) + p(t)$$

Shunting Model

$$\varepsilon\frac{dn(t)}{dt} = -n(t) + (b^+ - n(t))[^+W^1]p - (n(t) + b^1)[^-W^1]p$$

Two-Layer Competitive Network

Layer 1

$$\varepsilon(dn^1/dt) = -n^1 + (^+b^+ - n^1)[^+W^1]p - (n^1 + ^-b^1)[^-W^1]p$$

$$^+W^1 = \begin{bmatrix} 1 & 0 & \cdots & 0 \\ 0 & 1 & \cdots & 0 \\ \vdots & \vdots & \ddots & \vdots \\ 0 & 0 & \cdots & 1 \end{bmatrix}, \quad ^-W^1 = \begin{bmatrix} 0 & 1 & \cdots & 1 \\ 1 & 0 & \cdots & 1 \\ \vdots & \vdots & \ddots & \vdots \\ 1 & 1 & \cdots & 0 \end{bmatrix}$$

Layer 2

$$\varepsilon(dn^2/dt) = -n^2 + (^+b^2 - n^2)\{[^+W^2]f^2(n^2) + W^2a^1\} - (n^2 + ^-b^2)[^-W^2]f^2(n^2)$$

Learning Law

$$\frac{dw_{i,j}^2(t)}{dt} = \alpha\left\{-w_{i,j}^2(t) + n_i^2(t)n_j^1(t)\right\}$$

If we substitute the output of the Grossberg network as its input, then the network is recurrent. In 1983, Michael A. Cohen and Stephen Grossberg described a general principle for assessing the stability of a certain class of neural networks, by a system of coupled nonlinear differential equations, given as

$$\frac{du_j}{dt} = a_j(u_j)\left[b_j(u_j) - \sum_{i=1}^N c_{ji}\phi_i(u_i)\right], \quad j = 1, 2, \ldots, N \qquad (5.24)$$

5.3 CELLULAR NEURAL NETWORKS

In 1988, Leon O. Chua and Lin Yang proposed a novel class of information processing systems called cellular neural networks [Chua and Roska, 2004]. The basic circuit unit of cellular neural networks is called a cell. It contains linear and nonlinear circuit elements, which typically are linear capacitors, linear resistors, linear and nonlinear controlled sources, and independent sources. The structure of cellular neural networks is similar to that found in cellular automata; namely, any cell in a cellular neural network is connected only to its neighbor cells. The adjacent cells can interact directly with each other. Cells not directly connected together may affect each other indirectly because of the propagation effects of the continuous-time dynamics of cellular neural networks. A simple example of a two-dimensional cellular neural network is shown

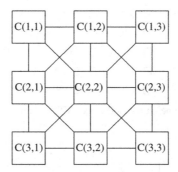

FIGURE 5.13 A two-dimensional cellular neural network. The circuit size is 3×3. The squares are the circuit units called cells.

in Figure 5.13. Actually, we can define a cellular neural network of any dimension. However, we will concentrate on the two-dimensional case in this chapter. The results can be easily generalized to higher dimension cases.

Now, consider an $M \times N$ cellular neural network, having $M \times N$ cells arranged in M rows and N columns. We call the cell on the ith row and jth column cell (i, j), and denote it by $C(i, j)$ as in Figure 5.14.

The $r -$ neighborhood of a cell $C(i, j)$ of a cellular neural network is defined by

$$N_r(i,j) = \left\{ C(k,l) : \max\{|k - i|, |l - j|\} \leq r, 1 \leq k \leq M; 1 \leq l \leq N \right\} \quad (5.25)$$

where r is a positive integer number. The Figure 5.15 shows a neighborhood of the cell located at the center. Usually, we call the r neighborhood a "$(2r + 1) \times (2r + 1)$ neighborhood." Note that the neighborhood defined above exhibits a symmetry property, that is, if $C(i, j) \in N_r(k, l)$, then $C(k, l) \in N_r(i, j)$, for all $C(i, j)$ and $C(k, l)$ in a cellular neural network.

A typical example of a cell $C(i, j)$ of a cellular neural network is shown in Figure 5.16, where the suffixes u, x, and y denote the input, state, and output, respectively. The node voltage v_{xij} of $C(i, j)$ is called the state of the cell and its initial condition is assumed to have a magnitude less than or equal to 1. The node

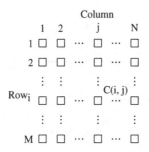

FIGURE 5.14 A two-dimensional cellular neural network. The circuit size is $M \times N$.

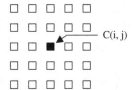

FIGURE 5.15   The neighborhood of cell $C(i, j)$ defined by (5.25) for $r = 2$.

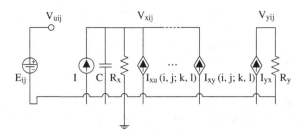

FIGURE 5.16   An example of a cell circuit.

voltage v_{uij} is called the input of $C(i, j)$ and is assumed to be a constant with magnitude less than or equal to 1. The node voltage v_{yij} is called the output.

According to Figure 5.16, each cell $C(i, j)$ contains one independent voltage source E_{ij}, one independent current source I, one linear capacitor C, two linear resistors R_x and R_y, and at most 2 m linear voltage-controlled current sources, which are coupled to its neighbor cells via the controlling input voltage v_{ukl}, and the feedback from the output voltage v_{ykl} of each neighbor cells. In particular, $I_{xy}(i, j; k, l)$ and $I_{xu}(i, j; k, l)$ are linear voltage-controlled current sources with the characteristics $I_{xy}(i, j; k, l) = A(i, j; k, l)v_{ykl}$ and $I_{xu}(i, j; k, l) = B(i, j; k, l)v_{ukl}$ for all $C(i, j) \in N_r(i, j)$. The only nonlinear element in each cell is a piecewise linear voltage-controlled current source $I_{yx} = (1/R_y)f(v_{xij})$ with characteristic $f(\cdot)$ as shown in Figure 5.17.

All of the linear and piecewise line-controlled sources used in the cellular neural network can be easily realized via operational amplifiers (op amps). Applying

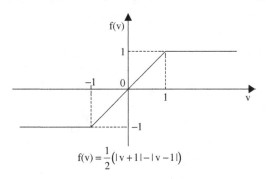

$$f(v) = \frac{1}{2}(|v + 1| - |v - 1|)$$

FIGURE 5.17   The characteristic of the nonlinear controlled source.

Kirchhoff's current law and Kirchhoff's voltage law, the circuit equations of a cell are easily derived as follows:

State equation:

$$C\frac{dv_{xij}(t)}{dt} = -\frac{1}{R_x}v_{xij}(t) + \sum_{C(k,l) \in N_r(i,j)} A(i,j;k,l)v_{ykl}(t)$$

$$+ \sum_{C(k,l) \in N_r(i,j)} B(i,j;k,l)v_{ukl}(t) + I, \quad 1 \le i \le M; \quad 1 \le j \le N \tag{5.26}$$

Output equation:

$$v_{yij}(t) = \frac{1}{2}\left(\left|v_{xij}(t) + 1\right| - \left|v_{xij}(t) - 1\right|\right), \quad 1 \le i \le M; \quad 1 \le j \le N \tag{5.27}$$

Input equation:

$$v_{uij} = E_{ij}, \quad 1 \le i \le M, \quad 1 \le j \le N \tag{5.28}$$

Constraint conditions:

$$\left|v_{xij}(0)\right| \le 1, \quad 1 \le i \le M, \quad 1 \le j \le N \tag{5.29}$$

$$\left|v_{uij}\right| \le 1, \quad 1 \le i \le M, \quad 1 \le j \le N \tag{5.30}$$

Parameter assumptions:

$$A(i,j;k,l) = A(k,l;i,j), \quad 1 \le i,k \le M, \quad 1 \le j, \quad 1 \le N \tag{5.31}$$

$$C > 0, \quad R_x > 0 \tag{5.32}$$

The inner cell is the cell that has $(2r + 1)^2$ neighbor cells, where r is defined in (5.25). All other cells are called boundary cells. All inner cells of a cellular neural network have the same circuit structures and element values. A cell neural network is completely characterized by the set of all nonlinear differential equations (5.26)–(5.32) associated with the cells in the circuit.

5.4 NEURODYNAMICS AND OPTIMIZATION

First, we describe the mathematical model of a nonlinear dynamic system [Haykin, 2009]. Let $x_1(t)$, $x_2(t)$, ..., $x_N(t)$ denote the state variables of a nonlinear dynamic system, where the continuous time t is the independent variable and N is the order of the system. The dynamics of a large class of nonlinear dynamic systems may be written as the following first-order differential equations:

$$\frac{d}{dt}x_j(t) = F_j(x_j(t)), \quad j = 1, 2, \ldots, N \tag{5.33}$$

where the function $F_j(\cdot)$ is, in general, a nonlinear function of its argument. For convenience of notation, the N state variables are collected into an N-by-1 vector $x(t)$, that is,

$$x(t) = [x_1(t), x_2(t), \ldots, x_N(t)]^T \tag{5.34}$$

which is called the state vector, or simply state of the system. Then, we can express Eq. 5.33 in a compact form as follows:

$$\frac{d}{dt}x(t) = F(x(t)) \tag{5.35}$$

where the nonlinear function $F(x(t))$ is

$$F(x(t)) = [F_1(x_1(t)), F_2(x_2(t)), \ldots, F_N(x_N(t))]^T \tag{5.36}$$

A nonlinear dynamic system for which the vector function $F(x(s))$ does not depend explicitly on time t is said to be autonomous. A constant vector $\bar{x}$ is said to be an equilibrium state of system (Eq. 5.35) if it satisfies

$$F(\bar{x}) = 0 \tag{5.37}$$

where 0 is the null vector. Clearly, the constant function $x(t) = \bar{x}$ is a solution of Eq. 5.35 since the velocity vector dx/dt vanishes at $\bar{x}$.

Now, we discuss some of the important issues involved in neurodynamics. Neurodynamics possesses the following general characteristics:

1. *A large number of degrees of freedom:* The human cortex is a highly parallel, distributed system that is estimated to possess about 10 billion neurons, with each neuron modeled by one or more state variables. It is generally believed that both the computational power and the fault-tolerant capability of such a neurodynamic system are the result of the collective dynamics of the system. The system is characterized by a very large number of coupling constants represented by the strengths of the individual synaptic junctions.

2. *Nonlinearity:* A neurodynamic system is inherently nonlinear. In fact, nonlinearity is essential for creating a universal computing machine.

3. *Dissipation:* A neurodynamic system is dissipative. It is therefore characterized by the convergence of the state space volume onto a manifold of lower dimensionality as time goes on.

4. *Noise:* Noise is an intrinsic characteristic of neurodynamic systems. In real-life neurons, membrane noise is generated at synaptic junctions.

5.5 STABILITY ANALYSIS OF RECURRENT NEURAL NETWORKS

For the purpose of studying the stability of recurrent neural networks, it is necessary to be familiar with the definition of stability, in the context of an autonomous nonlinear dynamic system (Eq. 5.35 with equilibrium state $\bar{x}$ [Khalil, 1992].

Definition 5.1 The equilibrium state $\bar{x}$ is said to be uniformly stable if, for any positive constant ε, there exists another positive constant $\delta = \delta(\varepsilon)$ such that the condition

$$\|x(0) - \bar{x}\| < \delta$$

implies that

$$\|x(t) - \bar{x}\| < \varepsilon$$

for all $t > 0$.

Definition 5.2 The equilibrium state $\bar{x}$ is said to be convergent if there exists a positive constant δ such that the condition

$$\|x(0) - \bar{x}\| < \delta$$

implies that

$$x(t) \rightarrow \bar{x} \text{ as } t \rightarrow \infty$$

Definition 5.3 The equilibrium state $\bar{x}$ is said to be asymptotically stable if it is both stable and convergent.

Definition 5.4 The equilibrium state $\bar{x}$ is said to be globally asymptotically stable if it is stable and all trajectories of the system converge to $\bar{x}$ as time t approaches infinity.

By making a comparison among the above definitions, we find that uniform stability means that a trajectory of the system can be made to stay within a small neighborhood of the equilibrium state $\bar{x}$ if the initial state $x(0)$ is close to $\bar{x}$. In addition, the convergence reveals that if the initial state $x(0)$ of a trajectory is close enough to the equilibrium state $\bar{x}$, then the trajectory described by the state vector $x(t)$ will approach $\bar{x}$ as time t approaches infinity. Furthermore, it is only when stability and convergence are both satisfied that we have asymptotic stability. Global asymptotic stability implies that the system will ultimately settle down to a steady state for any choice of initial conditions.

The definition of the positive definite function is also required when doing stability analysis. A function $V(x)$ is called positive definite if it satisfies the following conditions:

1. The function $V(x)$ has continuous partial derivatives with respect to the elements of the state x.
2. $V(\bar{x}) = 0$.
3. $V(x) > 0$ if $x \in \mathfrak{A} - \bar{x}$,, where $\mathfrak{A}$ is a small neighborhood around $\bar{x}$.

Incidentally, the definition of negative definite can be easily derived in the light of the above conditions.

An elegant approach for investigating the stability of dynamic systems is the direct method of Lyapunov, which is based on a continuous scalar function of the state, called a Lyapunov function. The Lyapunov's theorems on the stability analysis of the dynamic system (5.35) are stated as follows [Khalil, 1992]:

Theorem 5.1   The equilibrium state $\bar{x}$ is stable if, in a small neighborhood of $\bar{x}$, there exists a positive definite function $V(x)$ such that its derivative with respect to time is negative semidefinite in the region.

Theorem 5.2   The equilibrium state $\bar{x}$ is asymptotically stable if, in a small neighborhood of $\bar{x}$, there exists a positive definite function $V(x)$ such that its derivative with respect to time is negative definite in the region.

The scalar function $V(x)$ that satisfies the requirements of these two theorems is called a Lyapunov function for the equilibrium state $\bar{x}$.

To summarize, given that $V(x)$ is a Lyapunov function, then according to Theorem 5.1, the equilibrium state $\bar{x}$ is stable if

$$\frac{d}{dt} V(x) \leq 0, \quad \text{for} \quad x \in \mathfrak{A} - \bar{x} \tag{5.38}$$

Similarly, according to Theorem 5.2, the equilibrium state $\bar{x}$ is asymptotically stable if

$$\frac{d}{dt} V(x) < 0, \quad \text{for} \quad x \in \mathfrak{A} - \bar{x} \tag{5.39}$$

It should be noted that global stability of a nonlinear dynamic system generally requires that the condition of radial unboundedness holds, that is,

$$V(x) \rightarrow \infty \text{ as } \|x\| \rightarrow \infty$$

This condition is usually satisfied according to the Lyapunov functions constructed for neural networks with sigmoid activation functions.

5.5.1 Stability Analysis of the Hopfield Network

To facilitate the stability analysis of Hopfield network (5.28), we make the following three assumptions [Haykin, 2009]:

1. The matrix of synaptic weights is symmetric, that is,

$$w_{ji} = w_{ij}, \quad \text{for all i and j} \tag{5.40}$$

2. Each neuron has a nonlinear activation of its own, that is, $\phi_i(\cdot)$.
3. The inverse of the nonlinear activation function exists. According to (5.26), we may write

$$v = \phi_i^{-1}(x) \tag{5.41}$$

Let the sigmoid function $\phi_i(\cdot)$ be defined by the hyperbolic tangent function as

$$x = \phi_i(v) = \tanh\left(\frac{a_i v}{2}\right) = \frac{1 - \exp(-a_i v)}{1 + \exp(-a_i v)} \tag{5.42}$$

which has a slope of $a_i/2$ at the origin, that is,

$$\frac{a_i}{2} = \frac{d\phi_i}{dv}\bigg|_{v=0} \tag{5.43}$$

Thus, we refer to a_i as the gain of neuron i.

Based on (5.42), the inverse output–input relation of (5.42) can be further written as

$$v = \phi_i^{-1}(x) = -\frac{1}{a_i} \ln\left(\frac{1-x}{1+x}\right) \tag{5.44}$$

Let the standard form of the inverse output–input relation for a neuron of unity gain be denoted as

$$\phi^{-1}(x) = -\ln\left(\frac{1-x}{1+x}\right) \tag{5.45}$$

Then, Eq. 5.44 has an equivalent form presented as follows:

$$\phi_i^{-1}(x) = \frac{1}{a_i}\phi^{-1}(x) \tag{5.46}$$

Define the Lyapunov function of the Hopfield network shown in Figure 5.3 as

$$V = -\frac{1}{2}\sum_{i=1}^{N}\sum_{j=1}^{N} w_{ji}x_ix_j + \sum_{j=1}^{N}\frac{1}{R_j}\int_0^{x_j}\phi_j^{-1}(x)dx - \sum_{j=1}^{N} I_jx_j \qquad (5.47)$$

Differentiating V with respect to time t, we can obtain

$$\begin{aligned} \frac{dV}{dt} &= -\sum_{i=1}^{N}\sum_{j=1}^{N} w_{ji}x_i\frac{dx_j}{dt} + \sum_{j=1}^{N}\frac{1}{R_j}\phi_j^{-1}(x_j)\frac{dx_j}{dt} - \sum_{j=1}^{N} I_j\frac{dx_j}{dt} \\ &= -\sum_{j=1}^{N}\left(\sum_{i=1}^{N} w_{ji}x_i - \frac{1}{R_j}\phi_j^{-1}(x_j) + I_j\right)\frac{dx_j}{dt} \end{aligned} \qquad (5.48)$$

From (5.10) and (5.41), we get

$$\frac{dV}{dt} = -\sum_{j=1}^{N}\left(\sum_{i=1}^{N} w_{ji}\phi_i(v_i) - \frac{v_j}{R_j} + I_j\right)\frac{dx_j}{dt} \qquad (5.49)$$

Combining (5.12) with (5.49), we can further derive that

$$\frac{dV}{dt} = -\sum_{j=1}^{N} C_j\frac{dv_j}{dt}\frac{dx_j}{dt} \qquad (5.50)$$

Next, we apply (5.49) to (5.50) and find that

$$\begin{aligned} \frac{dV}{dt} &= -\sum_{j=1}^{N} C_j\frac{d\phi_j^{-1}(x_j)}{dt}\frac{dx_j}{dt} \\ &= -\sum_{j=1}^{N} C_j\frac{d\phi_j^{-1}(x_j)}{dx_j}\left(\frac{dx_j}{dt}\right)^2 \end{aligned} \qquad (5.51)$$

In accordance with Figure 2.11, we see that the hyperbolic tangent function is a monotonically increasing function. Thus, the inverse output–input relation $\phi_j^{-1}(x_j)$ is also a monotonically increasing function of the output x_j, which implies that

$$\frac{d\phi_j^{-1}(x_j)}{dx_j} \geq 0, \quad \text{for all } x_j \qquad (5.52)$$

Clearly,

$$\left(\frac{dx_j}{dt}\right)^2 \geq 0, \quad \text{for all } x_j \tag{5.53}$$

Therefore, for the energy function V defined in (5.31), we have

$$\frac{dV}{dt} \leq 0, \quad \text{for all t} \tag{5.54}$$

Additionally, from the definition of (5.47), we note that the function V is bounded. In summary, the energy function V is a Lyapunov function of the continuous Hopfield model. Furthermore, the model is stable in accordance with Theorem 5.1.

The time evolution of the continuous Hopfield model described by the system of nonlinear first-order differential equations given in (5.11) represents a trajectory in state space that seeks out the minima of the Lyapunov function V and comes to a stop at such fixed points. From Eq. 5.51, we note that the derivative dV/dt vanishes only if

$$\frac{dx_j}{dt} = 0, \quad \text{for all } j \tag{5.55}$$

Hence, we have

$$\frac{dV}{dt} < 0, \text{ except at a fixed point} \tag{5.56}$$

which reveals that the Lyapunov function V of a Hopfield network is a monotonically decreasing function of time. Consequently, the Hopfield network is asymptotically stable in the Lyapunov sense.

5.5.2 Stability Analysis of the Cohen–Grossberg Network

The Lyapunov function corresponding to Eq. 5.24 is defined as

$$V = \frac{1}{2}\sum_{i=1}^{N}\sum_{j=1}^{N} c_{ji}\phi_i(u_i)\phi_j(u_j) - \sum_{j=1}^{N}\int_0^{u_j} b_j(\lambda)\phi_j'(\lambda)d\lambda \tag{5.57}$$

where $\phi_j'(\lambda)$ is the derivative of $\phi_j(\lambda)$ with respect to λ. The following conditions should be held in order to ensure the validity of definition of Eq. 5.41.

1. The synaptic weights of the network are symmetric, that is,

$$c_{ji} = c_{ij} \tag{5.58}$$

2. The function $a_j(u_j)$ satisfies the nonnegativity condition, that is,

$$a_j(u_j) \geq 0 \tag{5.59}$$

3. The nonlinear input–output function $\phi_j(u_j)$ satisfies the monotonicity condition, that is,

$$\phi_j'(u_j) = \frac{d\phi_j(u_j)}{du_j} \geq 0 \tag{5.60}$$

With this background, we may now formally state the Cohen–Grossberg theorem.

Theorem 5.3 [Haykin, 2009]   Provided that the system of nonlinear differential equations (5.24) satisfies the conditions of symmetry, nonnegativity, and monotonicity, the Lyapunov function V of the system defined by Eq. 5.57 satisfies the condition

$$\frac{dV}{dt} \leq 0 \tag{5.61}$$

Once this basic property of the Lyapunov function V is in place, stability of the system follows from Theorem 5.1.

By comparing the general system of Eq. 5.24 with the system of Eq. 5.11 for a continuous Hopfield model, we may make the correspondences between the Hopfield model and the Cohen–Grossberg theorem that are summarized in Table 5.1. Applying the correspondences in Table 5.1 to Eq. 5.57, we can obtain a Lyapunov function for the continuous Hopfield model expressed as

$$V = -\frac{1}{2}\sum_{i=1}^{N}\sum_{j=1}^{N} w_{ji}\phi_i(v_i)\phi_j(v_j) + \sum_{j=1}^{N}\int_0^{v_j}\left(\frac{v_j}{R_j} - I_j\right)\phi_j'(v)dv \tag{5.62}$$

TABLE 5.1   Correspondences between the Cohen–Grossberg Theorem and the Continuous Hopfield Model

Cohen–Grossberg Theorem	Hopfield Model
u_j	$C_j(v_j)$
$a_j(u_j)$	1
$b_j(u_j)$	$-\dfrac{v_j}{R_j} + I_j$
c_{ji}	$-w_{ji}$
$\phi_i(u_i)$	$\phi_i(v_i)$

Note that the following equations hold:

1. $\phi_i(v_i) = x_i$.
2. $\int_0^{v_j} \phi_j'(v)dv = \int_0^{x_j} dx = x_j$.
3. $\int_0^{v_j} v\phi_j'(v)dv = \int_0^{x_j} v\,dx = \int_0^{x_j} \phi_j^{-1}(x)dx$.

By applying them to Eq. 5.62, we can further obtain a Lyapunov function that is identical with Eq. 5.47. This demonstrates that the Hopfield model can be seen as a special case of the Cohen–Grossberg theorem [Haykin, 2009].

The Cohen–Grossberg theorem is a general principle of neurodynamics with a wide range of applications.

EXERCISES

5.1. The Lyapunov function of a Hopfield network is written as

$$V = -\frac{1}{2}\left(7x_1^2 + 12x_1x_2 - 2x_2^2\right)$$

Point out the matrix of synaptic weight of the network.

5.2. Compute the equilibrium state of the following dynamic system:

$$\frac{dx}{dt} = \begin{bmatrix} -x_1 + x_2^2 \\ -x_2(x_1 + 1) \end{bmatrix}$$

Then, confirm the stability of the equilibrium state by choosing the Lyapunov function as

$$V(x) = x^T x$$

5.3. Given a nonlinear dynamic system as

$$\frac{dx}{dt} = \begin{bmatrix} x_2 - 2x_1\left(x_1^2 + x_2^2\right) \\ -x_1 - 2x_2\left(x_1^2 + x_2^2\right) \end{bmatrix}$$

study the stability of the origin, by making use of the following Lyapunov function:

$$V(x) = \alpha x_1^2 + \beta x_2^2$$

5.4. The discrete (time) gamma model of a neurodynamical system is described by the following pair of equations:

$$x_j(k) = \phi\left(\sum_{i<j}\sum_m w_{ji}^{(m)} x_i^{(m)}(k)\right) + K_j$$

and

$$x_j^{(m)}(k) = \left(1 - \mu_j\right) x_j^{(m)}(k-1) + \mu_j x_j^{(m-1)}(k-1)$$

where k denotes discrete time, $j = 1, 2, \ldots, N$, and $m = 1, 2, \ldots, M$.

1. Compare the discrete gamma model with the model described in (5.6).

2. Construct a signal flow graph for the recursive part of the gamma model.

3. Find the value of the control parameter μ_j for which the discrete gamma model is stable.

5.5. Consider a Hopfield network made up two neurons. The synaptic weight matrix of the network is

$$W = \begin{bmatrix} 0 & -1 \\ -1 & 0 \end{bmatrix}$$

The bias applied to each neuron is zero. The following are the four possible states of the network:

$$x^1 = [+1, +1]^T$$
$$x^2 = [-1, +1]^T$$
$$x^3 = [-1, -1]^T$$
$$x^4 = [+1, -1]^T$$

1. Demonstrate that states x^2 and x^4 are stable, whereas states x^1 and x^3 exhibit a limit cycle. Do this demonstration by using the stability condition and energy function, respectively.

2. Confirm the length of the limit cycle characterizing the states x^1 and x^3.

5.6. Define the Lyapunov function related to the cellular neural network described in Section 5.3 as

$$V = -\frac{1}{2}\sum_{(i,j)}\sum_{(k,l)} A(i, j; k, l) v_{yij}(t) v_{ykl}(t) + \frac{1}{2R_x}\sum_{(i,j)} v_{yij}^2(t)$$
$$- \sum_{(i,j)}\sum_{(k,l)} B(i, j; k, l) v_{yij}(t) v_{ukl} - \sum_{(i,j)} Iv_{yij}(t)$$

Try to perform the stability analysis of the network.

PART II

FUZZY SET THEORY
AND FUZZY LOGIC

■■■■ **CHAPTER 6**

Basic Fuzzy Set Theory

6.1 INTRODUCTION[1]

Uncertainty[2] is universal! Take, for example, a computer system whose task is to recognize trees in a visual image. Sources of uncertainty in this task include (but are not limited to) noise in the sensed imagery,[3] distortion due to pose and lens conditions, variability of the class of interest (what is a "tree"?), faithfulness of the features used to described a tree, missing features, spatial context (a tree in a forest versus a tree in New York City), temporal context (a tree in summer versus a tree in winter), the choice of recognition algorithm, and so on. If multispectral (or hyperspectral) imagery is available or multiple algorithms are applied to the decision-making aspect, then the problems of how to fuse compensatory or even conflicting information becomes important.

The historical framework for dealing with uncertainty has been probability theory. This is a powerful tool that has served science well in modeling situations where the primary source of uncertainty is randomness. In some instances, we argue that uncertainty takes other forms. Many times, instead of asking *whether* something is true, we ask *how much* it is true, that is, how much is a certain property exhibited in a particular instance. For example, we may want to know how much a particular object matches an ideal prototype.

Jim Bezdek [1993] in the inaugural editorial of the *IEEE Transactions on Fuzzy Systems* gave the following example: You are dying of thirst in a desert when you come across two bottles. One has a label that says "probability of being potable is

---

[1] Chapters 6–9 follow the pattern in a short introductory chapter in Xu *et al.* [2008] with permission from World Scientific Publishing & Imperial College Press. Several good applications of fuzzy set theory to bioinformatics can be found in that source.

[2] We use the term "uncertainty" in a very general sense. For the most part, we concentrate on the notions of vagueness, similarity, or preference as opposed to ambiguity.

[3] Noise can be thought of as a source of ambiguity: There is a true value but it is hard to distinguish which one is the correct value from a set of possible answers. Fuzzy measures in Chapter 9 also provide a characterization of ambiguity.

Fundamentals of Computational Intelligence: Neural Networks, Fuzzy Systems, and Evolutionary Computation, First Edition. James M. Keller, Derong Liu, and David B. Fogel.

0.91," while the other reads "membership in the class of potable liquids is 0.91." Jim asks the question "which one would you drink?" The probability bottle may contain something really tasty or it may contain acid. The fuzzy bottle, assuming a reasonable definition of membership, will probably not have the tastiest liquid, but shouldn't hurt you. Is one model of uncertainty better than the other? No! They measure different aspects of uncertainty and should be used as is appropriate to a particular problem.

In many cases, there is a lack of clear boundaries between classes of objects: When does an image object stop being a tree and become, say, a bush? In doing some landscaping, we planted a Crepe Myrtle, which turns out to be both a bush and a tree early in its life, that is, there are objects that "completely" belong to multiple classes. In these situations, alternative methodologies should be utilized to aid us in making automated evaluations. Fuzzy set theory and fuzzy logic provide a different way to view the problem of modeling uncertainty and offer a wide range of computational tools to aid decision making. Clearly, it is not our intention to diminish the vital role of probabilistic models in science and engineering. Fuzzy set theory and fuzzy logic provide complementary information to that which comes from a probabilistic view.

Many science and engineering problems are formulated in a deterministic manner. Most of these problems are defined by fixed objective functions and solved through optimization. Many dynamic processes are modeled using differential equations with deterministic behavior. However, there are at least three situations in which fuzziness should be considered: intrinsic fuzziness in real-life systems (e.g., biology), multiple states or roles of a real object (organisms that can be male or female as environmental conditions dictate), and fuzzy descriptions of phenomena, that is, when our knowledge is incomplete and/or vague. For example, our descriptions of many biological concepts often have difficulty fitting into a deterministic (crisp) explanation. As a result, our knowledge, concepts, and representations of biological and other domain terms may also be fuzzy. Hybrids of flowers or hybrids of types of engines in an automobile make crisp classification of species or vehicle type problematic. The auto industry just created a new class called Hybrid to solve the marketing issues.

Descriptions for similarity and typicality can be fuzzy. Consider defining a chair. How much does a king's throne resemble a three-legged stool? What properties do they (partially) share? How close is a given example to the prototypical chair? Such fuzziness could result from the limitations of classifications, natural language, or poor understanding of the underlying mechanism. Tolerance of fuzziness allows us to explore complex concepts effectively.

But we still ask the question: do we really need fuzziness? If the world were deterministic, the answer would be no. Boolean logic and probability theory would clearly suffice. This is the "balls in the urn" world; you know that when you put 12 red balls and 7 green balls into an urn and ask questions like "If you pick out 5 balls, what's the probability that they are all red?" If however, you put balls of varying radii into the urn and picked out five, what's the probability that they are all SMALL? Here the event, SMALL balls, is ambiguous. You can convert this into the former case if you set a threshold radius below which a ball is considered SMALL, but then two balls just on each side of the threshold will feel the same, but one will be SMALL and the other Not SMALL. So, establishing such a threshold doesn't match well with human intuition in ambiguous cases. It would be nice to have models and calculi that handle these situations.

Fuzzy set theory in general and fuzzy logic specifically are natural ways to model ambiguous events that occur in human-like reasoning. People have no trouble operating with phrases such as "large risk factor," "somewhat likely to be involved in cancer," "accelerate rapidly," and so on. As will be seen, rules containing such ambiguous clauses can be successfully handled in a fuzzy logic system.

The beauty (and also a danger, if we are not careful) of fuzzy set theory is that it offers a multitude of calculi for the fusion of partial support for a hypothesis under investigation, that is, flexible mechanisms to increase or decrease confidence in a decision as evidence unfolds. In his seminal text on computer vision, David Marr [1982] stated two principles to be followed in the design of intelligent (vision) algorithms. The first is called the principle of least commitment (PLC). He states it simply as "Don't do something that later must be undone." Hence, in a complex computing scenario, one where there are many decision-making steps, avoid making deterministic decisions for as long as is possible. It is very difficult, perhaps impossible, to recover from a wrong crisp decision early on. Keep your options open until the situation demands a final answer. While Marr was interested in computer vision, the PLC certainly applies to all complex automated decision-making problems. Clearly, the concept of assigning and maintaining degrees of membership (perhaps confidence in competing hypotheses) or more general linguistic labels in fuzzy set theory supports the PLC for complex decision-making applications such as bioinformatics, eldercare, landmine detection, and human activity recognition.

The second principle of Marr is called the principle of graceful degradation (PGD). By this he meant that algorithms should deliver a partial (reasonable) answer as input degrades. In other words, intelligent algorithms should encompass a degree of robustness and continuity. Here also, techniques that utilize membership degrees or other fuzzy constructs in the calculation of their response to input conditions have the potential to degrade much more gracefully than their crisp counterparts. Consider, for example, a simple stability assessment for an elder that relies on a single test value, say the amount of time it takes the elder to get up from a chair and walk 10 m, turn around, walk back, and sit down (similar to an actual test and good for illustrative purposes). The standard outcome is categorical (one of four discrete states) and the algorithm is crisp, based on time threshold. A slight difference in how the stopwatch was pressed can (and does) actually flip the screening result from a high-risk- to medium-risk category. If however, the output is a set of memberships in the various categories generated by putting trapezoidal functions around the thresholds [Klir and Yuan, 1995], such small perturbations can be ameliorated and a more realistic assessment of stability generated. Figure 6.1 shows a typical trapezoidal curve, along with other standard fuzzy membership functions. This simple example illustrates the point that fuzzy models embrace the concept of the PGD.

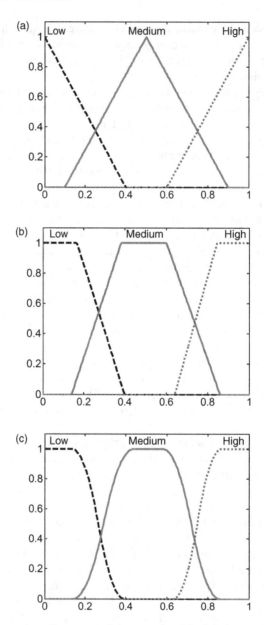

FIGURE 6.1 Examples of common fuzzy membership functions where $X = [0,1]$. (a) triangular, (b) trapezoidal, and (c) smooth quadratic functions. See Example 6.1 and Problems 6.1 and 6.2.

6.2   A BRIEF HISTORY

Concepts of vagueness and fuzziness have been contemplated in mathematics and science for quite a while. For example, in 1923 Bertrand Russell [1923] stated that "All traditional logic habitually assumes that precise symbols are being employed. It is therefore not applicable to this terrestrial life, but only to an imagined celestial existence." Like Russell, the philosopher Max Black [1937] was concerned with vagueness and imprecision in language, and the effect of these concepts on logic. In fact, he believed that all terms whose application involves using our senses are vague. Black, in 1937, actually came up with the concept that we now associate with membership functions. He even conducted a cognitive psychological experiment with a group of people that effectively constructed membership functions exemplifying vagueness of certain words. However, most people attribute the beginning of fuzzy set theory to Lotfi A. Zadeh's 1965 paper [Zadeh, 1965] that developed this topic in its current form. An excellent treatment of the history of fuzzy sets and fuzzy logic can be found in Seising [2005]. The journal *Fuzzy Sets and Systems* published a "40th Anniversary of Fuzzy Sets" in December 2005 that contains 14 position papers covering various aspects of the role and future prospects of fuzzy sets [Dubois, 2005].

The mathematical basis for formal fuzzy logic can be found in infinite-valued logics, first studied by the Polish logician Jan Lukasiewicz in the 1920s (see Borkowski [1970]). Lukasiewicz constructed a series of multivalued logical systems, generalizing from small finite numbers of truth values to those containing infinite sets of truth values. His work and calculation formulas are ingrained in modern fuzzy set theory and fuzzy logic, the genesis of which is credited to Zadeh in his seminal three-part treatise on the theory and applications of linguistic variables [Zadeh, 1975a, 1975b, 1976].

Perhaps the biggest boost to the visibility and perceived utility of fuzzy set theory came from the application of rule-based fuzzy systems to problems in control [Mamdani and Assilian, 1975; Mamdani, 1977; Sugeno, 1985; Takagi and Sugeno, 1985; Verbruggen and Babuska, 1999, Yurkovich and Passino, 1999]. In what has become commonplace now, sets of linguistically described rules were created and inserted into a variety of nonlinear control systems. The ease of design and the smoothness of the control surface from only a handful of rules made fuzzy controllers very popular in a variety of products from the automotive industry, consumer electronics markets, and so on. Fuzzy controllers are well suited for low-cost embedded systems.

While the big economic impact of fuzzy set theory and fuzzy logic centers on control, particularly in consumer electronics, there has been, and continues to be, much research and application of these technologies in pattern recognition, information fusion, data mining, and automated decision making [Keller *et al.*, 1996; Bezdek *et al.*, 1999]. There are national, multinational, and international fuzzy systems professional societies around the globe whose purposes are to foster research, development, and application of fuzzy set theory and fuzzy logic. Fuzzy systems comprise one of the core pillars of the IEEE Computational Intelligence Society.

The "fuzzy" chapters of this book cover the basics of fuzzy set theory and fuzzy logic along with a few more advanced topics, but there is a wealth of literature to explore. For example, the reader is referred to Klir and Yuan [1995] and Bezdek *et al.*

[1999] for more extensive development of the theory and selected applications. In Xu *et al.* [2008], fuzzy set theory and fuzzy logic are examined in several applications to bioinformatics.

6.3 FUZZY MEMBERSHIP FUNCTIONS AND OPERATORS

6.3.1 Membership Functions

Traditional set theory is based on binary, or two-valued, logic. Given a "universe" set X, a subset A of X can be defined in several ways. Suppose that X is the set of integers. The subset of prime numbers less than 10 can be specified by listing its members:

$$A = \{2, 3, 5, 7\} \tag{6.1}$$

or by providing defining properties:

$$A = \{x \in X | x \text{ is a positive integer less than } 10 \text{ and } x \text{ has only two distinct divisors: } 1 \text{ and } x\} \tag{6.2}$$

Alternatively, we define a subset A by its characteristic function, which is also denoted by the set name, $A: X \rightarrow \{0, 1\}$ from X into the binary set $\{0,1\}$ given by

$$A(x) = \begin{cases} 1, & \text{if } x \in A \\ 0, & \text{if } x \notin A \end{cases} \tag{6.3}$$

Zadeh [1965] simply defined a fuzzy subset of X as a function $A: X \rightarrow [0, 1]$, that is, a characteristic function from X into the interval [0,1]. The value $A(x)$ is called the membership of the point x in the fuzzy set A or the degree to which the point x belongs to the set A. For example, the fuzzy subset of "big positive integers" could be defined by

$$A(x) = \begin{cases} 1 - \dfrac{1}{\sqrt{x}}, & \text{if } x > 0 \\ 0, & \text{else} \end{cases} \tag{6.4}$$

All fuzzy set theory is based on the concept of a membership function. Where do these membership functions come from? In many cases, they are defined as in the two examples above: Common sense definitions that convey some linguistic expression. More generally, they come from expert knowledge directly or they can be derived from questionnaires, heuristics, and so on. This is a human-centric view and is certainly open to debate. In many cases, the membership functions take on specific functional forms such as triangular, trapezoidal, S-functions, pi-functions, sigmoids, and even Gaussians for convenience in representation and computation. Pi-functions and S-functions are constructed from quadratic functions "pieced together" to make smooth curves. Figure 6.1 displays several common fuzzy membership functions, with definitions in Example 6.1. Alternatively, membership functions (or the parameters of the specific equation forms) can be learned from training data, much as probability density functions are learned. Some fuzzy clustering algorithms naturally produce membership functions as their output. A neural network, given the proper input/output training data, also acts as a membership function for new input.

One of our favorite practical membership functions comes from the field of pain assessment. Figure 6.2a is a rendition of an analog version of a pain scale, in the spirit of McGrath *et al.* [1996] and Marquie *et al.* [2007]. A patient is asked to slide the bar to a position indicative of his or her level of pain. The color and width provide a guide. The corresponding scale on the right is effectively a membership in the fuzzy set "pain," and can be used as a guide to pain remediation treatment. The goal is to provide sufficient pain medication without overdosing. This is a continuous membership function. For young children, and in fact for many people, a discrete pain scale is preferred. You can find these in doctors' offices and clinics, and are usually made from a set of graphic "faces" going from happy (no pain) to crying (most pain) [Hicks *et al.*, 2001; Wong *et al.*, 2001]. A discrete membership array accompanies the set of figures. Figure 6.2b contains our "pirate pain scale" as an example.

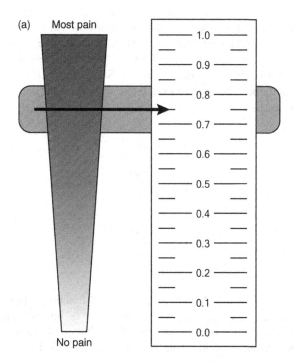

FIGURE 6.2   (*Continued*)

(b)

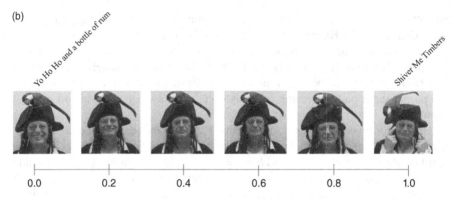

FIGURE 6.2 Practical membership function generators from a pain assessment instrument for medication dosage. In (a), the patient slides a marker to the place on the visual scale that best represents his or her level of pain. The nurse can read off the analog "pain membership" value from the corresponding scale. Part (b) is discrete version for younger children or others to pick a face to estimate a similar, although quantized, version of pain membership.

EXAMPLE 6.1   DEFINITIONS OF MEMBERSHIP FUNCTIONS

It is pretty obvious how to define triangular and trapezoidal fuzzy membership functions over a real-valued domain, that is, an interval subset of $\Re$ (piecewise linear functions—right?) (see Problem 6.1). S-functions are defined by three parameters (a,b,c) where $a < b < c$. The function $S(x; a,b,c)$ is required to be 0 up to a; a parabola that opens up from a to b with $S(b; a,b,c) = 1/2$; a parabola that "matches up" and opens downward from b to c with $S(c; a,b,c) = 1$. The equation for such an S-function is

$$S(x; a, b, c) = \begin{cases} 0, & x \leq a \\ \dfrac{(x-a)^2}{2(b-a)^2}, & a < x \leq b \\ \dfrac{-(x-c)^2}{2(b-c)^2} + 1, & b < x \leq c \\ 1, & x > c \end{cases} \tag{6.5}$$

How do we get that equation? Do Problem 6.3. A Z-function is just the "flip" of the S-function, that is, $Z(x;a,b,c) = 1 - S(x;a,b,c)$. A pi-function just pieces an S-function with a Z-function to look something like a Gaussian, although it actually reaches 0 and is made from parabolic sections. It has six parameters $a < b < c \leq d < e < f$ and is defined by

$$Pi(x; a, b, c, d, e, f) = \begin{cases} S(x; a, b, c), & x \le c \\ 1, & c < x < d \\ Z(x; d, e, f), & x \ge d \end{cases} \qquad (6.6)$$

Note that if $c < d$, a pi-function resembles a "soft" trapezoid. Of course, we can (and do) use properly scaled Gaussian functions as membership functions in many applications. They have the advantage of having derivatives of all orders, but they never actually reach 0.

Notice that all of these membership functions have at least one value where the function value is 1. A fuzzy set A whose membership function $A(x)$ "reaches" 1 is called *normal*. To be precise, define the height of A by $ht(A) = \sup_{x \in X}\{A(x)\}$. We use supremum (sup) instead of maximum to handle certain infinite domain cases, like the logistic membership function $A(x) = 1/(1 + e^{-x})$, where the domain X is the set of all real numbers $\mathfrak{R}$. $A(x)$ never actually reaches 1 but is asymptotic to 1. Its height is 1. If $Ht(A) < 1$, we say A is *subnormal*.

6.3.2 Basic Fuzzy Set Operators

Once fuzzy subsets of a universal set X are defined, definitions for the complement of a set, the union of two sets, and the intersection of two sets are required to actually generate a "set theory." In 1965, Zadeh proposed the following. Suppose $A: X \to [0, 1]$ is a fuzzy subset of X. The complement A^c of A is defined by

$$A^c(x) = 1 - A(x) \qquad (6.7)$$

Additionally, if $B: X \to [0, 1]$ is another fuzzy subset of X, Zadeh defined

$$(A \cup B)(x) = \max\{A(x), B(x)\} = A(x) \vee B(x) \qquad (6.8)$$

and

$$(A \cap B)(x) = \min\{A(x), B(x)\} = A(x) \wedge B(x) \qquad (6.9)$$

Why did Zadeh define the operators in this manner? Quite obvious, it was because these definitions revert back to the standard crisp definitions if the subsets are crisp. Hence, this forms a true extension of normal set theory. As can be found in the many textbooks on fuzzy set theory (see, for example, Klir and Yuan [1995] and Pedrycz and Gomide [1998]), all of the theorems of crisp set theory hold for this fuzzy set theory except two: the law of contradiction (LOC) and the law of excluded middle (LEM).

The LOC states that the intersection of a set and its complement must be empty ($A \cap A^c = \phi$), while the LEM requires that the union of a set and its complement must be the whole universe set ($A \cup A^c = X$). Since crisp set theory is formally equivalent to the first-order predicate logic, these two laws state that a proposition cannot be both true and not true simultaneously, and that either a proposition or its negation (complement) must be true. While these statements seem reasonable, they give rise to a paradox within classical logic, commonly called Russell's paradox. A simple version goes something like this: Russell's barber has a sign that states "I shave everyone, and only those, who do not shave themselves." Then who shaves the barber? If he shaves himself, then he cannot (he shaves only those who do not shave themselves); but if he does not shave himself, then he must (since he shaves everyone who don't shave themselves). Such a dilemma! As mentioned earlier, the Crepe Myrtle is both a tree and a nontree (it is also a bush). Hence, it is impossible to put it only into one set; it also naturally fits into the complement set. So, perhaps it is not that unreasonable to disobey the LOC and the LEM (Can you demonstrate this? See Problem 6.7).

EXAMPLE 6.2   FUZZY SET PROPERTIES

A Simple law: Show that standard fuzzy set theory satisfies the commutative law for union, that is, $(A \cup B) = (B \cup A)$. This will follow from what we know about real numbers. We have to show that for each $x \in X$, $(A \cup B)(x) = (B \cup A)(x)$. Just "follow your nose": $(A \cup B)(x) = A(x) \vee B(x) = B(x) \vee A(x) = (B \cup A)(x)$. The commutative law for union is a result of the fact that the maximum of two real numbers is commutative.

A little more complicated law: Show that standard fuzzy set theory satisfies the distributive law: $(A \cup (B \cap C)) = (A \cup B) \cap (A \cup C)$.

To verify this property, we have to show that for each $x \in X$, $(A \cup (B \cap C))(x) = ((A \cup B) \cap (A \cup C))(x)$.

First consider the left-hand side (LHS) of the conjectured equality.

Now, $(A \cup (B \cap C))(x) = A(x) \vee (B \cap C)(x) = A(x) \vee (B(x) \wedge C(x))$. Similarly, the right-hand side (RHS) becomes $((A \cup B) \cap (A \cup C))(x) = (A \cup B)(x) \wedge (A \cup C)(x) = (A(x) \vee B(x)) \wedge (A(x) \vee C(x))$. The proof is established by looking at all the possible configurations for the three values $A(x), B(x)$, and $C(x)$:

Case 1. $A(x) \leq B(x) \leq C(x)$. Here, the LHS becomes $A(x) \vee B(x) = B(x)$ and similarly the RHS becomes $(A(x) \vee B(x)) \wedge (A(x) \vee C(x)) = B(x) \wedge C(x) = B(x)$, that is, for case 1, LHS = RHS. $\sqrt{}$

Case 2. $A(x) \leq C(x) \leq B(x)$. You get the idea, right?

For all configurations, you verify that the operations on the LHS and RHS give the same result. How many configurations are there for three real numbers? (see Exercise 6.5). It's tedious but straightforward to verify this law for fuzzy set theory. The proof is like the truth table proof done for the crisp version of the law in a digital logic class (except that there the three variables can only take on values of 0 and 1).

Suppose that the membership function values for a particular element x in X are interpreted as the confidences that x possesses certain properties, for example, A(x) is the confidence that an image region x is LONG and B(x) corresponds to the confidence that x is STRAIGHT. Then the original Zadeh definitions of complement, union, and intersection produce confidences related to the linguistic concepts of NOT, OR, and AND: $A^c(x)$ is the confidence that x is NOT LONG; (A∪B)(x) gives the degree to which x is either LONG or STRAIGHT (or both); (A∩B)(x) computes the confidence that x is both LONG and STRAIGHT.

The 2012 World Congress on Computational Intelligence was held in Brisbane Australia. If A is the set of mammals, then the platypus at the Lone Pine Koala Sanctuary is an example of an object that has nonzero membership in both A and A^c. Of course, if B represents all reptiles, then B(platypus) > 0 and B^c(platypus) > 0[4]. So, nonbinary confidence values for concept membership are a normal occurrence of nature, and hence, should be considered in our computational models of nature.

The good news and the bad news in fuzzy set theory is that there are infinite numbers of ways to define complement, union, and intersection [Dubois and Prade, 1985; Klir and Yuan, 1995].[5] An alternative fuzzy set theory that is useful for fuzzy logic inference is generated by the operators,

$$(A\cup_b B)(x) = 1 \wedge (A(x) + B(x)) \tag{6.10a}$$

$$(A\cap_b B)(x) = 0 \vee (1 - (A(x) + B(x))) \tag{6.10b}$$

called the bounded sum and bounded difference, along with the standard complement,

$$A^c(x) = 1 - A(x) \tag{6.10c}$$

Each such extension of crisp set theory loses either LOC and LEM or two other properties (idempotency and distributivity) [Klir and Yuan, 1995]. For this choice, LOC and LEM are satisfied, while idempotency and distributivity are lost. Actually, there are infinite families of union, intersection, and complement operators that are extremely useful in multicriteria decision making where partially supported criteria are to be combined in disjunctive (OR) and/or conjunctive (AND) manners to reach an overall evaluation of an alternative.

One such infinite family of connectives is due to Yager [1980]. Here, complement, union, and intersection are given by

$$A^c(x) = (1 - A(x)^w)^{1/w}, \quad w \in (0, \infty) \tag{6.11a}$$

---

[4] Yes, a platypus has membership greater than zero in the set of reptiles. See http://usatoday30.usatoday.com/tech/science/genetics/2008-05-08-platypus-genetic-map_N.htm.

[5] Axioms have been developed for classes of operators that behave like intersections, unions, and complements (see Klir and Yuan [1995]). The intersections are usually called T-norms and the unions T-conorms. A rich mathematical diversity evolves from the satisfaction of subsets of these axiom sets.

$$(A\cup_w B)(x) = \min\{1, (A(x)^w + B(x)^w)^{1/w}\}, \quad w \in (0, \infty), \tag{6.11b}$$

$$(A\cap_w B)(x) = 1 - \min\{1, ((1 - A(x))^w + (1 - B(x))^w)^{1/w}\}, \quad w \in (0, \infty) \tag{6.11c}$$

For all choices of w, the value of the Yager union operator is greater than the standard union (max), while that for the intersection is less than the standard intersection (min). In other words, a Yager union operator is more optimistic than the maximum (in combining confidence), whereas each Yager intersection produces values that are more pessimistic than the minimum. The parameter w controls the degree of optimism or pessimism. In fact, the following limits hold:

$$\lim_{w\to\infty} (A\cup_w B)(x) = A(x) \vee B(x) \tag{6.12}$$

and

$$\lim_{w\to\infty} (A\cap_w B)(x) = A(x) \wedge B(x) \tag{6.13}$$

At the other end, that is, the limits as $w \to 0$, generate the drastic union and intersection, defined by

$$(A\cup_d B)(x) = \begin{cases} A(x), & \text{if } B(x) = 0 \\ B(x), & \text{if } A(x) = 0 \\ 1, & \text{else} \end{cases} \tag{6.14a}$$

$$(A\cap_d B)(x) = \begin{cases} A(x), & \text{if } B(x) = 1 \\ B(x), & \text{if } A(x) = 1 \\ 0, & \text{else} \end{cases} \tag{6.14b}$$

Fuzzy operators have been used extensively in multicriteria decision making [Bellman and Zadeh, 1970; Yager, 1988, 2004].

EXAMPLE 6.3   FUZZY DECISION TREES

There are many problems where several different criteria influence a final decision. Sometimes the criteria need to be conjunctively combined: They all need to be satisfied for decision confidence to be high. The satisfaction of other criteria may be disjunctively aggregated: any one of them being high will allow for a confident output. Some satisfaction values need to be negated to fit into a, say, maximization scheme. These aggregations of degrees of satisfaction of criteria can be organized in many different ways to match individual preferences about the interactions of

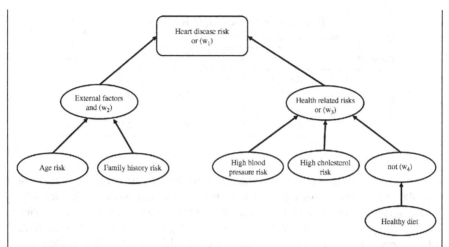

FIGURE 6.3 Fuzzy tree structure to demonstrate the utility of fuzzy operators for hypothetical heart disease risk.

the criteria. Because of the logical nature of fuzzy operators, a convenient model for analyzing alternatives with respect to a given set of criteria is a fuzzy decision tree. As a particularly simplistic illustration, consider an algorithm that might be used to assess the risk of heart disease for a person. Risk of heart disease is a complicated matter that is influenced by many factors. The simple tree shown in Figure 6.3 was made up, not to actually solve this problem, but to demonstrate how fuzzy operators can be connected within a multicriteria decision-making problem. Making it relate to heart disease is only to grab your attention (it certainly grabbed mine). In this hypothetical configuration, heart disease risk is determined to be high if either external factors or health-related risks are high. This is modeled by a union operator (OR). External factors are defined as the conjunction (AND) of age risk and family history risk. Why use an AND operator? While a family history of heart issues presents a risk, combining with increased age heightens the propensity (age by itself is only an indirect risk factor). The health-related risks are more obvious, and we list only three: high blood pressure, high cholesterol, and an unhealthy diet. Any one of these can trigger high heart disease risk, so they are connected by a disjunction (OR). We suppose that a "healthy diet index" can be ascertained so that we demonstrate the negation operator. Clearly, the decision tree shown in Figure 6.3 is only loosely based on medical science, but the example is meant to demonstrate the utility of fuzzy operators in multicriteria decision making more than focusing on reality. If you don't want to assess risk, think about how you would rate pizza parlors in your town or pick a new car. In Figure 6.3, we model each of the operators with the corresponding Yager connective (Eq. 6.11). The parameters for these four connectives will be labeled w_1 for the top disjunction, w_2 for the conjunction of external factors, w_3 for the disjunction of health-related factors, and w_4 for the complement.

TABLE 6.1   Heart Disease Risk Output for Various Weights for
the Input Values Displayed in the Text

Parameters	w_1	w_2	w_3	w_4	Risk
1	0.5	0.5	2.0	0.5	0.1
2	1.0	10.0	1.0	0.5	0.3
3	0.5	0.5	1.0	2.0	0.7
4	0.5	0.5	0.5	2.0	1.0

Obviously, there must be membership mappings to create the levels of satisfaction of the various base criteria in a fuzzy decision tree. Incidentally, these mappings effectively normalize the original domains for the underlying variables into the unit interval. Given the tree in Figure 6.3, suppose that we have determined the following fuzzy membership values for the leaf nodes: age risk $= 0.2$, family history risk $= 0.8$, high blood pressure risk $= 0.0$, high cholesterol risk $= 0.1$, and healthy diet $= 0.8$. That is, a particular patient is fairly young but with a family history of heart problems, while being pretty heart healthy. If the logical operators are the classical Boolean ones, then the risk of cancer would be zero for this set, assuming that the fuzzy memberships are hardened at a 0.5 threshold. The advantage of fuzzy set theory is that the operators that govern complement, disjunction, and conjunction can be tailored to reflect different user dispositions. Table 6.1 displays the heart disease risk output for a few choices of connection parameters. For example, using the parameters on line 3 of Table 6.1, we calculate the risk as

$$\text{Risk} = \min\{1, ((1 - \min\{1, ((1 - 0.2)^{0.5} + (1 - 0.8)^{0.5})^2\})^{0.5}$$
$$+ (\min\{1, ((\min\{1, (0^1 + 0.1^1)^{1/1}\})^1 + ((1 - 0.8^2)^{0.5})^1)^{1/1}\})^{0.5})^2\} = 0.7$$

The weights in case 1 of Table 6.1 produce operators that behave like Boolean logic (yes or no), and hence produce risk near zero. If the patient or doctor is more aggressive relative to assessing risk, Table 6.1 provides examples that produce low, moderate, and even complete risk for those same inputs. While we claim that this flexibility is an advantage of fuzzy set theory, some may argue that it confuses the situation. The message is that no one should use computational or logical operations on data without understanding how these operators combine the data. By studying fuzzy set connectives (as in Klir and Yuan [1995]), different degrees of aggressiveness can be quantified and produce meaningful trade-offs to a patient in this case, or for more general multicriteria decision-making processes. There are even models that allow you to weight each criteria's satisfaction differently within the logical operator aggregation. For example, you might want family history of heart-related problems to be considered much more strongly than age in the above example.

TABLE 6.2   Risk Output for Various Amounts of High Blood Pressure for $w_1 = 1.0$, $w_2 = 10.0$, $w_3 = 1.0$, $w_4 = 0.5$, and for the Input Values Stated in the Text

High blood pressure risk	0.0	0.1	0.2	0.3	0.4	0.5	0.6	0.7	0.8	0.9	1.0
Overall heart disease risk	0.3	0.4	0.5	0.6	0.7	0.8	0.9	1.0	1.0	1.0	1.0

Additionally, for a given choice of parameters, a "what-if" game can be played. In the above example, with the parameters as in case 2 of Table 6.1, we can examine the change in heart disease risk given by changing a patient's risk resulting from high blood pressure, as shown in Table 6.2.

6.4 ALPHA-CUTS, THE DECOMPOSITION THEOREM, AND THE EXTENSION PRINCIPLE

Suppose you want to take the average, or weighted average of a bunch of senor measurements that are uncertain. We might model the uncertainty by fuzzy numbers, that is, by normal, convex fuzzy sets over the real line. You know what normal fuzzy sets are; intuitively convex fuzzy subsets of the real numbers have membership function that "go up" for a while and then "go down," like triangles, trapezoids, pi-functions, and Gaussians, but nothing bimodal for example. We'll make this more precise shortly. So, how do we do arithmetic with fuzzy numbers? The story goes like this.

Let A be a fuzzy subset of X. For each $\alpha \in (0, 1]$, define ${}^{\alpha}A = \{x \in X | A(x) \geq \alpha\}$. The crisp set ${}^{\alpha}A$ is called the *α-cut* of A. The set ${}^{1}A$, that is, the set of x such that $A(x) = 1$, is called the *core* of A. Of course, ${}^{0}A$ is all of X. If we define the *strong α-level set* of A by ${}^{\alpha+}A = \{x \in X | A(x) > \alpha\}$, then ${}^{0+}A$ is the set of all x such that $A(x) > 0$, known as the *support* of A. Now we can formalize in the simple case what we mean by convex fuzzy subsets of the reals $\mathfrak{R}$. They are those fuzzy subsets all of whose α-cuts are crisp convex subsets of the real numbers (intervals for one dimension). See Klir and Yuan [1995] for a more general definition of a convex fuzzy set along with the theorem connecting that definition to α-cuts.

For each $\alpha \in (0, 1]$, define a fuzzy set $({}_{\alpha}A)(x) = \begin{cases} \alpha, & x \in {}^{\alpha}A \\ 0, & x \notin {}^{\alpha}A \end{cases}$. Note that for all α except 1, this is a subnormal fuzzy subset.

Why go through all this work? We can decompose a fuzzy set into the "union" of its α-cuts [Klir and Yuan, 1995].

Decomposition Theorem: Let A be a fuzzy subset of X. Then

$$A = \bigcup_{\alpha \in [0,1]} {}_{\alpha}A, \quad \text{where} \quad \left(\bigcup_{\alpha \in [0,1]} {}_{\alpha}A \right)(x) = \sup_{\alpha \in [0,1]} \{{}_{\alpha}A(x)\} \qquad (6.15)$$

This theorem (actually pretty straightforward to prove) states that if you know the α-cuts of A, then you know A itself. This is one of the two building blocks to perform fuzzy math. The other is the extension principle [Zadeh, 1975a, 1975b, 1976]. Simply put, given a function $f: X \rightarrow Y$ between two domains, we can "extend" f to be a function between fuzzy subsets of X and fuzzy subset of Y. Note that you already know how to extend f to a function on the crisp subsets of X by $f(A) = \{y \in Y | y = f(x) \text{ for some } x \in A\}$. The image of a crisp subset A of X is again a crisp subset f(A) of Y. There might be many values of x that map to a given y (think of $f(x) = x^2$ and $y = 1$), but all you need for crisp subsets is one such correspondence. Hence, for crisp subsets,

$$f(A)(y) = \begin{cases} 1, & \text{if } y = f(x) \text{ for some } x \in A \\ 0, & \text{else} \end{cases} \qquad (6.16)$$

If A is a fuzzy subset of X, then the extension principle states that for a fuzzy subset A of X,

$$f(A)(y) = \sup\{A(x) | y = f(x)\} \qquad (6.17)$$

that is, you find the "largest" membership in the set of elements of X that map to y. If the domain of f has a dimension greater than 1, say $f: X_1 \times X_2 \times \cdots \times X_n \rightarrow Y$, and you have fuzzy subsets $A_1, \ldots, A_n$ in their respective domains, the extended function becomes

$$f(A)(y) = \sup\{A_1(x_1) \wedge A_2(x_2) \wedge \cdots \wedge A_n(x_n) | y = f(x_1, x_2, \cdots, x_n)\} \qquad (6.18)$$

(we'll see more of this in Chapter 7).

Now we combine these two ingredients to get fuzzy number math. While the approach below can extend to many kinds of functions (see the problems at the end of the chapter), we restrict to those needed in computing the weighted average of fuzzy numbers where the weights themselves are fuzzy numbers, that is, addition, subtraction, multiplication, and division. To perform one of these binary operations on a pair of fuzzy numbers, the function under consideration looks like $\#: \Re \times \Re \rightarrow \Re$, where # is one of $\{+, -, \times, /\}$. Suppose A and B are two fuzzy numbers. By the extension principle (6.17), we could directly compute $(A\#B)(y) = \sup\{A(x_1) \wedge B(x_2) | y = x_1 \# x_2\}$. This is tedious at best, and in the continuous case involves solving a non-linear program for each value of y. However, for fuzzy numbers and the basic arithmetic operators, we have [Klir and Yuan, 1995]

$$^\alpha(A\#B) = (^\alpha A)\#(^\alpha B) \qquad (6.19)$$

Since the α-cuts of a fuzzy number are closed intervals, computing the α-cut of the extended arithmetic operation reduces to interval arithmetic, something that is easy to do. Then using the decomposition theorem, we finish this off by noting that

$$A\#B = \left(\bigcup_{\alpha \in [0,1]} \alpha(A\#B) \right) \qquad (6.20)$$

EXAMPLE 6.4   EXTENSION PRINCIPLE AND α-CUTS

A. Suppose $X = \{-3, -2, -1, 0, 1, 2, 3\}$, thought of as a subset of the integers. Let fuzzy subsets A and B of X be defined by their membership vectors $A = (0.0, 0.3, 0.8, 1.0, 0.8, 0.3, 0.0)$ and $B = (1.0, 0.9, 0.7, 0.5, 0.2, 0.0, 0.0)$.

According to the extension principle, $(A + B)(0) = \sup_{x+y=0}(A(x) \wedge B(y))$. Since X is a finite set, the "sup" is just the maximum value, that is, $(A + B)(0) = \vee_{x+y=0}(A(x) \wedge B(y))$. Consider the following table of values with x and y coming from X and $x + y = 0$:

x	y	A(x)	B(y)
1	−1	0.8	0.7
2	−2	0.3	0.9
3	−3	0.0	1.0
−1	1	0.8	0.2
−2	2	0.3	0.0
−3	3	0.0	0.0

Clearly, the max of the mins of the columns 3 and 4 is 0.7 and so, $(A + B)$ $(0) = 0.7$. Note that $(A + B)(0)$ is NOT $A(0) + B(0)$, since that would result in a value of 1.5, not a legal option for a fuzzy set. Exercise 6.12 will ask you to compute all of the values of $A + B$ over all of the integers and not just X itself.

Can we compute other extensions? What about Max(A,B)? Be careful, it's not the same as the union of A and B. For example, $Max(A, B)(-1) = \sup_{(x \vee y) = -1}(A(x) \wedge B(y))$. The value of $Max(A,B)(-1)$ is computed to be 0.8 from row 3 of the following table:

x	y	A(x)	B(y)
−1	−1	0.8	0.7
−1	−2	0.8	0.9
−1	−3	0.8	1.0
−2	−1	0.3	0.7
−3	−1	0.0	0.7

B. Using this approach, you can do these tedious calculations on finite sets to get extensions of many functions, such as A^2, exp(A), ln(A), and so on.

Thankfully, Eqs. 6.19 and 6.20 make it possible to generate approximations to complex domains, like intervals of the real line.

$$\text{Let} \quad A(x) = \begin{cases} 0, & \text{if } x < -4 \text{ or } x > 3 \\ \dfrac{x+4}{4}, & \text{if } -4 \le x \le 0 \\ \dfrac{-x+3}{3}, & \text{if } 0 < x \le 3 \end{cases}$$

$$B(x) = \begin{cases} 0, & \text{if } x < -2 \text{ or } x > 0 \\ x+2, & \text{if } -2 \le x \le -1 \\ -x, & \text{if } -1 < x \le 0 \end{cases}$$

Both A and B are triangular fuzzy numbers, and hence, their α-cuts are closed intervals. To invoke the decomposition theorem, we use the fact that $^{\alpha}(A + B) = (^{\alpha}A) + (^{\alpha}B)$ along with interval arithmetic to compute the right hand side of this expression. Now, for $0 < \alpha \le 1$, we have $^{\alpha}A = [4\alpha - 4, 3 - 3\alpha]$ and $^{\alpha}B = [\alpha - 2, -\alpha]$ (calculate the intersection of the line $y = \alpha$ with the lines defining A and B). Hence, $^{\alpha}(A + B) = (^{\alpha}A) + (^{\alpha}B) = [5\alpha - 6, 3 - 4\alpha]$. For a fixed α, say $\alpha = 0.5$, we have $^{0.5}A = [-2, 1.5]$, $^{0.5}B = [-1.5, -0.5]$, and thus, $^{0.5}(A + B) = (^{0.5}A) + (^{0.5}B) = [-3.5, 1]$. Now, we could have computed $(A + B)(1)$ directly from the extension principle by analyzing various subintervals between -4 and 3 and finding that in the interval $[2,3]$, the "sup of the mins" is equal to 0.5 and is less than that for all other subintervals (can you do it?). The solution is actually a constrained optimization problem for every value (ouch) where the constraints are given by the arithmetic expression linking the domain variables. However, by computing the α-cut intervals for several values of α, and plugging them into Eqs. 6.19 and 6.20, a good approximation can be gotten to the extended functional equation. Exercise 6.13 will have you explore this in more detail.

6.5   COMPENSATORY OPERATORS

Of course, the meaningfulness of the results of an analysis as in Example 6.3 depends on the faithfulness of the model and the accuracy of assessing the input values. This problem is not specific to fuzzy set theory, but is inherent to all computational paradigms. The discussion here makes no overt claims to accurately model cancer risk, but is only used to demonstrate the flexibility of fuzzy connectives in decision processes.

In Figure 6.3, we might conjecture that the external factors and health-related risks might better be combined in a compensative manner, that is, more like an average than a union. Besides modeling negation (NOT), disjunction (OR), and conjunction (AND), fuzzy set theory admits mechanisms to model compensatory connections, that is, aggregation operators where a high value in matching one criterion can compensate to some extent for a low value for another criterion. The simplest of these is called the generalized mean. If $a_1, a_2, \ldots, a_n$ are the degrees of satisfaction of n criteria, the generalized mean is defined as

$$h_\alpha(a_1, a_2, \cdots, a_n) = \left(\frac{a_1^\alpha + a_2^\alpha + \cdots + a_n^\alpha}{n} \right)^{1/\alpha} \tag{6.21}$$

where α is a fixed real number. For $\alpha = 1$, this equation implements the arithmetic average, for $\alpha = -1$, we have the harmonic average, and for α converging to 0, Eq. 6.21 produces the geometric mean, the nth root of the product of the values. All instantiations of the generalized means produce values between the minimum and maximum of the degrees of satisfaction of the individual criteria. Additionally,

$$\lim_{\alpha \to -\infty} h_\alpha(a_1, \ldots, a_n) = \min\{a_1, \ldots, a_n\} \tag{6.22}$$

and

$$\lim_{\alpha \to \infty} h_\alpha(a_1, \ldots, a_n) = \max\{a_1, \ldots, a_n\} \tag{6.23}$$

This high degree of coverage makes fuzzy set connectives appealing for multicriteria decision making.

In Krishnapuram and Lee [1992a, 1992b], Yager unions and intersections, along with generalized means, were used in hierarchical decision networks, and a gradient descent-based training algorithm was created to learn the parameters of the connectives in the network from a set of input/output training data. However, there were fairly cumbersome tests to decide if a node should be a union, intersection, or mean (and to flip between them). A more general class of connectives, called fuzzy hybrid operators, combine all three types of linguistic connectives into a single equation. The typical arithmetic and multiplicative hybrid operators are given by

$$A \oplus_\gamma B = (1 - \gamma)(A \cap B) + \gamma(A \cup B) \tag{6.24}$$

$$A \oplus_\gamma B = (A \cap B)^{1-\gamma} \cdot (A \cup B)^\gamma \tag{6.25}$$

where γ is between 0 and 1 and controls the amount of "mixing" of the union and intersection components, that is, if γ is close to 0, the hybrids acts like an intersection, near 1 produces a union-like response, and for γ around 0.5, the hybrid takes on the characteristics of a generalized mean.

Zimmermann and Zysno [1980] proposed a hybrid operator for multicriteria aggregation that was modeled after the compensatory nature of human aggregation. This hybrid operator (γ model) is an example of Eq. 6.25 and is given by

$$Y = \left(\prod_{i=1}^{n} (a_i)^{\delta_i} \right)^{1-\gamma} \left(1 - \prod_{i=1}^{n} (1 - a_i)^{\delta_i} \right)^\gamma \tag{6.26}$$

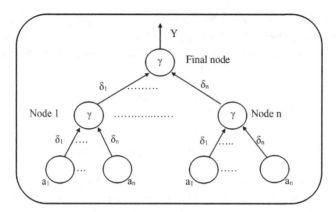

FIGURE 6.4 A fuzzy aggregation network of multiplicative hybrids used in Parekh and Keller [2007].

where $a_i \in [0,1]$ are the criteria satisfactions to be aggregated, $0 \le \gamma \le 1$ is the mixing coefficient, and $\sum_{i=1}^{n} \delta_i = n$. Here, δ_i are weights associated with each criterion a_i and n is the number of criteria being aggregated.

Krishnapuram and Lee [1992a, 1992b] also developed a backpropagation algorithm to learn the parameters of operators of this type in a network-based decision application. While the algorithm converged, the derivatives were quite messy and as with all such algorithms, convergence could only be guaranteed to a local minimum of a least-squares fitness function. Keller *et al.* [1994] extended the approach to additive hybrid networks. In Parekh and Keller [2007], particle swarm optimization [Eberhart and Kennedy, 1995; Clerc, 2004] was used to train these aggregation networks. Figure 6.4 displays such a network. The advantage of swarm optimization is that many potential solutions (here, the list of all node parameters) are randomly generated and through individual particle memory and communication between particles, large areas of the optimization search space can be covered while still moving quickly to a (usually very good) local optimum of the fitness function. Additionally, in this case, no derivatives were necessary, since each particle contains all the node parameters and evaluation is performed directly at each time step.

EXAMPLE 6.5   LEARNING FUZZY AGGREGATION NETWORK PARAMETERS

As an example, synthetic data were used to verify this approach. Parameters for the multiplicative hybrid operators were randomly generated and assigned to each node. Then, a table of 1000 input values was randomly generated and corresponding outputs were calculated from successive applications of Eq. 6.25. The training data consisted of 800 data points and the test data had 200 data points. Table 6.3 shows a sample of the training and test data from one

TABLE 6.3   Sample of the 800 Training and 200 Testing Input/Output Data for Learning the Parameters of the Nodes in the Network of Figure 6.4, Where Each Node Has Two Inputs

Sample of Training Data						
Node 1		Node 2				
a1	a2	a1	a2	Y	Y'	SSE
0.425	0.590	0.655	0.861	0.364	0.363	0.00175
0.768	0.452	0.629	0.668	0.209	0.211	
0.532	0.053	0.521	0.548	0.018	0.018	
0.235	0.868	0.722	0.892	0.490	0.492	
0.673	0.925	0.428	0.829	0.542	0.541	
Sample of Test Data						
0.467	0.538	0.518	0.990	0.423	0.420	0.00052
0.771	0.678	0.617	0.999	0.621	0.622	
0.810	0.344	0.392	0.109	0.005	0.005	
0.997	0.644	0.235	0.630	0.242	0.244	
0.272	0.032	0.821	0.528	0.009	0.008	

TABLE 6.4   Actual and Recovered Parameters Corresponding to Table 6.3

	Parameter	Actual	Recovered
Node 1	δ_1	0.440	0.446
	δ_2	1.559	1.553
	γ	0.255	0.341
Node 2	δ_1	0.161	0.163
	δ_2	1.838	1.836
	γ	0.180	0.198
Final Node	δ_1	0.786	0.816
	δ_2	1.213	1.183
	γ	0.0846	0.028

experiment while Table 6.4 shows the original and recovered hybrid parameters. With an easy and effective training mechanism, such fuzzy aggregation networks are attractive tools for hierarchical confidence fusion. Besides the ability to approximate input/output training data, an additional advantage of these networks is that after training, each node can be associated with a linguistic connective (disjunction, conjunction, mean), based on the corresponding value of γ, and the weights give an indication of the importance of the particular criteria toward the fused result.

6.6 CONCLUSIONS

In this chapter, a small slice of the rich theory and application potential of basic fuzzy sets as a means to model and manipulate uncertainty was presented. Our hope is that this quick look will inspire you, the reader, to explore the theory and the numerous applications to real problem domains with a view to incorporate these principles into your own research and development endeavors.

EXERCISES

6.1. A trapezoid membership function is defined by four parameters $a < b \leq c < d$. (If $b = c$, you have a triangular function). Define the equations for trapezoid and triangular fuzzy sets over a fixed interval of reals $[r,s]$. Note that these membership functions do not need to be symmetric about the "center." What happens if $b < r$? What about $s < c$?

6.2. Let $X = [0, 10]$. Define and draw the graphs of the Trap(x; 1,2,3,5), Trap(x; 3,5,7,9), Trap(x; 8,9,9,10), Trap(x; −2,−1,3,5), and Trap(x; 8,9,11,12).

6.3. **A.** Derive the equation for the S-function of Example 6.1. *Hint:* There are two parabola pieces and for each piece you know two points, the value of the function at two of the parameters. But wait you say, a parabola has three coefficients, so you need at least three equations to find them. Think about the derivative.

   B. Show that a symmetric S-function (where b is the midpoint of a and c) has a well-defined derivative at all points in its domain (even at the "join" points).

   C. Let $X = [0, 10]$. Define and graph the functions S(x; 5,7,9), Z(x; 2,3,6), Pi(x; 1,2,3,3,4,5), and Pi(x; 4,5,6,7,8,10).

6.4. Suppose $X = \{-3, -2, -1, 0, 1, 2, 3\}$. Let fuzzy subsets A and B of X be defined by their membership vectors $A = (0.0, 0.3, 0.8, 1.0, 0.8, 0.3, 0.0)$ and $B = (1.0, 0.9, 0.7, 0.5, 0.2, 0.0, 0.0)$

   A. Using Zadeh's original definitions, compute

$$A^c$$

$$A \cup B$$

$$A \cap B$$

   B. What is $A \cup B$ if $A \cup B(x) = 1 \wedge (A(x) + B(x))$?

6.5. Finish the proof of the distributive law in Example 6.2.

6.6. Show that DeMorgan's laws hold for the standard fuzzy set theory definitions, that is, that $(A \cup B)^c = A^c \cap B^c$ and $(A \cap B)^c = A^c \cup B^c$.

6.7. Show that LEM and LOC do not hold for the standard fuzzy set theory definitions, that is, show that $A \cup A^c = X$ and $A \cap A^c = \phi$ are not true in general for fuzzy set theory. Note that $X(x) = 1$ and $\phi(x) = 0$ for all $x \in X$. *Hint:* Think about how you show that a statement is *not* a theorem.

6.8. Consider the operators:

$$(A \cup B)(x) = 1 \wedge (A(x) + B(x))$$
$$(A \cap B)(x) = 0 \vee (A(x) + B(x) - 1)$$
$$A^c(x) = 1 - A(x)$$

A. Show that intersection and complement satisfy the law of contradiction.

B. Is it true that intersection is idempotent: $A \cap A = A$? (prove or give a counterexample)

6.9.

$$\text{Let} \quad A(x) = \begin{cases} 0, & \text{if } x < -3 \quad \text{or} \quad x > -1 \\ x + 3, & \text{if } -3 \le x \le -2 \\ -x - 1, & \text{if } -2 < x \le -1 \end{cases}$$

$$B(x) = \begin{cases} 0, & \text{if } x < 1 \quad \text{or} \quad x > 3 \\ x - 1, & \text{if } 1 \le x \le 2 \\ -x + 3, & \text{if } 2 < x \le 3 \end{cases}$$

A. Compute $^{0.7}A$.

B. Using the standard definitions, sketch a picture of B^c.

$$\text{C. Now, let } C(x) = \begin{cases} 0, & \text{if } x < 2 \quad \text{or} \quad x > 4 \\ x - 2, & \text{if } 2 \le x \le 3 \\ -x + 4, & \text{if } 3 < x \le 4 \end{cases} \qquad \text{Compute } B \cap C$$

6.10. Suppose $X = \{a, b, c, d, e\}$. Let A be the fuzzy set with memberships (0.5, 0.4, 0.7, 0.8, 1) for a, b, c, d, and e, respectively. (A common way you might see in the literature to write this is $A = 0.5/a + 0.4/b + 0.7/c + 0.8/d + 1/e$). List all nonempty α-cuts of A.

6.11. Let A be any fuzzy subset of X. Show that $^{\alpha}A \subseteq {}^{\beta}A$ if $\beta < \alpha$

6.12. For the fuzzy sets defined in Example 6.4 (A), generate all of the nonzero values of $(A + B)$ and Max(A,B).

6.13. For the fuzzy sets defined in Example 6.4(B), compute the α-cuts, $^{\alpha+}(A + B)$, for several values of $0 < \alpha \leq 1$ and sketch the graphs of A, B, and $(A + B)$. Does this match your intuition? Why or why not?

(*Hint:*To get a nice graph, you can use the strong α-cut $^{\alpha+}(A + B)$ for $\alpha = 0$, to define the support of the extension; then connect the endpoints.) Is $(A + B)$ a triangular fuzzy number?

6.14. Repeat Exercise 6.13 for $(A - B)$ and $(A \times B)$.

CHAPTER 7

Fuzzy Relations and Fuzzy Logic Inference

7.1 INTRODUCTION

There are times when domain knowledge, and hence, the decision functions, about a particular problem can best be described in terms of linguistic rules. For example, in the heart disease risk example, we might have rules like

IF   The External Factor Risk is SOMEWHAT LOW and
        The Health Related Risk is LOW
THEN
        The Overall Heart Disease Risk is LOW.

Traditional crisp expert systems including those that manipulate numeric confidences or probabilities have been around for many years [Ignizio, 1991; Giarratano and Riley, 2004]. Fuzzy logic extends this approach by modeling linguistic propositions, rules, and the inference procedure directly with fuzzy sets. In this chapter, we describe the background necessary to understand and construct fuzzy logic inference systems for decision-making problems and control applications.

Fuzzy logic begins with the concept of a linguistic variable [Zadeh, 1975a, 1975b; Zadeh, 1976]. A linguistic variable is, as its name suggests, a variable whose values are words. For example, the linguistic variable "Age" might take as values "infant," "young," "adult," "middle," "senior," and "very old." With any linguistic variable, there is an underlying domain, X that will be used to create the meanings for the linguistic values. In our simple example above, the underlying domain is the real numbers between 0 and 1 since risk factors are normalized to that range. Each linguistic value has a fuzzy subset of X that serves as its definition. An example will be given at the end of this chapter.

Fundamentals of Computational Intelligence: Neural Networks, Fuzzy Systems, and Evolutionary Computation, First Edition. James M. Keller, Derong Liu, and David B. Fogel.
© 2016 by The Institute of Electrical and Electronics Engineers, Inc. Published 2016 by John Wiley & Sons, Inc.

7.2 FUZZY RELATIONS AND PROPOSITIONS

Once we have this fundamental concept of a linguistic variable, we can build the machinery necessary for fuzzy logic inference. We will provide mechanisms for making deductions that are all based on the concept of fuzzy relations. A crisp relation is simply a mapping between two domains, X and Y, for example, $R : X \to Y$. Actually, R may be restricted to a subset of X. Alternatively, we can write R as a crisp subset of the cross-product domain $R : X \times Y \to \{0, 1\}$, where $R(x, y) = 1$ if, and only if $R(x) = y$, that is, if x is "related" to y by R. For example, suppose X is the set of cities in the United States and Y is the set of airports. Then R may be the mapping that relates cities to airports when the airport is within 25 miles of the city center. So, the pair (New York, LaGuardia) is in R, but so is the pair (New York, JFK). Hence, relations don't need to be functions. Certainly, there are cities in the United States that are related to no airports with respect to this definition. Now, just as we extended crisp sets to fuzzy sets by introducing continuous memberships, we do the same for relations, since they are also sets. That is, a fuzzy relation is simply a fuzzy subset of the cross-product domain $R : X \times Y \to [0, 1]$, where now we interpret $R(x,y)$ as the strength of the relation between x and y. We could convert the above example into a fuzzy relation by creating the relation strength between a city and an airport with an S function defined over driving distance from the airport to the city center. Then, R(New York, LaGuardia) and R(New York, JFK) would have different relational strengths. There are many examples of direct applications of fuzzy relations, and we will concentrate on the main use, that of providing an engine for logical inference in a fuzzy rule-based system.

In what follows, let $X, X_1, X_2, \ldots, X_n$ and Y be domains, $U, U_1, U_2, \ldots, U_m$ and $V, V_1, \ldots, V_m$ be linguistic variables, and $A, A_1, \ldots, A_n$ and $B, B_1, \ldots, B_m$ be the fuzzy sets that model linguistic values over respective domains. An atomic proposition in fuzzy logic is a statement of the form "U is A," where U is the name of a linguistic variable and A is the name of a linguistic value, that is, it is the name of a fuzzy subset of the domain X; think of something like "AGE is Young."

To make the language richer, atomic propositions can have values that contain "hedges" like NOT, SOMEWHAT, MORE_OR_LESS, VERY, RATHER, QUITE; the list can go on and on. Either you can define all such values directly, perhaps a tedious job, or you can posit that a hedged linguistic value has a fuzzy set that is a hedged function of the base value. The hedge NOT is the easiest; just use your favorite complement operator on the original fuzzy set: $NOTYoung(x) = Young^c(x)$. Other hedges are defined by functions that carry the linguistic semantics of the words. For example, VERY Young(x) might be defined by squaring the values of Young(x) while MORE_OR_LESS Young uses the square root. Using these functions, it's easy to see that $VERYYoung(x) \le Young(x) \le MORE\_OR\_LESSYoung(x)$ for all actual ages x. Of course, many other formulations can be conceived; the important point is to preserve the common semantic relationships between hedged values.

The conjunctive proposition between two fuzzy sets can be written as follows [Klir and Yuan 1995]:

$$U_1 \text{ is } A_1 \text{ and } U_2 \text{ is } A_2 \tag{7.1}$$

where U_i are linguistic variables over domains X_i and where $A_i(x_i)$ are linguistic values represented by fuzzy sets on those domains. Here, an example is "AGE is Young and HEALTH is Good." Note that the challenge is to create the appropriate domains for the definitions of the linguistic values of a given linguistic variable; AGE is straightforward, but HEALTH is more problematic. Using such linguistic variables requires care in the definition of suitable scales to characterize the fuzzy sets that specify the meanings of linguistic values. Exercise 7.9 explores this issue in more detail.

The result of this operation is a fuzzy relation of the cross-product domain based on U_1 and U_2, which is called the cylindrical closure of the fuzzy sets A_1 and A_2. A fuzzy relation so referenced is just a fuzzy subset of $X_1 \times X_2$. The cylindrical closure can be viewed as the intersection of the extension of each fuzzy set to the cross-product domain $X_1 \times X_2$, that is, a fuzzy subset, $A_1 \times A_2$, of $X_1 \times X_2$ where

$$A_1 \times A_2(x_1, x_2) = A_1(x_1) \wedge A_2(x_2) \tag{7.2}$$

Here we use minimum as the intersection operator, but note that any fuzzy intersection operator could be used.

EXAMPLE 7.1

Suppose $X_1 = \{1,2,3,4\}$, $X_2 = \{@, \#, \&\}$. For $A_1 = \text{SMALL} = 1.0/1 + 0.8/2 + 0.0/3 + 0.0/4$ and $A_2 = \text{LARGE} = 0.0/@ + 0.6/\# + 1.0/\&$, the meaning of the compound proposition U_1 is A_1 and U_2 is A_2 is given by the cylindrical closure of A_1 and A_2 in $X_1 \times X_2$. Since the domains are finite, this fuzzy relation (using the min operator) is viewed as the 4×3 matrix:

$$A_1 \times A_2 = \begin{pmatrix} 0.0 & 0.6 & 1.0 \\ 0.0 & 0.6 & 0.8 \\ 0.0 & 0.0 & 0.0 \\ 0.0 & 0.0 & 0.0 \end{pmatrix}$$

The condition proposition, or fuzzy implication, between two fuzzy propositions is written as

$$\text{IF } U \text{ is } A \text{ THEN } V \text{ is } B \tag{7.3}$$

where U and V are linguistic variables that have elements $x \in X$ and $y \in Y$, respectively, and where $A(x)$ and $B(y)$ are linguistic values represented by fuzzy sets on those elements. The definition of an implication proposition is a fuzzy relation R between X and Y, based on U and V, that can take many forms in combining the input fuzzy sets (see Klir and Yuan [1995] for many possibilities). Three common definitions used in many fuzzy rule systems are as follows:

The Lukasiewicz implication (Zadeh's original implication operator):

$$R_Z(x, y) = \min(1, 1 - A(x) + B(y)) \tag{7.4}$$

Correlation min implication[1]:

$$R_{cm}(x, y) = \min(A(x), B(y)) \tag{7.5}$$

Correlation product implication:

$$R_{cp}(x, y) = A(x)*B(y) \tag{7.6}$$

EXAMPLE 7.2

Suppose $X = \{1,2,3,4\}$, $Y = \{a,b,c,d,e\}$. For $A = \text{SMALL} = 1.0/1 + 0.8/2 + 0.0/3 + 0.0/4$ and $B = \text{MEDIUM} = 0.0/a + 0.5/b + 1.0/c + 0.5/d + 0.0/e$, consider the rule: IF U is SMALL then V is MEDIUM. First note that we can express both fuzzy sets as "vectors" over their respective domains: $A = (1.0, 0.8, 0.0, 0.0)$ and $B = (0.0, 0.5, 1.0, 0.5, 0.0)$. For $x = 2$ and $y = d$, we have

$$R_z(2, d) = \min(1, 1 - A(2) + B(d)) = \min(1, 1 - 0.8 + 0.5) = 0.7$$

$$R_{cm}(2, d) = \min(A(2), B(d)) = \min(0.8, 0.5) = 0.5, \text{and}$$

$$R_{cp}(2, d) = A(2)*B(d) = 0.8*0.5 = 0.40$$

The full fuzzy relations can be visualized as matrices (of size 4×5) generated by the above equations. They are formally equivalent to "fuzzy outer products" $A^t \cdot B$, where the matrix arithmetic is interpreted to be one of the defining relation equations. Hence,

$$R_Z = \begin{pmatrix} 1.0 \\ 0.8 \\ 0.0 \\ 0.0 \end{pmatrix} \cdot \begin{pmatrix} 0.0 & 0.5 & 1.0 & 0.5 & 0.0 \end{pmatrix} = \begin{pmatrix} 0.0 & 0.5 & 1.0 & 0.5 & 0.0 \\ 0.2 & 0.7 & 1.0 & 0.7 & 0.2 \\ 1.0 & 1.0 & 1.0 & 1.0 & 1.0 \\ 1.0 & 1.0 & 1.0 & 1.0 & 1.0 \end{pmatrix}$$

---

[1] Note that modeling of logical implication (If A Then B) as Not A OR B creates a fuzzy relation. With Correlation Min (A AND B), we derive a fuzzy relation, although strictly speaking it is not an implication. The same is true for correlation product. It establishes a relationship between the antecedent and consequent that is useful for inference procedures. This is a fundamental difference between traditional logic and fuzzy logic.X

Note: The details of the calculation of element (2,d) are shown above. Similarly,

$$R_{cm} = \begin{pmatrix} 1.0 \\ 0.8 \\ 0.0 \\ 0.0 \end{pmatrix} \cdot \begin{pmatrix} 0.0 & 0.5 & 1.0 & 0.5 & 0.0 \end{pmatrix} = \begin{pmatrix} 0.0 & 0.5 & 1.0 & 0.5 & 0.0 \\ 0.0 & 0.5 & 0.8 & 0.5 & 0.0 \\ 0.0 & 0.0 & 0.0 & 0.0 & 0.0 \\ 0.0 & 0.0 & 0.0 & 0.0 & 0.0 \end{pmatrix},$$

and

$$R_{cp} = \begin{pmatrix} 1.0 \\ 0.8 \\ 0.0 \\ 0.0 \end{pmatrix} \cdot \begin{pmatrix} 0.0 & 0.5 & 1.0 & 0.5 & 0.0 \end{pmatrix} = \begin{pmatrix} 0.0 & 0.5 & 1.0 & 0.5 & 0.0 \\ 0.0 & 0.4 & 0.8 & 0.4 & 0.0 \\ 0.0 & 0.0 & 0.0 & 0.0 & 0.0 \\ 0.0 & 0.0 & 0.0 & 0.0 & 0.0 \end{pmatrix}$$

This is a very simple example and is intended only to demonstrate the hand calculation of an implication proposition for finite domains. Other than visualization, there is no inherent reason why these matrices need to be formed *a priori*. In some cases, it is faster to compute the values needed on the fly (as we will see later). For "fun," combine this example with Example 7.1 and write out the implication relations for the proposition: IF U_1 is SMALL and U_2 is LARGE Then V is MEDIUM. Note that the resulting relations will be $4 \times 3 \times 5$ matrices (see Exercise 7.3).

7.3 FUZZY LOGIC INFERENCE

Note that a fuzzy implication proposition is just a (fuzzy) rule. The compositional rule of inference or generalized modus ponens can now be described to combine a fuzzy rule and a linguistic proposition.

The compositional rule of inference is [Zadeh, 1973] as follows:

Rule: If U is A then V is B
Fact: U is A'
Conclusion: V is B'

where the conclusion expression is the composition operation:

$$B'(y) = A'(x) \circ R(x, y) \tag{7.7}$$

Here, $R(x, y)$ is the chosen translation of the fuzzy implication. The composition operation is defined as

$$B'(y) = A'(x) \circ R(x, y) = \sup_{x \in X} \min\{A'(x), R(x, y)\} \tag{7.8}$$

where "sup" is the supremum of the set, that is, the least element that is greater than or equal to each element in the set. Note that for finite domains X and Y, the compositional rule of inference looks like matrix multiplication where min replaces multiplication and sup replaces summation.

EXAMPLE 7.3

Consider the rule from Example 7.2, IF U is SMALL then V is MEDIUM, and suppose that A' is SMALL. Using the three translations of implication given above (Eqs. 7.4–7.6), the conclusion of the inference is

$$B'_Z = \begin{pmatrix} 1.0 & 0.8 & 0.0 & 0.0 \end{pmatrix} \circ \begin{pmatrix} 0.0 & 0.5 & 1.0 & 0.5 & 0.0 \\ 0.2 & 0.7 & 1.0 & 0.7 & 0.2 \\ 1.0 & 1.0 & 1.0 & 1.0 & 1.0 \\ 1.0 & 1.0 & 1.0 & 1.0 & 1.0 \end{pmatrix}$$

$$= \begin{pmatrix} 0.2 & 0.7 & 1.0 & 0.7 & 0.2 \end{pmatrix}$$

$$B'_{cm} = \begin{pmatrix} 1.0 & 0.8 & 0.0 & 0.0 \end{pmatrix} \circ \begin{pmatrix} 0.0 & 0.5 & 1.0 & 0.5 & 0.0 \\ 0.0 & 0.5 & 0.8 & 0.5 & 0.0 \\ 0.0 & 0.0 & 0.0 & 0.0 & 0.0 \\ 0.0 & 0.0 & 0.0 & 0.0 & 0.0 \end{pmatrix}$$

$$= \begin{pmatrix} 0.0 & 0.5 & 1.0 & 0.5 & 0.0 \end{pmatrix}$$

$$B'_{cp} = \begin{pmatrix} 1.0 & 0.8 & 0.0 & 0.0 \end{pmatrix} \circ \begin{pmatrix} 0.0 & 0.5 & 1.0 & 0.5 & 0.0 \\ 0.0 & 0.45 & 0.8 & 0.45 & 0.0 \\ 0.0 & 0.0 & 0.0 & 0.0 & 0.0 \\ 0.0 & 0.0 & 0.0 & 0.0 & 0.0 \end{pmatrix}$$

$$= \begin{pmatrix} 0.0 & 0.5 & 1.0 & 0.5 & 0.0 \end{pmatrix}$$

Here, all three outputs *could be* recognized as MEDIUM, although with the Zadeh translation (Eq. 7.4), B'_Z is less certain than those for correlation min and correlation product. Why would anyone want to consider the Zadeh translation if it doesn't precisely satisfy crisp modus ponens? Part of the answer is that it uses the Lukasiewicz multivalued logic implication, and hence is a true translation of "Not A or B." Neither correlation min nor correlation product translates the logical equivalence of implication. Also, in operation, the answer lies in the

following situation. What should happen to the above inference if A' is NOT
SMALL? Students of symbolic logic know that nothing should be concluded from
((IF U is A then V is B) and U is NOTA). So, when A' is NOT SMALL, we have

$$B'_Z = \begin{pmatrix} 0.0 & 0.2 & 1.0 & 1.0 \end{pmatrix} \circ \begin{pmatrix} 0.0 & 0.5 & 1.0 & 0.5 & 0.0 \\ 0.2 & 0.7 & 1.0 & 0.7 & 0.2 \\ 1.0 & 1.0 & 1.0 & 1.0 & 1.0 \\ 1.0 & 1.0 & 1.0 & 1.0 & 1.0 \end{pmatrix}$$

$$= \begin{pmatrix} 1.0 & 1.0 & 1.0 & 1.0 & 1.0 \end{pmatrix}$$

$$B'_{cm} = \begin{pmatrix} 0.0 & 0.2 & 1.0 & 1.0 \end{pmatrix} \circ \begin{pmatrix} 0.0 & 0.5 & 1.0 & 0.5 & 0.0 \\ 0.0 & 0.5 & 0.8 & 0.5 & 0.0 \\ 0.0 & 0.0 & 0.0 & 0.0 & 0.0 \\ 0.0 & 0.0 & 0.0 & 0.0 & 0.0 \end{pmatrix}$$

$$= \begin{pmatrix} 0.0 & 0.2 & 0.2 & 0.2 & 0.0 \end{pmatrix}$$

$$B'_{cp} = \begin{pmatrix} 0.0 & 0.2 & 1.0 & 1.0 \end{pmatrix} \circ \begin{pmatrix} 0.0 & 0.5 & 1.0 & 0.5 & 0.0 \\ 0.0 & 0.45 & 0.8 & 0.45 & 0.0 \\ 0.0 & 0.0 & 0.0 & 0.0 & 0.0 \\ 0.0 & 0.0 & 0.0 & 0.0 & 0.0 \end{pmatrix}$$

$$= \begin{pmatrix} 0.0 & 0.2 & 0.2 & 0.2 & 0.0 \end{pmatrix}$$

The fuzzy set where all membership values are 1 represents "UNKNOWN," that is,
all elements of the output domain have maximum possibility of being the "true"
value. Correlation min and product produce output sets with very small memberships (in this example), indicating low correlation between antecedent and fact, but
this is not logic!

Rarely will rules only have one antecedent clause. Rules with multiple antecedent
clauses pose no conceptual problem. The compositional rule of inference with
multiple antecedent clauses is the following:

Rule: IF U_1 is A_1 and U_2 is A_2 and . . . and U_n is A_n THEN V is B
Fact: U_1 is A'_1 and U_2 is A'_2 and . . . and U_n is A'_n
Conclusion: V is B'

The first step in this case is to find the cylindrical closure, $A_1 \times A_2 \times \cdots \times A_n$ of
the n antecedent clauses, that is, the intersection of the extensions of all these fuzzy

sets to the domain $X_1 \times X_2 \times \cdots \times X_n$. Once computed, the above definition for implication can be applied to the rule:

$$IF\langle U_1, U_2, \ldots, U_n \rangle \text{ is } A_1 \times A_2 \times \cdots \times A_n \text{ Then V is B} \qquad (7.9)$$

This produces a fuzzy relation R between $X_1 \times X_2 \times \cdots \times X_n$ and Y, that is, a fuzzy subset of $X_1 \times X_2 \times \cdots \times X_n \times Y$. Finally, the fuzzy conclusion can be drawn with the compositional rule of inference as

$$B'(y) = A_1' \times \cdots \times A_n'(x_1, \ldots, x_n) \circ R(x_1, \ldots, x_n, y), \quad \text{for each y in Y} \qquad (7.10)$$

The compositional rule of inference with several rules takes the following form:

Rule 1: IF U_1 is A_{11} and . . . and U_n is A_{1n} THEN V is B_1
Rule 2: IF U_1 is A_{21} and . . . and U_n is A_{2n} THEN V is B_2

$$\vdots$$

Rule k: IF U_1 is A_{k1} and . . . and U_n is A_{kn} THEN V is B_k
Fact: U_1 is A_1' and U_2 is A_2' and . . . and U_n is A_n'
Conclusion: V is B'

Each rule is translated as above to form $R_i(x_1, \ldots, x_n, y)$ and then the compositional rule of inference is applied to that rule with the fact to obtain

$$B_i'(y) = A_{i1}' \times \cdots \times A_{in}'(x_1, \ldots, x_n) \circ R_i(x_1, \ldots, x_n, y) \qquad (7.11)$$

These partial conclusion expressions are usually aggregated into a single output fuzzy set by either

$$B'(y) = \sum_{i=1}^{k} B_i'(y) \qquad (7.12)$$

or

$$B'(y) = \max_{i=1}^{k} \left\{ B_i'(y) \right\} \qquad (7.13)$$

Note that the first expression may exceed 1 for particular values of y and hence, not formally be a fuzzy subset of Y. However, it is easy to normalize. In fact, this formula is popular in those cases like fuzzy control where the output fuzzy set needs to be converted to a single numeric value. This process is known as defuzzification. The most common form is centroid defuzzification:

$$\bar{y} = \frac{\sum_{y \in Y} y \cdot B'(y)}{\sum_{y \in Y} B'(y)} \qquad (7.14)$$

although many other defuzzification techniques are available (Figure 7.1).

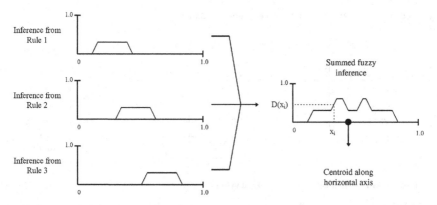

FIGURE 7.1   Example of using Eq. 7.12 for aggregation and Eq. 7.14 for defuzzification.

7.4   FUZZY LOGIC FOR REAL-VALUED INPUTS

While the above development handles the general case of fuzzy inference, in most applications of fuzzy rule-based systems, the inputs are not actually fuzzy sets themselves, but crisp values in their respective domains. For example, in a fuzzy system to perform classification, the inputs may be values of features extracted from the objects to be classified. The rules may contain propositions like "Feature 1 is LOW," "Class 1 confidence is HIGH," and so on, indicating the uncertainty in the decision process. However, in application, given an object to classify, Feature 1 is normally a real number $\tilde{x}_1$. The standard method to convert it to a fuzzy set for the inference process is to create a (crisp) set that is 1 for $\tilde{x}_1$ and zero everywhere else in the domain X_1. This makes the firing of the rules particularly simple [Klir and Yuan 1995]. To see this, recall that for finite domains, the result of generalized modus ponens for a single rule is given by

$B'(y) = A'_1 \times \cdots \times A'_n(x_1, \ldots, x_n) \circ R(x_1, \ldots, x_n, y)$ for each y in Y (7.10). The cylindrical closure is usually defined by the min operator and so, if correlation min is also used to encode the rule relation R, we have

$$
\begin{aligned}
B'(y) &= A'_1 \times \cdots \times A'_n(x_1, \ldots, x_n) \circ R(x_1, \ldots, x_n, y) \\
&= \text{Max}_{(x_1, \ldots, x_n)} \big[\big(A'_1(x_1) \wedge A'_2(x_2) \wedge \\
&\quad \cdots \wedge A'_n(x_n)\big) \wedge \big(A_1(x_1) \wedge A_2(x_2)\big) \wedge \cdots \wedge A_n(x_n) \wedge B(y) \big]
\end{aligned}
\tag{7.15}
$$

Now, if each input clause is modeled by a singleton fuzzy set, zero membership everywhere except the numeric input value, this complex looking equation will be identically zero except for the n-tuple $\tilde{x}_1, \ldots, \tilde{x}_n$ producing the output

$$
B'(y) = (A_1(\tilde{x}_1) \wedge A_2(\tilde{x}_2) \wedge \cdots \wedge A_n(\tilde{x}_n) \wedge B(y))
\tag{7.16}
$$

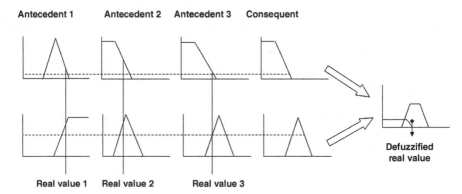

Antecedent 1   Antecedent 2   Antecedent 3   Consequent

Real value 1   Real value 2   Real value 3

FIGURE 7.2 Graphical illustration of generalized modus ponens firing of rules using Correlation-Min encoding, followed by aggregation and defuzzification (see text for details).

The interpretation is that you let each input numeric value fire the corresponding antecedent clause, take the minimum firing strength, and use that minimum to "chop off" the consequent clause fuzzy set B. Figure 7.2 shows graphically Eq. 7.16 in action with two rules, each having three antecedent clauses. In this case, at each iteration the input will be a triple of real values (x_1, x_2, x_3). For each rule, the membership of x_i in Antecedent$_i$ is computed (the vertical lines). The minimum of those three values (denoted by the dashed line) is used in the minimum operator with each value of the rule consequent (thus the comment about "chopping" the consequent). This very fast operation is done for each rule, the chopped consequents are added and then the centroid is computed (defuzzification). That value can be used as a crisp classifier confidence or it can be sent to a controller and a new cycle begins. Sure takes some of the mystery out of fuzzy logic controllers! The point is that this can be done very fast and there is no need to precompute and store large multidimensional rule matrices. Most of the early fuzzy logic controllers were built with this scheme.

Figure 7.3 describes a generic numeric-based fuzzy logic system. If the system is to be used for classification, then the outputs don't go back to a "physical plant," but are interpreted as confidence in class labels. An example of such a classification system for landmine detection built on fuzzy rules can be found in Gader *et al* [2001], although many other applications of this type exist.

Alternatively, a simple triangular set, a pi-function, or any membership function can be centered at x_1 to explicitly model the uncertainty in the feature extraction (see Figure 6.1). This process of converting measured crisp inputs into fuzzy sets for inference purposes is known as fuzzification. It may seem artificial at times, but the rule clauses describe the uncertainty and variability in the problem domain. For more complex situations, the input values themselves may be fuzzy sets, and hence, the computational simplification described above cannot be used. However, new methods to increase processing speed are being developed. For example, Harvey [2008]

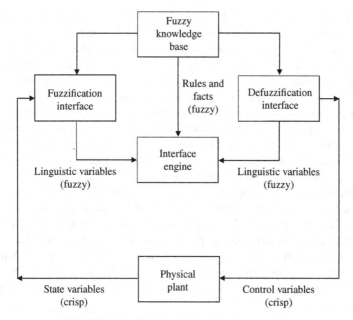

FIGURE 7.3 "Standard" fuzzy logic controller.

describes a fast method to do general modus ponens using a graphics processing unit (GPU).

The system of inference described above is referred to as a Mamdani–Assilion or MA fuzzy rule system [Mamdani, 1977]. An alternative formulation, denoted as a Takagi–Sugeno–Kang (TSK) system [Takagi and Sugeno, 1985; Sugeno and Kang, 1988], only modifies a function described in the consequent clause. It was developed for control applications where the output of the rule firing should be a function of a set of crisp input values. Instead of a general fuzzy set B of Y, the output of each rule is a specific function of the real inputs. The antecedent part of each rule, R_i, is matched as in the MA approach, but the output then becomes

$$y_i = A'_{i1} \times \cdots \times A'_{in}(x_1, \ldots x_n) \cdot f_i(x_1, \ldots x_n) \tag{7.17}$$

The weighted average of this set of k values is used as the system output.[2] One of the main motivations to recast fuzzy inference in this way is that stability theory for fuzzy controllers could be developed [Passino, 1998; Verbruggen and Babuska, 1999]. Both methods, as well as other formulations, can be used to produce similar results for specific application domains. The choice is really in the description of the consequent clause, as will be demonstrated in the example below.

---

[2] Here also, like with Correlation Min and Correlation Product, the rule in a TSK system is not a direct translation or generalization of logical implication. Rather the rule describes a relationship between the fuzzy antecedents and a function describing the desired output.

You might recall Theorem 3.1 from Chapter 3, the universal approximation theorem for one hidden layer nonlinear perceptrons. It said that given a continuous function defined on a compact subset D of Euclidean n space, $\Re^n$, mapping into Euclidean m space, $f : D \to \Re^m$, there exists a one hidden layer perceptron that uniformly approximates f on D. The same turns out to be true for fuzzy systems. The earliest mention we can find for such a result is in a 1990 paper presented in the Proceedings of the North American Fuzzy Information Processing Society meeting [Cao, 1990].[3] In 1992, several variations of the universal approximation theorem started to surface, including Buckley [1992], Wang [1992] and Kosko [1992]. A nice survey can be found in Kreinovich *et al.* [1998]. Because of the flexibility of defining operators and aggregation functions in a fuzzy system, this theorem has been reproved numerous times for different configurations. The bottom line is that fuzzy rule-based systems fall into the category of universal approximators for real (vector)-valued functions just like multilayer perceptrons. You need to be cautious because just like with perceptrons, proofs don't tell you how to build the fuzzy systems, only that one exists for any level of approximation accuracy. A potential added value over neural networks is that the fuzzy system has the connection to the semantics of the problem, possibly adding to the understandability of the approximation process.

7.5   WHERE DO THE RULES COME FROM?

Fuzzy logic systems are quite powerful and have been used in many applications from nonlinear control to classification. However, a common question is often asked: Where do the rules come from? Much like our discussion of membership functions, sometimes the rules come from experts. The person who has controlled a complex piece of equipment can linguistically describe his or her reactions to a variety of input conditions. Think about a simple task of balancing a broom in the palm of your hand. While you may not be able to solve the equations for motion in your head, at any instant, you can see roughly the angle the broom makes with the desired vertical orientation, say in terms of words like "big," "medium," "small," and so on, and you can "feel" the rate at which the broom is moving, either up or down and quantized in a similar way to the angle. (Note that here we only consider the broom falling away from us, though both directions should be taken into account.) It wouldn't take long for you to come up with rules like "IF the broom's angle is MEDIUM and the broom is falling away from me SLOWLY THEN push my hand forward FAIRLY FAST". Balancing the "inverted pendulum" on a motorized cart was one of the early demonstrations of fuzzy logic rule-based control.

In cases where training data are available, fuzzy rules can be learned, many times through the use of clustering algorithms or other computational intelligence techniques like neural networks, evolutionary computation, swarm intelligence, and so on. (see, for example, Pedrycz and Gomide [1998], Zurada *et al.* [1994], and Fogel

---

[3] However, the manuscript contained no proofs due page limitations (author comment in the paper).

and Robinson [2003)]. Learning the rules (and their membership functions) is treated as an optimization problem; the performance of the rule system on the training data is the function that needs to be maximized. There are many tools available to manually build or learn fuzzy rule systems. In fact, in the example given below, Matlab contains a "neuro-fuzzy" implementation of the TSK model (called ANFIS [Math Works, 1995]) that supports learning both the antecedent membership functions and the consequent function parameters. One important point to remember about fuzzy rule-based systems versus statistically trained classifiers is that important conditions that are improbable (i.e., ones that don't happen frequently enough to be captured in the training data) can be easily "hand" encoded into rules that will only contribute to output in those infrequent cases.

Fuzzy rule-based systems clearly generalize standard expert systems. In the inverted pendulum example, an equivalent crisp rule control system could be developed by quantizing the angle and the rate of motion into small intervals and building a rule for all pairs of intervals, the output of each rule being a set velocity of cart motion. How fine does the quantization need to be? That depends on how smooth you desire the control. Fuzzy rule systems, by their very construction, are great interpolators and so few rules are usually needed compared to crisp expert systems. Should fuzzy rules always be used? If the data are purely symbolic in nature, then fuzzy logic certainly doesn't apply. Probabilities can be associated with crisp rules and uncertainty can be updated along with rule firing. Bayesian networks, or more generally, belief networks, offer alternative ways to encode and manage probabilistic uncertainty in hierarchical frameworks. The choice of model should always be dictated by the form of the problem, the nature of the uncertainty, the ease of use of the particular formulation, and the meaningfulness of the results.

EXAMPLE 7.4

We close with a simple example of a fuzzy inference system that comes from Wang *et al.* [2006]. It involves softening the output of the short physical performance battery (SPPB) test, a series of timed physical activities that have been created to evaluate, discriminate, and predict physical functional performance for both research and clinical purposes, primarily for physically impaired older adults. The original scoring system of the SPPB test uses crisp time boundaries to assign the subject to discrete classes of performance. The crisp (and somewhat arbitrary) nature of the thresholds can easily produce anomalies. The SPPB test measures balance, gait, strength, and endurance. Although it is a timed performance test, each subtask score is an integer value in the range of 0–4. A score of 0 indicates the inability to complete the task in a nominal time frame, while categories 1–4 are assigned to the corresponding quartiles of time needed to perform the action.

The original scoring for the SPPB standing test is as given in Table 7.1 [Guralnik *et al.*, 1994]:

TABLE 7.1   Scoring Performance on Tests of Standing Balance

Score	Side-by-Side Stand	Semi-Tandem Stand	Full-Tandem Stand
0	$t < 10\,s$	Not attempted	Not attempted
1	$t = 10\,s$	$t < 10\,s$	Not attempted
2	$t = 10\,s$	$t = 10\,s$	$t < 3\,s$
3	$t = 10\,s$	$t = 10\,s$	$3\,s <\, =t < 10\,s$
4	$t = 10\,s$	$t = 10\,s$	$t = 10\,s$

In Wang *et al.* [2006], rules and the membership functions for the linguistic values were constructed manually with input from nurses. Here is a simple set of fuzzy rules for Standing Test Performance:

1. If (Side-by-Side_Stand_Time is SHORT) then (Standing_Test_Performance is VERY_POOR)
2. If (Side-by-Side_Stand_Time is LONG) and (Semi-Tandem_Stand_Time is SHORT) then (Standing_Test_Performance is POOR)
3. If (Side-by-Side_Stand_Time is LONG) and (Semi-Tandem_Stand_Time is LONG) and (Full-Tandem_Stand_Time is ShortSHORT) then (Standing_Test_Performance is OK)
4. If (Side-by-Side_Stand_Time is LONG) and (Semi-Tandem_Stand_Time is LONG) and (Full-Tandem_Stand_Time is MEDIUM) then (Standing_Test_Performance is GOOD)
5. If (Side-by-Side_Stand_Time is LONG) and (Semi-Tandem_Stand_Time is LONG) and (Full-Tandem_Stand_Time is LONG) then (Standing_Test_Performance is EXCELLENT)

Membership functions were modeled either by triangles and trapezoids or smooth curves, in this case, chosen heuristically to reflect common sense. As an example, the membership functions for short, medium, and long for the linguistic variable Full-Tandem Stand are shown in Figure 7.4.

The system was implemented in Matlab [Math Works, 1995] using both an MA fuzzy set output and a functional TSK output format (the output function for each rule is just the class label value, 0–4). Figure 7.5 displays one implementation of an MA system response when Side-by-Side Stand is 10 s, Semi-Tandem Stand is 10 s, and Full-Tandem Stand is 9 s. Here, the defuzzified output is 3.2, close to the crisp output of 3 in this case. The fuzzy system provides a smoother transition from one category to the next as the times change. The goal is to do frequent passive monitoring of elders to detect gradual changes in their physical performance.

Figure 7.6 shows a similar configuration for a TSK version of the rule base, with smooth membership functions. Particularly with small rule bases, the performance is not overly sensitive to the form of the precise definition of the membership functions.

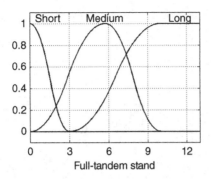

FIGURE 7.4 Membership functions for Full-Tandem Stand, used in the fuzzified scoring rule-based SPPB system.

Finally, Figure 7.7 shows the complete output surfaces obtained by varying two of the three input values across their entire respective ranges. The figure clearly shows the smoothness of the output function to small changes.

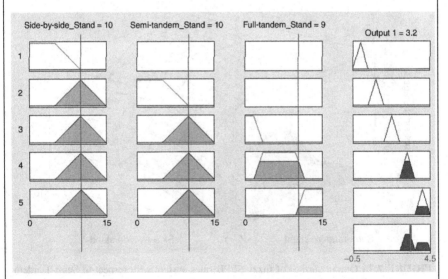

FIGURE 7.5 Matlab implementation of simple MA rule system for PSSB scoring. The antecedents of each of the five rules are activated by a singleton fuzzy set (measured value represented by the vertical lines). The outputs of each rule, according to Eq. 7.16, is shown as a darkened area. Note that for these inputs, only rules 4 and 5 produce nonzero output sets. The bottom of the fourth column shows the aggregation and defuzzification.

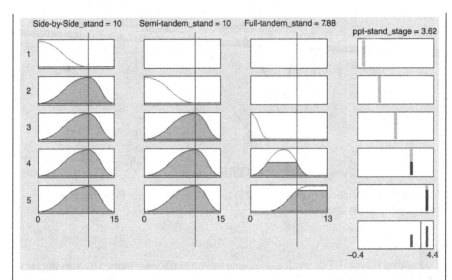

FIGURE 7.6 Matlab implementation of simple TSK rule system for fuzzy PSSB scoring. Here, the membership functions for antecedent values are pi-functions, but the interpretation is the same as in Figure 7.5.

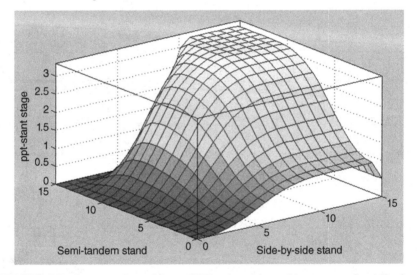

FIGURE 7.7 Output surface of fuzzy SPPB rules system with respect to Semi-Tandem Stand time and Side-by-Side Stand time using the TSK system described in the text.

7.6 CHAPTER SUMMARY

Just as fuzzy set theory extends traditional crisp set theory, fuzzy logic represents an extension of classical logic. Logical propositions are translated into fuzzy relations where the diversity of underlying fuzzy operators provides many possible interpretations. IF-THEN rules form the basis of a fuzzy logic inference system. Generalized modus ponens allow for deductions to be made from propositions that only approximately match the antecedents of a fuzzy rule. Systems of fuzzy rules can be built or learned to perform control functions (the original big pay-off), but also to work as a pattern classifier. These systems in many cases behave like statistical classifiers, but can also encode human knowledge directly into the structure in a linguistically pleasing manner.

EXERCISES

7.1. In a fuzzy logic control application, explain what is meant by the terms "fuzzification" and "defuzzification." Why are they necessary?

7.2. Let $Y = \{-3, -2, -1, 0, 1, 2, 3\}$. Suppose that the output of three rules are as follows:

$$B_1' = (0, .1, .6, .6, .6, .2, .1)$$
$$B_2' = (0, 0, 0, .2, .7, .7, .2)$$
$$B_3' = (0, 0, 0, 0, 0, 1, 0)$$

Compute the crisp output generated by a fuzzy control system, using centroid defuzzification.

7.3. Let $X = \{1, 2, 3\}$; $Y = \{a, b\}$; $Z = \{w, x, y, z\}$; and

let	$A = 1.0/1 + 0.4/2 + 0.1/3$
	$B = 0.2/a + 0.8/b$
	$C = 0.0/w + 0.4/x + 0.8/y + 1.0/z$
Consider the rule:	IF U is A THEN W is C

A. Translate this rule using Zadeh's original translation formula (i.e., the Lukasiewicz implication).

B. Show the result of inference if the input is
  i. U is A;
  ii. U is NOT A;
  iii. U is A' where $A' = 0.6/1 + 1.0/2 + 0.0/3$.

C. Translate the following rule using Correlation-Min encoding:
  If U_1 is A and U_2 is B THEN W is C

7.4. If the exact fuzzy antecedent is applied to a fuzzy rule that is translated using Zadeh's original formula (i.e., the Lukasiewicz implication), the result is not exactly the consequent fuzzy set. However, what nice property does this translation of modus ponens possess? Can you prove it (not just give an example)?

7.5. Let $X_1 = \{1, 2, 3\}$; $X_2 = \{w, x, y, z\}$; $Y = \{a, b\}$; and

let	$A_1 = (0.0, 1.0, 0.6)$	(fuzzy subset of X_1)
	$A_2 = (0.0, 0.4, 0.8, 1.0)$	(fuzzy subset of X_2)
	$B = (0.2, 0.8)$	(fuzzy subset of Y)
Consider the rule:	IF V is A_1 THEN W is B	

A. Translate this rule using Correlation-Product encoding.

B. Show the result of inference (for the translation in part A) if the input is
  i. V is A_1
  ii. V is NOT A_1

7.6. Let $X = \{a, b, c, d, e, f\}$; $Y = \{1, 2, 3, 4\}$; $Z = \{s, t\}$; and

let	$A = 0.0/a + 0.0/b + 1.0/c + 1.0/d + 0.0/e + 1.0/f$
	$B = 1.0/1 + 0.7/2 + 0.2/3 + 0.0/4$
	$C = .9/s + 0.5/t$
Consider the rule:	IF V is A THEN W is B

A. Translate this rule using Zadeh's original translation formula (i.e., the Lukasiewicz implication).

B. Show the result of the compositional rule of inference (for the translation in part A) when the input is "V is A."

C. Based on the result of part B, state and prove a theorem about the "firing" of rules that have a crisp antecedent.

7.7.[4] You are to design a Takagi–Sugeno fuzzy rule-based system to perform a 2-input, 1-output control function. Based on a considerable experiment in which a skilled human controlled the plant, it is decided that the linguistic variable for the first antecedent should have two linguistic values, while that of the second needed three values, and with linear functions for the output. While you believe that the membership functions should be triangular (though not necessarily symmetric), their actual definitions are uncertain. You decide to adopt an evolutionary computation approach to optimize the complete rule base. (For simplicity, assume only two rules in your system.)

A. Sketch a diagram of the input domains for the antecedent linguistic variables and their corresponding output functions.

B. Define an evolutionary algorithm data structure to optimize the membership functions and function parameters.

C. Describe a good fitness function, and how it is used to evaluate each candidate solution.

---

[4] Problems 7.7 and 7.8 ask you think about using evolutionary computation approaches to optimize a fuzzy rule base. You may need to delay solving these until you study those chapters.

7.8.[5] You are to design a fuzzy rule-based system to balance a broomstick: a 2-input, 1-out control function. The inputs are the angle the broom makes with the vertical axis and the rate of change in that angle between instances. The output is the force you apply to your hand to bring the broom into balance. (Actually, this is called the inverted pendulum problem and the output is voltage to a cart motor.) Based on a considerable data collection experiment in which a skilled human balanced the broom, it is decided that the linguistic variable for the first antecedent should have two linguistic values, while that of the second needed three values, and the output linguistic variable needed four values. While you believe that the membership functions should be triangular (though not necessarily symmetric), their actual definitions are uncertain. You decide to adopt a PSO approach to optimize the complete rule base.

A. Sketch a diagram of the input and output domains for the linguistic variables.

B. Define a particle structure to optimize the membership functions.

C. Describe a good fitness function, and how it will be evaluated for each particle.

7.9. Right after Eq. 7.1, we mentioned how it is a challenge to define appropriate domains to describe the values (create fuzzy sets) for some linguistic variables. Take the example of HEALTH. How could you define a domain on which you could define values like good, average, poor, and so on, so that some (hopefully) measured index could be used to activate a rule with HEALTH in one of its antecedent clauses? If you don't like HEALTH, think of a different variable that doesn't directly map into an obvious numeric range, like AGE does. In many applications, standard nominal ranges (0–10, 0–100) are used and a person is asked to pick a number—like the faces in the membership function for pain in Chapter 6 (Figure 6.2). A nice extension of this would be to ask a person to draw a fuzzy set over the nominal range that they feel represents their HEALTH condition—might be multimodal. Of course, using this type of fuzzification requires a full generalized modus ponens since the input is no longer a singleton.

---

[5] This is a "naïve physics" version of control of the inverted pendulum problem. Section 12.5.1 provides a good description of a few variations of this problem along with an experiment using evolutionary computation to learn a neural network controller. Chapter 12 of Klir and Yuan [1995] contains a nice description of a fuzzy controller for the simple inverted pendulum. You should solve this problem before looking there (or the many other references on this problem).

CHAPTER 8

Fuzzy Clustering and Classification

8.1 INTRODUCTION TO FUZZY CLUSTERING

Unsupervised learning is the area of pattern recognition comprising theories and algorithms that attempt to search for "natural structure" in unlabeled data. The principal tool in unsupervised learning is clustering where the goal is to group objects that are similar to each other into one set and have objects that are not similar placed into different groups. Most clustering techniques work with sets of feature vectors in Euclidean d-space, in which each vector of measured features represents an object in some real problem. Approaches that utilize numeric feature vectors are called object-based methods. This will be the main focus of this chapter, with the fuzzy C-means (FCM) and the possibilistic C-means (PCM) as the primary examples. However, any set of objects for which a dissimilarity measure can be developed can be used for clustering. Such methods are referred to as relational clustering approaches. For example, in Wilbik and Keller [2012], a distance metric was created to measure distance between linguistic summaries of human activities. Clustering algorithms have then been applied to sets of such summaries [Wilbik *et al.*, 2012]. If you have a set X of objects, then as long as you have a method to assess the distance or even the dissimilarity (doesn't have to be a metric) between pairs of objects, you are in business to do clustering. There are relational variations of many clustering approaches, including the FCM and PCM [Bezdek *et al.*, 1999], although we won't discuss them here.

There are really three distinct problems in clustering. Given a data set, X, the first question is whether there are any clusters in the data. Many times we forget this step and dive right into finding them since we have intuition as to what we think might be present in the data. Intuition is sometimes hard to come by when the problem dimensionality is large. There are visual methods to create a two-dimensional image to augment our intuition. The VAT and iVAT families of algorithms, for example, supply such visualizations [Bezdek and Hathaway, 2002; Havens and Bezdek, 2012]. The second question deals with actually finding the clusters once you believe there are some, the main topic of this chapter. All clustering algorithms will do their best at finding clusters. But, are the clusters good? The final problem in clustering addresses

Fundamentals of Computational Intelligence: Neural Networks, Fuzzy Systems, and Evolutionary Computation, First Edition. James M. Keller, Derong Liu, and David B. Fogel.
© 2016 by The Institute of Electrical and Electronics Engineers, Inc. Published 2016 by John Wiley & Sons, Inc.

this question. Various cluster validity measures have been proposed to create a numeric index of goodness. This way, different sets of algorithm parameters or different algorithms, producing different clusters in the data, can be tried with the validity measure deciding the optimal choice.

Object-based clustering focuses on vectors of real numbers. We denote the set by $X = \{x_1, x_2, \ldots, x_n\}$ where $x_k \in \mathfrak{R}^d$. What constitutes natural structures in X? The easy answer is that they are the ones we like, but of course, this glib answer is hard to define when the data are of high dimensionality. Hence, assumptions must be made, and with each assumption, we place constraints on the groupings allowed by automated techniques. For example, it is completely reasonable to assert that points in a real-valued feature space that are "close" to each other end up in the same cluster. How do we define "close?" Usually, we pick a distance metric in $\mathfrak{R}^d$. Certainly, the choice of distance measure strongly influences the resultant grouping of data, that is, it defines the geometry of clusters of feature vectors. The standard Euclidean distance $d^2(\mathbf{x}, \mathbf{y}) = (\mathbf{x} - \mathbf{y})^t(\mathbf{x} - \mathbf{y})$, the dot product of the difference between the two vectors, favors groups of vectors that are hyperspherical. Different choices of distance functions or even dissimilarity measures give rise to alternative definitions of closeness of objects for clustering approaches.

Clustering is based on the concept of a C-partition of the data set X. A partition of n data points into C clusters, $A_1, \ldots, A_C$, is defined by a partition matrix $U = \{u_{ik}\}$, where $0 \leq u_{ik} \leq 1$ is the degree that data point x_k belongs to cluster A_i, usually subject to the constraint that the total degree of a data point belonging to all clusters is 1, that is,

$$\sum_{i=1}^{C} u_{ik} = 1 \text{ for all } k \qquad (8.1)$$

For simplicity, we will also call u_{ik} the degree of membership of data point x_k in cluster A_i. Note that in the crisp case, each x_k will be assigned to one and only one cluster A_i, that is, a crisp partition satisfies the Laws of Contradiction and Excluded Middle. In other words, for each $k = 1, \ldots, n$, $u_{ik} = 1$ for some i between 1 and C and $u_{jk} = 0$ for all other cluster indices j. Fuzzy partitions ease this binary constraint and allow the degrees of membership to be in the unit interval and spread across the clusters. This spreading constraint is rooted in Eq. 8.1. We will see later that possibilistic partitions relax the constraint even further.

In the following sections, we develop some standard clustering algorithms and display results on three simple data sets.[1] The collections of points are in $\mathfrak{R}^2$ so that we can visualize what should happen, but always think about data in some higher dimensional space where intuition fails us. Table 8.1 contains a version of the "famous" butterfly data, created by Jim Bezdek in his classic book [Bezdek, 1981] to illustrate the advantages of the fuzzy C-means. It is a 15 point two-dimensional data set, shown graphically in Figure 8.1. Visually, you can imagine that there are two small symmetric triangular looking clusters with a bridge point (#8) between them.

---

[1] JK is grateful to his good friend, Professor Mihai Popescu, who graciously provided results of various algorithms depicted in this chapter.

TABLE 8.1   Fifteen-Point Butterfly Data Set

Point Index	x_i	y_i
1	−3	−2
2	−3	0
3	−3	2
4	−2	−1
5	−2	0
6	−2	1
7	−1	0
8	0	0
9	1	0
10	2	−1
11	2	0
12	2	1
13	3	−2
14	3	0
15	3	2

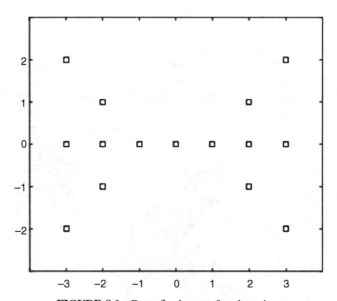

FIGURE 8.1   Butterfly data set for clustering.

Figures 8.2 and 8.3 contain other configurations of two-dimensional points, called Clouds 1 and Clouds 2. They both contain 1200 two-dimensional vectors. How many clusters do you see?

You probably have no trouble guessing that Clouds 1 contains three clusters. But what about Clouds 2? Are there still three clusters? It is harder to tell; might there be only two clusters or possibly four? Of course, we "know" because we made up the

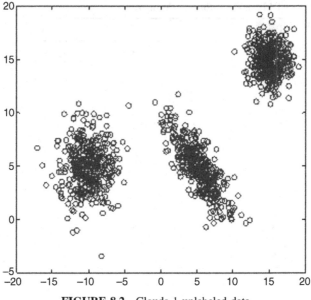

FIGURE 8.2   Clouds 1 unlabeled data.

data. The quotes in the previous sentence indicate that if we create labeled data, say from a mixture of probability distributions, there is no guarantee that clustering algorithms are going to agree with that labeling. If these points are randomly generated, and we compute all pairwise (Euclidean) distances, we can display the

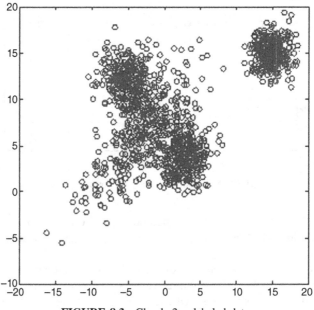

FIGURE 8.3   Clouds 2 unlabeled data.

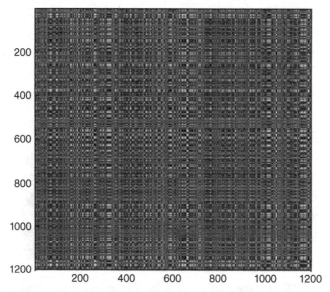

FIGURE 8.4   Distance matrix shown as an image for Clouds 1.

distance matrices as images. Figures 8.4 and 8.5 show the distance matrices for Clouds 1 and Clouds 2 (would you be able to do this for 10 million points?).

Big help, right? This might be all you have if, instead of two-dimensional vectors, the data sets were in dozens, hundreds, thousands, or higher numbers of dimensions, as is the case in real applications. Can you use these distance matrices to give you a clue that there might be structure in data sets? One way to get a

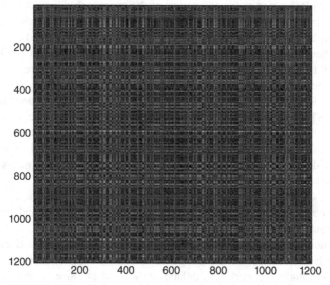

FIGURE 8.5   Distance matrix shown as an image for Clouds 2.

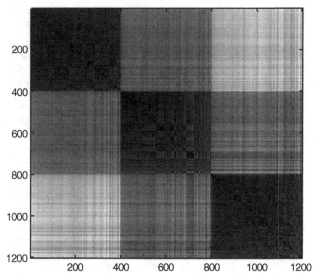

FIGURE 8.6 The VAT image of the distance matrix for Clouds 1.

glimpse of the geometric structure comes from the VAT/iVAT families of algorithms [Bezdek and Hathaway, 2002; Havens and Bezdek, 2012]. Essentially, VAT creates the minimal spanning tree (MST) of the data set and then reorders the points according to their indices in the MST. Euclidean distance is replaced by a geodesic distance to help iVAT generate a much cleaner version of the resultant image. Based on this reordering, dark blocks along the diagonal indicate that many points are close together in Euclidean space. The contrast between the dark diagonal blocks and the off-diagonal blocks hint at the separation of the groupings. Figure 8.6 is the VAT image for Clouds 1, while Figure 8.7 is the iVAT image for Clouds 2.

Does this help? Certainly, Figure 8.6 would lead us to believe that there is strong evidence that three pretty distinct clusters might exist in the data (look again at Figure 8.2). For Clouds 2, the situation is far less clear. If you stare at Figure 8.7, you might guess that there are four clusters (or not), but that they are somewhat jumbled up. Recall that VAT/iVAT construction is based on Euclidean distance to form the MST, so they inherently impose a geometry on feature space. Real clusters may not conform to this constraint, leading to misinformation from the visualization. Note also that this visualization technique does not produce clusters, but only provides evidence for whether there is some structure, along with a guess as to how many clusters to look for.

Actually, Clouds 1 and Clouds 2 are synthetic data sets generated as Gaussian mixtures. The generating parameters are given in Tables 8.2 and 8.3 and the "labeled" data are shown in Figures 8.8 and 8.9.

It is certainly much easier to understand if the data are labeled! Even with the component labels, Clouds 2 is a much more difficult data set to analyze. That is the challenge for clustering: start with unlabeled data and produce meaningful labels.

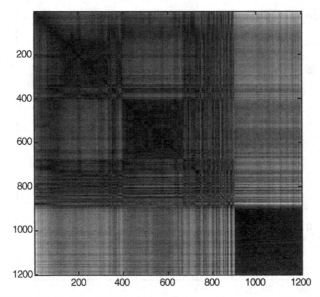

FIGURE 8.7 The iVAT image of the distance matrix for Clouds 2.

TABLE 8.2   Parameters for Clouds 1 Data Set

Mean Vector	Covariance Matrix	Probability
$\mu_1 = \begin{pmatrix} -10 \\ 5 \end{pmatrix}$	$\Sigma_1 = \begin{pmatrix} 4 & 0 \\ 0 & 4 \end{pmatrix}$	1/3 (circles)
$\mu_2 = \begin{pmatrix} 15 \\ 15 \end{pmatrix}$	$\Sigma_2 = \begin{pmatrix} 2 & 0 \\ 0 & 2 \end{pmatrix}$	1/3 (triangles)
$\mu_3 = \begin{pmatrix} 5 \\ 5 \end{pmatrix}$	$\Sigma_3 = \begin{pmatrix} 5 & -4 \\ -4 & 5 \end{pmatrix}$	1/3 (pluses)

TABLE 8.3   Parameters for Clouds 2 Data Set

Mean Vector	Covariance Matrix	Probability
$\mu_1 = \begin{pmatrix} -5 \\ 12 \end{pmatrix}$	$\Sigma_1 = \begin{pmatrix} 4 & 0 \\ 0 & 4 \end{pmatrix}$	1/4 (circles)
$\mu_2 = \begin{pmatrix} 15 \\ 15 \end{pmatrix}$	$\Sigma_2 = \begin{pmatrix} 2 & 0 \\ 0 & 2 \end{pmatrix}$	1/4 (pluses)
$\mu_3 = \begin{pmatrix} 3 \\ 4 \end{pmatrix}$	$\Sigma_3 = \begin{pmatrix} 3 & 0 \\ 0 & 3 \end{pmatrix}$	1/4 (crosses)
$\mu_4 = \begin{pmatrix} -3 \\ 6 \end{pmatrix}$	$\Sigma_4 = \begin{pmatrix} 18 & 15 \\ 15 & 18 \end{pmatrix}$	1/4 (triangles)

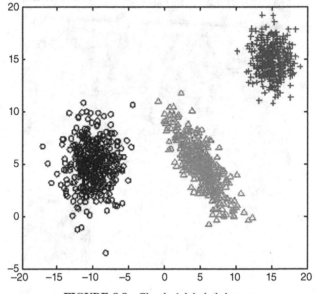

FIGURE 8.8 Clouds 1 labeled data set.

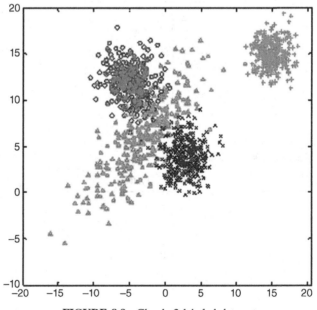

FIGURE 8.9 Clouds 2 labeled data set.

In the next two sections, we will explore one of the main approaches to produce soft clusters of object data, the fuzzy C-means. Results on our three sets of two-dimensional feature vectors will be used to illustrate the variations of the algorithms, including the most popular crisp clustering algorithm as a limit case. Following this, in Section 8.4, the possibilistic C-means algorithm will be introduced and studied.

8.2 FUZZY C-MEANS

As we mentioned above, the first task in unsupervised learning is to decide if there are any clusters in a given data set X, and if so, how many should we look for. Most clustering algorithms, including the one in this section, require the number of clusters as an input. Their job, then, is to search for the "best" partition of X, where the choice of what is meant by "best" partition varies from approach to approach. The fuzzy C-means [Bezdek, 1981; Bezdek *et al.*, 1999] is a scheme to partition X into a predefined number of clusters taking into account the uncertainty of cluster assignment. It produces a fuzzy partition of X that allows sharing of objects between clusters. In the FCM, each cluster is represented by a cluster center (or exemplar or prototype). Let v_i be the prototype of cluster A_i and let V be the set of all C cluster centers. The objective of the FCM is to minimize the following criterion function:

$$ J(U, V) = \sum_{k=1}^{n} \sum_{i=1}^{C} (u_{ik})^m d^2(x_k, v_i) \qquad (8.2) $$

subject to the constraint that $\sum_{i=1}^{C} u_{ik} = 1$ for all k. A fuzzy partition that minimizes J(U,V) is the "best" one for FCM. This is a generalized least-squares criterion function whose goal is to minimize the weighted sum of squared distances between the data and the set of cluster centers. The constraint is necessary to guard against the trivial solution, that is, setting all cluster memberships to zero, an obvious minimum of Eq. 8.2. Here, the parameter m is called the fuzzifier. This is because larger values of m favor more "fuzzy" partitions, that is, more similar degrees of membership of a data point in all clusters. Performing this minimization leads to two equations, one for memberships and one for cluster centers, expressing necessary conditions for a minimum.

The theorem [Bezdek, 1981] that gives rise to the FCM algorithm expresses necessary conditions on the partition membership values and the prototype vectors. We will state and prove it for distance measures that are based on Euclidean inner products, that is, where $d^2(\mathbf{x}, \mathbf{y}) = (\mathbf{x} - \mathbf{y})^t A(\mathbf{x} - \mathbf{y})$ for some positive definite d × d matrix A (like the inverse of a covariance matrix), but there are many extensions to other distance measures. After the proof, we'll see that the FCM generalizes a very basic and popular crisp clustering algorithm, the hard C-means (HCM).[2]

---

[2] The machine learning community likes to call this algorithm simply the K-means.

Theorem 8.1   Let $X = \{x_1, x_2, \ldots, x_n\}$, where $x_k \in \Re^d$; $U = \{u_{ik}\}$, $i = 1, \ldots, C$ and $k = 1, \ldots, n$, where $0 \le u_{ik} \le 1$ is a fuzzy C partition of X; $V = \{v_1, \ldots, v_C\}$ be a set of prototypes; and $m > 1$ be a fuzzifier. For each $k = 1, \ldots, n$, define $I_k = \{i | 1 \le i \le C \text{ and } d_{ik}^2 = d^2(x_k, v_i) = 0\}$. Then the pair (U, V) is a minimum of $J(U,V)$ only if

(a) $I_k = \phi$, then

$$u_{ik} = \frac{(1/d(x_k, v_i))^{2/(m-1)}}{\displaystyle\sum_{j=1}^{C} (1/d(x_k, v_j))^{2/(m-1)}} \tag{8.3a}$$

(b) $I_k \neq \phi$, then $u_{ik} = 0$ for all $i \notin I_k$ and

$$\sum_{i=1}^{C} u_{ik} = 1 \tag{8.3b}$$

Additionally, for point prototypes, that is, where each cluster is represented by a single vector in the feature space, these prototypes must have the form

$$v_i = \frac{\displaystyle\sum_{k=1}^{n} (u_{ik})^m x_k}{\displaystyle\sum_{k=1}^{n} (u_{ik})^m} \tag{8.4}$$

Proof: Assume that $I_k = \phi$. The objective is to find necessary conditions for a minimum of $J(U,V)$. First, fix V. Using LaGrange multipliers, write

$$J(U, V) = \sum_{k=1}^{n}\sum_{i=1}^{C} u_{ik}^m \, d^2(x_k, v_i) - \sum_{k=1}^{n} \lambda_k \left(\sum_{i=1}^{C} u_{ik} - 1 \right)$$

Then

$$\frac{\partial J}{\partial u_{rs}} = m \cdot u_{rs}^{m-1} d^2(x_s, v_r) - \lambda_s$$

Setting this partial equal to 0 and solving we get

$$u_{rs} = \left(\frac{\lambda_s}{m \cdot d^2(x_s, v_r)} \right)^{1/(m-1)}$$

Put this into the constraint (obtained by differentiating with respect to λ_s) to obtain

$$\sum_{i=1}^{C} \left(\frac{\lambda_s}{m \cdot d^2(x_s, v_i)} \right)^{1/(m-1)} = 1, \quad \text{or} \quad \lambda_s = \frac{m}{\left(\sum_{i=1}^{C} (1/d^2(x_s, v_i))^{1/(m-1)} \right)^{m-1}}$$

Hence,

$$u_{rs} = \left(\frac{m / \left(\sum_{i=1}^{C} (1/d^2(x_s, v_i))^{1/(m-1)} \right)^{m-1}}{m \cdot d^2(x_s, v_r)} \right)^{1/(m-1)} = \frac{1}{\left(\sum_{i=1}^{C} ((d^2(x_s, v_r))/(d^2(x_s, v_i)))^{1/(m-1)} \right)}$$

Now, in the event that $I_k \neq \phi$ for some x_k, we place nonzero memberships in those clusters where the distance is 0 (using the constraint $\sum_{i=1}^{C} u_{ik} = 1$) and zero memberships in the clusters where the distance is nonzero, thus providing a minimum value for the least-squares criteria function. This condition is called a singularity (although not the singularity event!). In practice, this situation is not common given the finite representation of real numbers in a computer. However, it can happen and needs to be addressed in any algorithm that implements the theorem.

To determine the necessary conditions on the prototypes, note that

$$\frac{\partial J(U, V)}{\partial v_i} = \sum_{k=1}^{n} u_{ik}^m \frac{\partial d^2(x_k, v_i)}{\partial v_i}$$

In the case where $d^2(\mathbf{x}, \mathbf{y}) = (\mathbf{x} - \mathbf{y})^t A(\mathbf{x} - \mathbf{y})$, where A is a $d \times d$ positive definite symmetric matrix, we have

$$\frac{\partial J(U, V)}{\partial v_i} = \sum_{k=1}^{n} u_{ik}^m \frac{\partial d^2(x_k, v_i)}{\partial v_i} = \sum_{k=1}^{n} u_{ik}^m \frac{\partial (x_k - v_i)^t A(x_k - v_i)}{\partial v_i} = -2 \sum_{k=1}^{n} u_{ik}^m A (x_k - v_i)$$

Setting this partial equal to the zero vector and knowing that A is invertible, we obtain

$$v_i = \frac{\sum_{k=1}^{n} (u_{ik})^m x_k}{\sum_{k=1}^{n} (u_{ik})^m}$$

the desired expression for the prototype update. QED.

Note that Eqs. 8.3 and 8.4 are coupled in the way that the partition memberships are needed to compute the prototypes and the prototypes are required to update the memberships. The FCM algorithm performs an iterative technique called alternating optimization (AO) in which cluster memberships and cluster centers are alternately updated in each iteration. The algorithm can be summarized as follows:

Let $X = \{x_1, x_2, \ldots, x_n\}$, where $x_k \in \mathfrak{R}^d$ is the set of vectors to be clustered.

```
Initialization: Set
        C, the number of clusters desired
        m, the fuzzifier
        ε, the convergence threshold
        V⁽⁰⁾ = {v₁⁽⁰⁾,...,v_C⁽⁰⁾} an initial set of cluster centers
    //Note: The v_i⁽⁰⁾ can be chosen randomly from X or through
      other mechanisms//
Set t = 0
REPEAT
    DO FOR each k = 1, . . . , n
        IF d(x_k,v_i) = 0 for some subset of clusters, i.e., I_k ≠ φ
        THEN
            Set u_jk⁽ᵗ⁾ = 0 for j ∉ I_k and u_jk⁽ᵗ⁾ > 0 for j ∈ I_k, as in Eq. 8.3b
        ELSE
            Compute u_ik⁽ᵗ⁾ from Eq. 8.3a.
        ENDIF
    END FOR
    Set t ← t + 1
    Using U⁽ᵗ⁻¹⁾, estimate V⁽ᵗ⁾ from Eq. 8.4.
```

$$\text{UNTIL } \sum_{i=1}^{C} \left\| v_i^{(t)} - v_i^{(t-1)} \right\| < \epsilon$$

```
where ‖*‖ is any vector norm (like Euclidean).
```

```
//Note: There are other stopping criteria, including number
of iterations, but this is the most common.//
```

This algorithm initializes by choosing C cluster centers. Alternatively, the algorithm can be initialized by randomly assigning values to the partition matrix subject to the constraint that the memberships for each vector sum to 1. Then, the computation steps need to be reversed inside the FOR loop (first estimate the cluster centers, and then update the partition matrix). How many initializations are there? Well, if you randomly choose prototypes from the data, there are $\binom{n}{C}$ such initializations—that's a bunch since n is usually much larger than C. If the initialization is done on the partition matrix, there is a "computationally infinite" number of starting points for FCM (not actually infinite, but effectively so). If you only run the FCM, or almost any other clustering algorithm, only one time, you risk misleading yourself with the result. Later, we'll show an example where different initializations lead to dissimilar final partitions.

Note that $m = 2$ is the default value for the fuzzifier because it makes Eq. 8.3a particularly simple. What happens as $m \to \infty$? Look at Eq. 8.3a. If the inverse distances are bigger than 1, then a large value of m in the exponent will drive them toward 1 from above, and if they started as fractions, they will be moved toward 1

TABLE 8.4   HCM Memberships for Butterfly Data

Index k	1	2	3	4	5	6	7	8	9	10	11	12	13	14	15
u_{k1}	0	0	0	0	0	0	0	1	1	1	1	1	1	1	1
u_{k2}	1	1	1	1	1	1	1	0	0	0	0	0	0	0	0

from below by the exponentiation. Hence, all terms in the calculation will become close, and so the cluster memberships will get closer to each other, that is, the memberships will be fuzzier. In the limit, all memberships will converge to 1/C, except possibly for the common cluster center if it lands on an element of X because of Eq. 8.3b—highly unlikely for real data. This is the most ambiguous partition possible.

If the cluster memberships are required to be binary at each step, that is, the clustering algorithm is to build a crisp C-partition of X, then the above AO algorithm reduces to the crisp or hard C-means with the two steps in the UNTIL loop: (i) Assign each vector to the cluster with the closest cluster center. (ii) Compute new cluster centers as the means of the vectors assigned to the respective clusters [Theodoridis and Koutroumbas, 2009]. In fact, as m → 1, the FCM partition converges to a HCM partition when d is the Euclidean distance. The HCM criterion is to find a set of cluster centers such that the sum of the distances of all points to their closest cluster center is minimized. Can you write that criterion function?

The hard 2-means was run on the butterfly data, using randomly chosen initial cluster centers. The final cluster memberships are shown in Table 8.4. Note that the bridge point (#8) is assigned to cluster 1 for this run. The final cluster center coordinates, $\{(-2.3, 0.0)^t, (2.0, 0)^t\}$, are shown in Figure 8.10. The bridge point

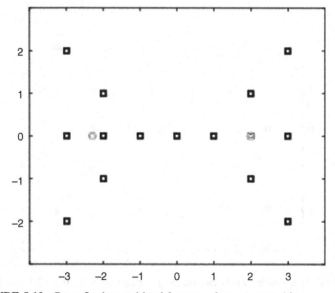

FIGURE 8.10   Butterfly data and hard 2-means cluster centers (shown as disks).

TABLE 8.5   FCM Memberships for Butterfly Data with C = 2 and m = 2

Index k	1	2	3	4	5	6	7	8	9	10	11	12	13	14	15
u_{k1}	0.13	0.03	0.13	0.05	0.00	0.05	0.12	**0.50**	0.88	0.95	1.00	0.95	0.87	0.97	0.87
u_{k2}	0.87	0.97	0.87	0.95	1.00	0.95	0.88	**0.50**	0.12	0.05	0.00	0.05	0.13	0.03	0.13

had to be assigned to one of the clusters, in this case the right cluster. The crisp assignment required by the HCM clearly affects the locations of the cluster centers.

One application of clustering is to use it in conjunction with a classification problem where the distribution of data within each class is not well understood. The idea is to cluster training data of only one class to look for substructure within the group. The resultant cluster centers can actually be used as multiple prototypes for the given class in a multiprototype pattern recognition approach—for test vectors, compute their distance to all prototypes of all classes and put them into the class of the closest prototype. You could also cluster all the training data and then examine the results (human intervention) to assign real class labels to the clusters, and then just use the cluster centers as single prototypes of each category. So, getting good estimates of the cluster centers is very important. Since they are normally averages of vectors, weighted by cluster member-ships, their locations can vary depending upon the treatment of points in overlapped areas and especially by outliers (perhaps caused by sensor failure or faulty feature extraction). Section 8.4 will address the outlier issue in more depth. An example of combining clustering and prototype classification can be found in Banerjee *et al.* [2013]. Think about this algorithm when you study Section 8.5.

Next, we ran the fuzzy 2-means on these data with m = 2. Table 8.5 shows the fuzzy memberships, while Figure 8.11 provides the locations of the final cluster

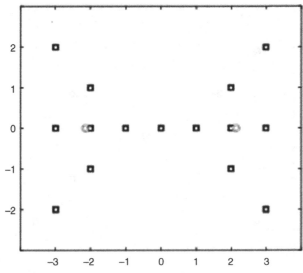

FIGURE 8.11   Butterfly data and fuzzy 2-means cluster centers $\{(-2.1, 0.0)^t, (2.1, 0.0)^t\}$ for m = 2.

TABLE 8.6   FCM Memberships for Butterfly Data with C = 2 and m = 1.25

Index k	1	2	3	4	5	6	7	8	9	10	11	12	13	14	15
u_{k1}	1.00	1.00	1.00	1.00	1.00	1.00	1.00	**0.50**	0.00	0.00	0.00	0.00	0.00	0.00	0.00
u_{k2}	0.00	0.00	0.00	0.00	0.00	0.00	0.00	**0.50**	1.00	1.00	1.00	1.00	1.00	1.00	1.00

TABLE 8.7   FCM Memberships for Butterfly Data with C = 2 and m = 6

Index k	1	2	3	4	5	6	7	8	9	10	11	12	13	14	15
u_{k1}	0.57	0.63	0.57	0.62	0.86	0.62	0.59	**0.50**	0.41	0.38	0.14	0.38	0.43	0.37	0.43
u_{k2}	0.43	0.37	0.43	0.38	0.14	0.38	0.41	**0.50**	0.59	0.62	0.86	0.62	0.57	0.63	0.57

centers. The bridge point now is shared equally between the two clusters. Because of the equal membership values for the bridge point, the cluster centers are symmetric.

As stated above, the FCM converges to the HCM as m approaches 1. As a demonstration, we ran the fuzzy 2-means on the butterfly set with m = 1.25. Table 8.6 corresponds to fuzzy memberships for this case. Memberships except for the bridge point are actually crisp. That vector is still shared between the two clusters with final memberships of 0.5 in each. The locations of the final cluster centers are in the same place as for m = 2.

On the other hand, as m gets larger, the memberships move toward 1/C. Table 8.7 displays the results of the FCM on the butterfly data with C = 2 and m = 6 (not infinity by a long shot, but fairly large for a FCM fuzzifier). You can easily see the "blurring" of vector membership. How would it look for m = 10?

Now let's see how these basic clustering algorithms do on the Gaussian cloud data sets. Clouds 1 should be no problem since the bunches of data are fairly compact and well separated (Hmm, might those criteria be used in judging the goodness of a final partition—the cluster validity issue?). Of course, you need to pick the right value of C. Figure 8.12 shows the output of the HCM with C = 3. The crisp membership symbols are generated from the final partition. No surprises here. The final partition of the HCM with C = 4 is displayed in Figure 8.13. The HCM did its job; it found a partition that at least was a local minimum of the HCM criterion. To get four clusters from Clouds 1, the HCM had to split a cluster. Given that we're talking about the standard Euclidean norm, it seems perfectly reasonable to split the elliptical cluster in half, favoring hyperspherical (circular in $\Re^2$) groupings.

Initialization is a big issue in clustering. What is shown in the figures is one run of the algorithm with a random initialization of the cluster centers (sampled from the data set). We'll show later what can happen for different initializations. Sensitivity to the choice of initialization is a significant question for clustering since the whole purpose of clustering is data exploration—you don't know the labels. A common practice is to

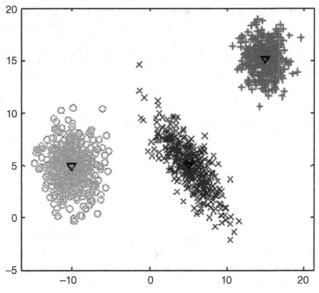

FIGURE 8.12 Results of HCM on Clouds 1 with C = 3.

run the clustering algorithm multiple times and see if the results are stable. Of course, then you have to define what you mean by stable. Essentially, this means that you have to compare the partition matrices or the cluster centers, or both. This is not as easy as it sounds.

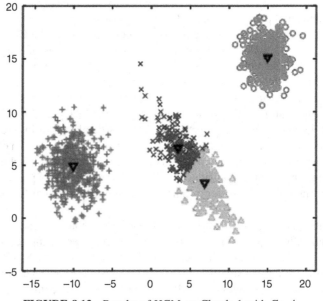

FIGURE 8.13 Results of HCM on Clouds 1 with C = 4.

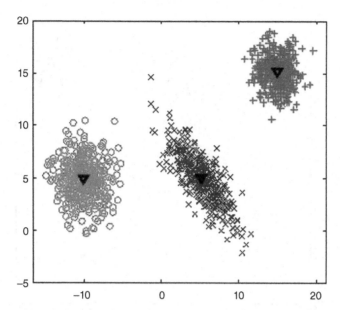

FIGURE 8.14   Results of FCM on Clouds 1 with $C = 3$ and $m = 2$.

How does FCM do on Clouds 1? Like HCM, the fuzzy clustering algorithm finds the same final crisp partition, as seen in Figure 8.14 with $C = 3$ and $m = 2$. Pretty hard to miss when the clusters are so well behaved. The final fuzzy partition is "hardened" to form the displayed crisp partition by assigning each vector to the cluster that has the highest membership.

If all data sets looked like Clouds 1, there would be little need for fuzzy (or other more generalized) clustering. The HCM algorithm would be sufficient. Clouds 2 is still a nice collection of points, being generated as a mixture of Gaussians. However, three of the components are close together and there is overlap in the distributions. Using a random initialization, the HCM doesn't handle the situation well according to what our intuition and knowledge of Clouds 2 construction tells us. Figure 8.15 shows the final HCM partition for $C = 4$. What we called component 4 is divided and combined with the two flanking clusters, while the component in the upper right area is split to form two clusters. Actually, with $C = 5$ (an intuitively wrong value), it's hard to argue with the breakdown of the data as seen in Figure 8.16, if we take a Euclidean viewpoint of the world.

The basic FCM using Euclidean distance does a little better in the final crisp partition on Clouds 2 as displayed in Figure 8.17, but it still doesn't match what we know about the class structure. Why? Euclidean distance in the criterion function imposes geometric constraints that are not satisfied by the actual data set. We will soon see how to accommodate more general distance metrics into the FCM structure.

Before leaving this section, we give a simple example that shows the importance of initialization. Figure 8.18 shows four seemingly (well) separated clusters. Do you

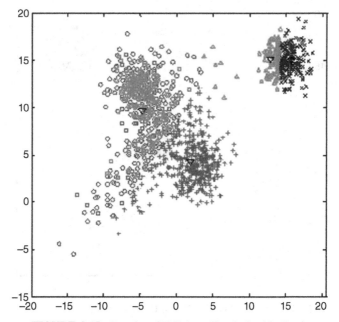

FIGURE 8.15   Results of HCM on Clouds 2 with C = 4.

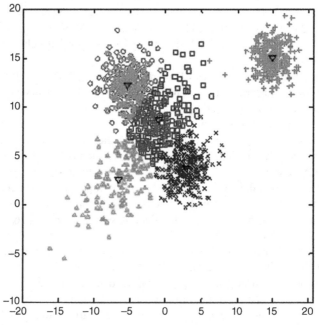

FIGURE 8.16   Results of HCM on Clouds 2 with C = 5. Is this reasonable for Euclidean distance?

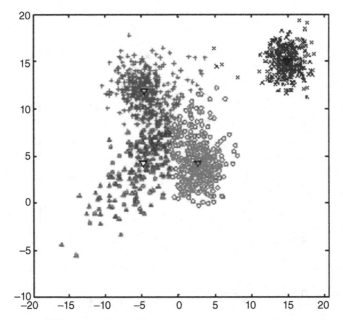

FIGURE 8.17 Results of FCM on Clouds 2 with C = 4 and m = 2.

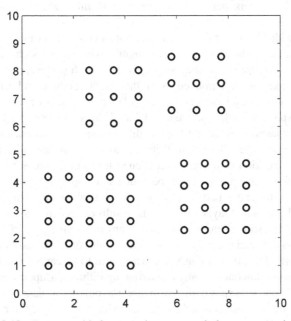

FIGURE 8.18 Data set with four seeming separated clusters to test initialization.

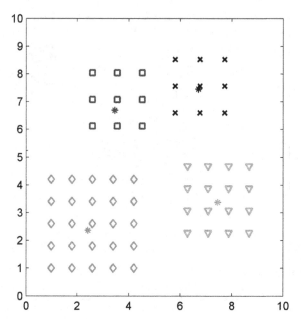

FIGURE 8.19 An intuitively correct hardened partition of the data in Figure 8.17 found by FCM with C = 4 and m = 2. Note that the cluster centers don't line up with the centers of the squares.

agree? You would think that it would be easy to find them with any clustering algorithm.

The FCM with C = 4 and m = 2 was run with two different initializations. Each initialization filled the beginning partition matrix with numbers chosen as uniform random with the constraint that they sum up to 1 for each vector. After convergence, the final partitions were hardened with the usual method and displayed with corresponding final cluster centers. Figure 8.19 shows a crisp partition that seems to match our expectations. But notice that the final cluster centers aren't exactly in the centers of the squares. This reflects the fact that all vectors contribute to the weighted average in the prototype calculation. With a second random initialization, the FCM finalizes on the partition displayed, in hardened form, in Figure 8.20. The different sizes and positions of what seems to be four natural groups stump the algorithm. Would you get the same partitions if you implemented and ran the FCM, say 100 times, on these data? Maybe not since the results depend heavily on initialization. It would be an interesting exercise to see how many different crisp partitions, and their frequencies, could be generated from ensembles of runs of HCM and FCM on the data from Figure 8.18. The point is that you always need to be careful (skeptical?) when using the partitions generated by any clustering algorithm. Unsupervised methods are by definition ways to explore the structure on unlabeled data. The better you understand the assumptions embedded in clustering techniques, the better you will be at interpreting their results.

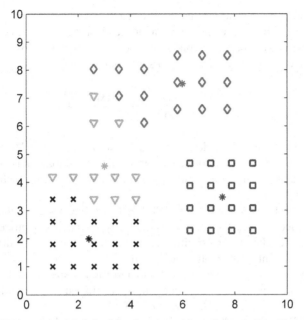

FIGURE 8.20 An unexpected final hardened partition of the data in Figure 8.17 from a second initialization by FCM with C = 4 and m = 2.

8.3 AN EXTENSION OF THE FUZZY C-MEANS

As we have seen from the examples above, the act of clustering, that is, running a clustering algorithm, enforces a belief about the geometry of feature space. Euclidean distance tends to produce clusters that are roughly hyperspherical in shape, and hence it determines the retrieved geometry of feature space. It didn't matter much with Clouds 1 since the groups of points had similar sizes and were well separated. But Clouds 2 proved to be more of a challenge (Figure 8.17). If the clusters are assumed to be hyperellipsoidal in structure as with the Clouds data, they can be modeled using a cluster-specific Mahalanobis distance:

$$d^2(\mathbf{x}_k, \mathbf{v}_i) = (\mathbf{x}_k - \mathbf{v}_i)^t \Sigma_i^{-1}(\mathbf{x}_k - \mathbf{v}_i) \tag{8.5}$$

Here, Σ_i represents the estimated "fuzzy" covariance matrix for the ith cluster [Gustafson and Kessel, 1979; Bezdek *et al.*, 1999; Theodoridis and Koutroumbas, 2009]. Note that this basic choice fits the assumptions of the FCM theorem, that is, it is an inner product norm distance. However, the distance metric should be cluster specific to capture the spread of the data within each group, which means that there are additional parameters that need to be estimated during each step of the resulting algorithm. Using this family of distance metrics, the resulting clusters will assume hyperellipsoidal shapes. As in the theorem for the FCM, the membership value and cluster center updates are the same (prove this! Problem 8.4). However, now necessary conditions on the cluster-specific fuzzy covariance matrices need to be determined for use in the AO algorithm, after the cluster centers are computed. In a similar fashion to the necessary conditions for cluster centers in Theorem 8.1, the fuzzy covariance approximation equations are found to be

$$\Sigma_i = \frac{\sum_{k=1}^{n} (u_{ik})^m (\mathbf{x}_k - \mathbf{v}_i)(\mathbf{x}_k - \mathbf{v}_i)^t}{\sum_{k=1}^{n} (u_{ik})^m} \tag{8.6}$$

Exercise 8.5 challenges you to derive Eq. Eq. 8.6 as a corollary to Theorem 8.1. We note that these necessary conditions are determined in a fashion that is essentially the same as its counterpart in the expectation–maximization (EM) approach to Gaussian mixture decomposition.

An extended FCM algorithm, called GK-FCM, incorporates into the FCM algorithm of Section 8.2 the estimation of fuzzy covariance matrices using Eq. 8.6. Also, it scales the cluster-specific Mahalanobis distance by the dth root of the determinant of Σ_i. This scaling attempts to normalize the clusters at each step to have the same hypervolume, thereby accommodating different sized clusters. Hence, the GK-FCM is more conducive to find hyperellipsoidal-shaped clusters of different sizes [Gustafson and Kessel, 1979; Bezdek *et al.*, 1999].

Applying GK-FCM to Clouds 2 (C = 4 and m = 2) and then hardening the final partition by assigning each point to the cluster with highest membership results in the distribution shown in Figure 8.21. It does find four clusters that look similar to the original distributions from Figure 8.9, although it tended to squeeze component 4.

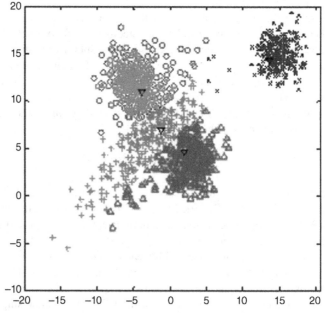

FIGURE 8.21   GK-FCM (C = 4, m = 2) hardened results on Clouds 2.

Many striking examples of the utility of this algorithm can be found in the clustering literature.

Gath and Geva [1989] proposed an extension to FCM that incorporated what they called exponential distance and an estimate of prior probabilities of clusters. Their algorithm, fuzzy maximum likelihood estimation (FMLE), is close in both form and results to Gaussian mixture decomposition (GMD) using the EM algorithm [Theodoridis and Koutroumbas, 2009]. In dealing with data that can be thought of as Gaussian mixtures, most clustering algorithms (certainly HCM, FCM, GK-FCM, FMLE, and GMD) are similar. The differences lie in the interpretation of the underlying structure and assumptions made on the distributions. You should use the approach that seems most natural for the problem you have.

8.4 POSSIBILISTIC C-MEANS

As we saw in Sections 8.2 and 8.3, the FCM clustering algorithm shares the soft memberships for each point across the clusters, although requiring them to sum to 1. This does ameliorate the problem of crisp assignment of vectors to particular clusters when the features possess ambiguity. When vectors are of high dimension, this fuzzy partition is useful not only in finding strong elements in a cluster (close to binary memberships) but also for detecting objects that lie in an overlapped region when memberships in multiple clusters are close to equal. This idea was used in Pal *et al.* [2005] to find, for example, proteins that had been incorrectly annotated in the Gene Ontology [Lord, 2003]. However, there is no reason that memberships of a given feature vector should always sum to 1. This is a definition within crisp clustering and is required in the FCM to avoid the trivial solution (all memberships equal zero) in minimizing the FCM criterion function. As mentioned in Chapter 6, a Crepe Myrtle is both a bush and a tree. In bioinformatics, there are proteins that belong completely to different families. So, the digital representation of this type of protein, or the Crepe Myrtle, is not consistent with the laws of excluded middle and contradiction, that is, under crisp algorithms, such objects must be placed entirely in one and only one cluster, certainly contrary to reality. The opposite is true for outliers, that is, for objects (or feature vectors) that do not belong to any of the groups. Sometimes, faults in sensors can give rise to completely bogus feature values. Such vectors should not be grouped into any cluster. Alternatively, if clusters in feature space are used to represent a normal condition, then identifying outliers is a means of creating an alert to signal an abnormal circumstance. The FCM spreads the degree of belonging across clusters better than does the HCM (recall Tables 8.5 and 8.6 for the butterfly data), but still does not capture the conditions of high membership in multiple clusters or low membership in all clusters. Crisp approaches have no choice other than to dump anomalies into a single group, and fuzzy algorithms can at best force their memberships close to 1/C. If the number of points being clustered is not huge, this can result in making noticeable changes in the cluster prototypes. Remember the multiprototype classifier technique discussed in Section 8.2. True outliers can drastically affect performance of that algorithm.

Krishnapuram and Keller [1993, 1996] found a way to relax the sum constraint while avoiding the trivial solution. It was done by modifying the FCM criterion function, resulting in a clustering technique called the possibilistic C-means. Their criterion function is

$$J(U, V) = \sum_{k=1}^{n} \sum_{i=1}^{C} u_{ik}^m d^2(x_k, v_i) + \sum_{i=1}^{C} \eta_i \sum_{k=1}^{n} (1 - u_{ik})^m \qquad (8.7)$$

where η_i are the appropriately chosen or estimated values [Krishnapuram and Keller, 1993, 1996].

The first term is the same weighted least-squares construct as in FCM, whereas the second term has the effect of trying to keep cluster memberships high. In Krishnapuram and Keller [1993], necessary conditions to minimize Eq. 8.7 with respect to the partition matrix entries are shown to be

$$u_{ik} = \frac{1}{1 + \left(d^2(x_k, v_i)/\eta_i\right)^{1/(m-1)}} \qquad (8.8)$$

for a minimum of $J(U,V)$. The necessary condition on the cluster centers is identical to Eq. 8.4. The proof mirrors that of the FCM theorem except that the zero distance case is unnecessary (see Problem 8.6). The PCM theorem and associated AO algorithm effectively decouples each cluster in computing membership values; hence, they possess a natural symmetry in membership. The values u_{ik} in Eq. 8.8 are called typicalities in the PCM since they measure how typical a particular input point is to each cluster, that is, how close the point is to each cluster prototype. Since all clusters are decoupled in the PCM formulation, the cluster centers have the freedom to seek out dense regions of feature space, and so the PCM algorithm is actually a mode-seeking technique. In the PCM algorithm, the parameters η_i have a significant influence on the final partition matrix; they determine the distance from the cluster center at which the membership crosses 0.5 (the 3dB point). In Krishnapuram and Keller [1996], it is recommended that if the data are not too noisy, FCM be run first for a few iterations to get a reasonable starting partition and then estimate these parameters as

$$\eta_i = \frac{\sum_{k=1}^{n} u_{ik}^m d^2(x_k, v_i)}{\sum_{k=1}^{n} u_{ik}^m} \quad \text{or} \quad \eta_i = \frac{\sum_{u_{ik} > \alpha} d^2(x_k, v_i)}{\sum_{u_{ik} > \alpha} 1}, \quad \text{for some } 0 < \alpha \le 1 \quad (8.9)$$

after running the FCM. The first form of the constant is simply the weighted average of all distances of vectors to the ith cluster center, whereas the second choice computes the average distance of "good" ith cluster points, that is, points that belong to the α-level cut of the cluster. Of course, now you have to decide what level cut to use, but it does add flexibility. For data sets where there might be more noise, the estimates in Eq. (8.9) can be done at the end of each iteration of the PCM algorithm. In the case of very noisy data, running FCM may not (probably won't) result in good

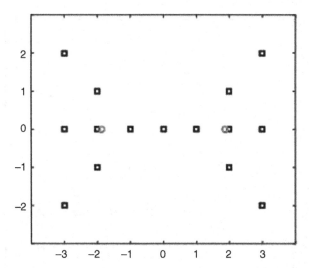

FIGURE 8.22   The butterfly data and final PCM cluster centers. Symmetric membership functions are produced by the PCM.

starting points for the PCM. In Krishnapuram and Keller [1993, 1996], there are several heuristics given for running the PCM.

You might ask what is "possibilistic" about the PCM algorithm. Actually, nothing in the true sense of possibility theory [Zadeh, 1978; Dubois and Prade, 2001]. The truth is that we were looking for a name for this approach to partitioning and "fuzzy" was already used. Hence, we picked possibilistic C-means just to distinguish it from the sum-to-one fuzzy partitions of FCM. So, please don't get confused over trying to attribute deeper significance to the name; it was just a name of convenience that, as names tend to do, stuck.

Figure 8.22 shows the final PCM cluster centers $\{(-1.9, 0.0)^t, (1.9, 0)^t\}$ for the butterfly data. These fuzzy clusters are completely symmetric, possible because PCM eliminated the constraint that the memberships for each point need to sum to 1 across the clusters. In this case, the bridge point obtained low and equal typicalities of 0.22 in both clusters and the cluster centers moved to symmetric positions. Table 8.8 gives the PCM typicalities for the butterfly data.

The PCM is very robust to outliers [Krishnapuram and Keller, 1996]. To demonstrate this capability, a 16th point was added to the butterfly data set. This point $(0, 9)^t$ can be considered an outlier since it is far from either of the two more recognizable clusters. This outlier causes serious problems for the hard 2-means, as seen in Figure 8.23.

TABLE 8.8   PCM Typicalities for Butterfly Data with C = 2 and m = 1.25

Index k	1	2	3	4	5	6	7	**8**	9	10	11	12	13	14	15
u_{k1}	0.16	0.44	0.16	0.50	0.98	0.50	0.57	**0.22**	0.11	0.06	0.06	0.06	0.03	0.04	0.03
u_{k2}	0.03	0.04	0.03	0.06	0.06	0.06	0.11	**0.22**	0.57	0.50	0.98	0.50	0.16	0.44	0.16

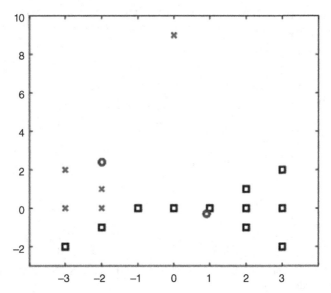

FIGURE 8.23   Final crisp partition and cluster centers of the HCM on the butterfly + outlier data.

The FCM does better, generating a 0.5 membership for both the bridge point and the outlier. The final cluster centers $\{(-2.0, 0.40)^t, (2.0, 0.40)^t\}$ are shifted vertically (refer to Figure 8.11) due to the 0.5 outlier membership in both clusters, as seen in Figure 8.24.

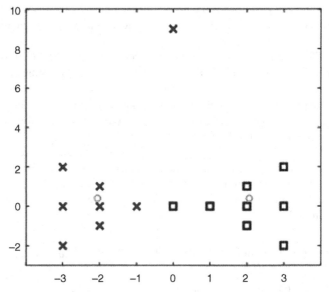

FIGURE 8.24   Final harden partition and cluster centers of the FCM on the butterfly + outlier data.

Using the final FCM cluster centers as initialization for the PCM, as suggested in Krishnapuram and Keller [1993], the possibilistic 2-means generates the same final cluster centers (up to the display precision), as shown in Figure 8.22, because it produces a typicality of 0.01 in each cluster for the outlier point. That is the whole point of how PCM can be used for outlier detection. The final typicalities are only slightly different from Table 8.8 for the other points.

The PCM tends to search for dense regions of feature space and can have the property that more than one cluster center, and hence the clusters themselves, end up identical. This has been cited as a bad property by some [Barni *et al.*, 1996], but Krishnapuram and Keller argue that it is a good trait in situations where the exact number of clusters is unknown. A larger number of clusters than "expected" can be specified as C and then identical clusters can be pruned. In fact, in Yang and Lai [2011], the authors propose to exploit this by having all points initially serve as cluster centers and then constructing an automatic merging algorithm (AM-PCM). The AM-PCM is a deterministic algorithm once the parameters are set, and so the key is to study the parameter setting. The fact that all points (or some gridding of feature space) are used as initial cluster centers gives rise to the major challenge of using this algorithm. In a report [Plodpradista, 2012], the author does an in-depth comparison between the PCM and the AM-PCM from the standpoint of studying the parameters of both algorithms, as well as through extensive controlled testing on numerous data sets. The final analysis there is summarized in Table 8.9.

In some sense, PCM searches for the correct number of clusters by overspecifying and looking at the coincident clusters after convergence, whereas AM-PCM starts with n clusters and performs merging. This situation is a classical issue in clustering since most clustering approaches require knowledge of C. This problem, known as cluster validity, is normally attacked by running the algorithm on a data set several times, varying the number of clusters. A number (called a validity measure) is calculated from the final output of each run. While there are numerous examples of

TABLE 8.9   Comparison of PCM and AM-PCM

PCM		AM-PCM
Pros	Pros	
	Great for large data sets	Deterministic (no random initialization)
	Inexpensive resource usage	Good for small data sets of high dimension
	Searches for modes in the data	Parameters relatively insensitive in presence of some noise in the data
		No need to specify the number of clusters
Cons	**Cons**	
	Random initialization can cause results to vary	Gridding is complicated on high-dimensional data
	Can produce coincident clusters (really a con? Sometimes)	Resource intensive for large data sets
	Fuzzifier has definite effect on results	Can leave cluster centers on outlier points in noisy data

validity measures, most (for object data clouds) are functions of the closeness of points assigned to, or shared by, a particular cluster and the separation of distinct clusters. The measure is usually maximized or minimized (depending on its form) when the "correct" number of compact, well-separated clusters is found. We direct the interested reader to Theodoridis and Koutroumbas [2009] for the general concepts of cluster validity as well as for several examples of such measures for both crisp and fuzzy clustering algorithms.

While the algorithms considered above require object data (vectors in $\mathfrak{R}^d$), some data sets have the property that only relational information is known. A distance, or more generally a dissimilarity, is calculated (or given somehow) between objects, but these numbers are not derived from vectors of real features. For example, Wilbik and Keller [2012] developed a metric between linguistic protoform summaries of the form "Q As are P" where the quantifier Q and summarizer P are modeled as fuzzy sets. So, the distance is between sentences; there is no vector space representation. Another common example is in the comparison of objects that are annotated by terms from an ontology, like proteins described by expressions in the Gene Ontology [Popescu *et al.*, 2006]. The representation there is a bag-of-words and the dissimilarity measures are constructed accordingly. Hence, the data, and algorithms that handle it, are called relational since all that is known is the relationship between objects. Most clustering algorithms have relational duals, although the development is beyond the scope of this text.

8.5 FUZZY CLASSIFIERS: FUZZY K-NEAREST NEIGHBORS

Clustering methods are used to look for structure in sets of unlabeled vectors. In many applications, we need to assign known labels to test data. In these cases, we assume that we have training sets of patterns that represent the various classes (labels) under consideration. The task now is to find a mapping, called a classifier, from the set of new samples into the set of class labels. As with clustering, there are crisp, probabilistic, fuzzy, and possibilistic classification models. In fact, most crisp classifiers can be extended to produce labels that contain uncertainty, be it probabilistic, fuzzy, or possibilistic. Many classifier methods can be extended to utilize fuzzy or possibilistic labels on the training data themselves. You saw a fuzzy classifier in Chapter 7; a set of fuzzy IF-THEN rules can be crafted for some pattern recognition problem where the number of features is modest and the outputs are (fuzzy) class assignments. Chapter 9 will introduce a very powerful classifier framework, fuzzy integrals, but again for situations where the input dimensionality is fairly low. This section defines a simple, yet powerful family of algorithms, referred to as k-nearest neighbor (k-NN) algorithms, to dynamically build a classification mapping that can be used for multiclass and high-dimensionality feature sets.

Let's suppose that you are a beer lover (well, if you don't like beer, switch to wine, tea, coffee, cigars, etc.). As you try new brews, you keep track of whether you like them or not, or even better, you assign each one either memberships or typicalities in the Like/Don't_Like classes. Now FlatBranch Brewery is coming out with a new edition and you wonder whether or not you will like it before buying a growler. We might tell you to find the closest beer from your list and make your decision based on its label. But how do you decide on the closest beer? Being a good pattern recognition person, when you tried each entry, you recorded important beer parameters like the International Bitterness Units (IBU), Alcohol By Volume (ABV), Maltiness Index, and so on. Good old Euclidean distance can come to the rescue, that is, we represent each beer by a vector of numeric features and then find the closest vector to the brew under consideration. If you liked the closest one (enough), you will probably like the new one; otherwise, don't spend your hard earned money. If you wanted a little more evidence, you might find the closest three and look for the majority opinion. If instead of picking three closest, you chose two or four, you might end up with a tie considering only the crisp labels. In such a case, a simple tie breaker is to then choose the class of the closest point (reverting back to the 1-NN). This simple example is the essence of the crisp and fuzzy k-nearest neighbors classifier.

To formalize this pattern recognition approach, let $X = \{x_1, x_2, \ldots, x_n\}$ where $x_j \in \Re^d$ is the set of feature vectors, except that we now assume that each x_j has class labels u_{ij}, $i = 1, \ldots, C$. The set X is called the labeled training data for the classifier. In the crisp case, $u_{ij} = 1$ for only one class i, and is zero elsewhere. This signifies that x_j represents an object sampled from class i. By relaxing the binary constraint to $\sum_{i=1}^{C} u_{ij} = 1$, for each j, or even just requiring $0 \le u_{ij} \le 1$, we produce fuzzy or possibilistic labels, respectively, on the training data. You might ask how training data could have soft labels. After all, training data are supposed to be pure representatives of their respective classes. That may be true in theory, but remember that objects are represented by vectors of features extracted from them. Sometimes these features aren't a completely faithful class mapping, and sometimes the classes have ambiguous representatives. Consider separating the handwritten digits "1" and "7". If you collected 10,000 samples of each from many people, you would see some very carefully written and some where you (the expert human) couldn't tell the difference. Even though a digit may have the label "1", features extracted from it may be in a region that has a population of "7"s and vice versa. Forcing binary memberships will only confuse matters. So, it is reasonable to postulate that training data should have soft labels.

The goal of the k-NN algorithm is to use the labeled training data to decide the class membership of a new test point x. The classical crisp k-NN goes like this. Given an input vector x to classify, find the closest k-neighbors in the training data (pick your favorite distance or dissimilarity measure). Assign x to the class with the majority label among the k-nearest neighbors. Ties are broken arbitrarily. The simplest case is when k = 1, called the nearest neighbor or the 1-NN algorithm—like our first beer predictor. Perhaps surprisingly, Cover and Hart [1967] proved that in the limit, the error rate of the 1-NN converged to a value that is no more than twice the optimal Bayes error rate. Of course, we can't realize the limit case (no matter how many pubs we visit) and so very little theory can be applied to predict the finite performance of the k-NN. However, its simplicity and reasonable (sometimes, very good) results have made it a very attractive numeric feature vector classifier.

In the "more sophisticated" beer predictor, a tie was broken by picking the label (likeableness) of the closest point. The idea of using distances is a natural idea to make the k-NN more sensitive to the actual data distribution. There have been numerous extensions to the k-NN algorithm that make use of the individual distances, $d_j(x) = d(x, x_j)$ for $j = 1, \ldots, k$ [Dasarathy, 1991]. Here, we show how both the distances and fuzzy (or possibilistic) labels are combined to create class labels for the test vector. Like the crisp counterpart, the fuzzy k-NN (FKNN) algorithm is simple in concept. Let $x_1, x_2, \ldots, x_k$ be the k-nearest neighbors of a test vector x. The goal is to compute the membership of x in each class. The formula for the membership of x in the ith class, $u_i(x)$, as given in Keller *et al.* [1985] is

$$u_i(x) = \frac{\sum_{j=1}^{k} u_{ij}\left(1/\left(d(x, x_j)^{2/(m-1)}\right)\right)}{\sum_{j=1}^{k}\left(1/\left(d(x, x_j)^{2/(m-1)}\right)\right)} \tag{8.10}$$

where $m > 1$ is a constant (like the fuzzifier in the FCM). Hence, $u_i(x)$ is proportional to a weighted average of the inverse distances of x to each of its k-nearest neighbors. The weights correspond to fuzzy ith class labels of neighbors. The denominator is used to scale the memberships for all classes so that they sum to 1. In the crisp case, this is an inverse distance weighted k-NN [Dasarathy, 1991]. Just be careful for the unlikely event that one of the distances is actually zero (handle that case separately). As with fuzzy clustering, if it is necessary to compute a crisp label, x can be assigned to the class with the largest computed membership.

From the form of Eq. 8.10, it's not hard to concoct many other variations that can be called a fuzzy k-NN classifier or even a possibilistic k-NN classifier—why force the confidences add up to 1 (see Problem 8.9).

As a simple two-class example, consider the classification of a two-dimensional feature vector with coordinates $(0.0, 0.0)^t$ shown as a star in Figure 8.24. The feature values and class labels for the six nearest neighbors in the training data are given in Table 8.10 and shown graphically in Figure 8.25.

Clearly, the crisp 6-NN would result in a tie. That's why most often in two-class problems, an odd number of neighbors is specified. We are using an even number of neighbors in this example only to make this point. How many neighbors should you use? The obvious, but unhelpful, answer is "enough, but not too many." Unfortunately, it is problem dependent and so there is no definitive answer. If you have a good training set, you could do a "leave-one-out" cross-validation experiment with a range of values for k

TABLE 8.10   Training Data for k-NN Classifier

x_1	x_2	Crisp Label	Soft Label for Class 1	Soft Label for Class 2
0.25	0.0	1	0.625	0.375
0.0	0.5	1	0.625	0.375
−0.5	0.0	1	0.625	0.375
0.0	−0.5	2	0.0	1.0
1.0	0.0	2	0.0	1.0
−1.0	0.0	2	0.0	1.0

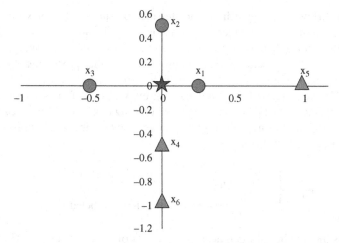

FIGURE 8.25   Graphical rendition of the k-NN example. Class 1 vectors are disks while class 2 points are triangles. The test point is the star.

and look for where the classification accuracy levels out. That knee in the performance curve should give you a good estimate for the correct number of neighbors to use.

Since the crisp 6-NN resulted in a tie, we can look at a fuzzy 6-NN result, here using Eq. 8.10. If the class memberships are crisp, that is, 1 in the designated class and 0 in the other, then for m = 2, Eq. 8.10 produces

$$u_1(x) = \frac{4 + 2 + 2}{4 + 2 + 2 + 2 + 1 + 1} = 0.67$$

and

$$u_2(x) = \frac{2 + 1 + 1}{4 + 2 + 2 + 2 + 1 + 1} = 0.33$$

Hence, class 1 receives the higher membership, and so if crisp assignments are required, the star would become a circle. However, consider the soft labels for the training data in Table 8.10. Here, for the sake of this example, the three vectors from class 1 have membership 0.625 in their class (with membership 0.375 in class 2), while those from class 2 are strong representatives with full class 2 membership (0 in class 1).

Now the calculations yield

$$u_1(x) = \frac{4(0.625) + 2(0.625) + 2(0.625) + 2(0) + 1(0) + 1(0)}{4 + 2 + 2 + 2 + 1 + 1} = 0.42$$

and

$$u_2(x) = \frac{4(0.375) + 2(0.375) + 2(0.375) + 2(1) + 1(1) + 1(1)}{4 + 2 + 2 + 2 + 1 + 1} = 0.58$$

Hence, the fuzzy 6-NN algorithm generates membership values for x that places it more in class 2 due to the weaker confidence in the typicality of the class 1 points. A crisp assignment would thus make the star a square. However, an advantage of a fuzzy classifier is that these memberships can be used in further processing if required.

How the training data receive fuzzy labels is problem dependent, but in Keller *et al.* [1985], we suggested using a version of the k-NN itself to assign those labels, that is, reflecting the fact that even for labeled training data, the feature vectors can lie in areas of more or less uncertainty. For a training point x_j, find the n closest neighbors in the training set and define

$$u_{ij} = \begin{cases} 0.51 + \left(\dfrac{n_i}{n}\right) \cdot 0.49, & \text{if } x_j \text{ is labeled class i} \\[2mm] \left(\dfrac{n_i}{n}\right) \cdot 0.49, & \text{if } x_j \text{ is not labeled class i} \end{cases} \tag{8.11}$$

where n_i is the number of the n-nearest neighbors of x_j labeled class i. This ensured that the membership of a training vector in its assumed class will always be over 0.5, even if the feature representation puts it squarely in a group of vectors from other classes. Is this reasonable? Certainly, it seemed like a good idea at the time, but you might want a different behavior in the soft label assignment. There is no reason that a class memberships of a particular training sample need to sum to 1 as above. Here too, Problem 8.9 will ask you to think about this issue. Bezdek [1981] was the first person that we know of to suggest that training data could (should) have soft labels.

This simple family of variations on the same theme has great utility in real applications. In Gader *et al.* [1995, 1997], k-NN algorithms were used to provide desired outputs for training data to be used in neural network classifiers for hand-written characters, and then for word recognition using dynamic programming or fuzzy integration (see Chapter 9). They trained multilayer perceptrons to approximate fuzzy sets instead of crisp sets. For example, using handwritten data from the U.S. Post Office, there was considerable confusion between some characters, like "U' and "J," and the extracted features showed this uncertainty. They assigned fuzzy desired outputs on both "U" and the "J" nodes for each of two training patterns that corresponded to how well the given pattern represented both a "U" and a "J." Actually, they used a variant of the k-NN that assigned possibilistic memberships by giving the ground truth class a membership of 1 regardless of the location of the feature vector. As a result, a multilayer perceptron learned and represented the training data more realistically by incorporating the uncertainty associated with the patterns. They compared networks trained with standard and supervised approaches on U.S. Postal Service data sets. For isolated character recognition, networks trained with crisp outputs achieved superior character recognition rates: 89.2% for the crisp networks and 83.8% for the fuzzy networks. However, the fuzzy networks achieved word recognition rates of about 76%, compared to 70% achieved with crisp networks. We can interpret this result as an affirmation of Marr's principle of least commitment [Marr, 1982]:

Don't do something that may later have to be undone.

In the case of handwritten word recognition, if we use crisp class memberships, then the neural network makes irreversible decisions concerning the class membership of an input segment. If the identity is not correct, the mistake will lead to an erroneous output in word recognition. Fuzzy class memberships are not definite; the network represents multiple class memberships and can more accurately make the final decision at a later stage of processing.

Frigui and Gader [2009] developed a possibilistic k-NN classifier to assign confidence to a large two-class problem, vehicle-based landmine detection from ground-penetrating radar volumes (cross-track, downtrack, depth). The feature vectors were built from edge histogram descriptors from areas in the 3D space that were identified by an inexpensive optimistic prescreening algorithm. They compared the possibilistic k-NN with several other high-performance approaches. Their algorithm outperformed all other algorithms involved in a large-scale cross-validation experiment that used a diverse data set covering over $41,807 \, m^2$ acquired from four outdoor test sites at different geographic locations over the course of several years.

The concept of nearest neighbors normally refers to distance in a metric space. That need not be the case. For example, in Bondagula and Xu [2007], comparisons of amino acid sequences don't lend themselves to standard Euclidean space feature vectors. In secondary structure prediction (how a protein folds in a three-dimensional space), sequence alignment of portions of the protein to large databases of known structures is used to match an unknown protein structure. Based on the search, numeric similarities to known states of the database structures are calculated from a statistical measurement. The resulting dissimilarities are not true distance values, but were used in the fuzzy k-NN algorithm to form one subset of features for a neural network predictor. Together with another set of structure-related features, the resulting algorithm outperformed two other state-of-the-art approaches on large data sets, the only substantial difference being the addition of the fuzzy k-NN features.

As a final note, you might wonder what if anything the fuzzy k-NN has to do with clustering. Well, at least in Gader et al. [1995, 1997], the k-NN soft labels were used to drive a self-organizing feature map (SOFM), a neural-based clustering algorithm. It can also be considered the extreme case of the multiprototype classification that is related to clustering. Pretty big stretch, but there is a connection.

8.6 CHAPTER SUMMARY

Searching for structure in data has been a problem for a long time and continues to hold an important role as data sets become larger and larger and of higher and higher dimensionality. Because it is unsupervised, in real applications where there is no actual ground truth, clustering approaches using crisp, fuzzy, probabilistic, possibilistic, neural, evolutionary, swarm, and hybrid frameworks abound. There are thousands of clustering algorithms in the literature, some very general and others specific to particular problem domains, some with strong mathematical grounding and others very heuristic. There is no way to do justice to this enormous field. We presented a few of the basic clustering approaches that have their roots in fuzzy set theory. Then, we presented a single family of techniques in the field of fuzzy classifier design, the k-NN, and showed the versatility and usefulness of these simple algorithms. As with the other chapters, our hope is that these sections whet your appetite to study further.

EXERCISES

8.1. What is the basic concept underlying all clustering procedures?

8.2. Let $X = \left\{ \begin{pmatrix} -2 \\ 1 \end{pmatrix}, \begin{pmatrix} -2 \\ -1 \end{pmatrix}, \begin{pmatrix} 0 \\ 0 \end{pmatrix}, \begin{pmatrix} 0 \\ 10 \end{pmatrix}, \begin{pmatrix} 2 \\ 1 \end{pmatrix}, \begin{pmatrix} 2 \\ -1 \end{pmatrix} \right\}$ and let $v_1 = \begin{pmatrix} -2 \\ 0 \end{pmatrix}$ and $v_2 = \begin{pmatrix} 2 \\ 0 \end{pmatrix}$.

Execute one complete iteration of the fuzzy 2-means algorithm with $m = 2$ and one complete iteration of the possibilistic 2-means with $m = 2$ and each η is equal to 1.

8.3. a. Apply the crisp 2-means algorithm using Euclidean distance to the data set:

$$X = \left\{ \begin{pmatrix} 0 \\ 0 \end{pmatrix}, \begin{pmatrix} 0 \\ 1 \end{pmatrix}, \begin{pmatrix} 5 \\ 4 \end{pmatrix}, \begin{pmatrix} 5 \\ 5 \end{pmatrix}, \begin{pmatrix} 4 \\ 5 \end{pmatrix}, \begin{pmatrix} 1 \\ 0 \end{pmatrix}, \begin{pmatrix} 2 \\ 2 \end{pmatrix}, \begin{pmatrix} 3 \\ 3 \end{pmatrix} \right\},$$

with $v_1(1) = \begin{pmatrix} 0.5 \\ 0 \end{pmatrix}$ and $v_2(1) = \begin{pmatrix} 3.5 \\ 3.75 \end{pmatrix}$

b. Now run the crisp 3-means beginning with the final two cluster centers from (a) and a third center that is the point of X furthest from its cluster center. Do you think there are two or three clusters in these data? This is usually done computationally by using a validity index.

8.4. Prove that the necessary condition for the cluster centers in an FCM algorithm with Mahalanobis distances takes the same form as in Theorem 8.1.

8.5. For the mathematically hardy, prove that Eq. 8.6 represents the necessary conditions on the covariance matrices when using cluster-specific Mahalanobis distances.

8.6. Develop the necessary conditions for a minimum of the possibilistic C-means criterion function.

8.7. What are the similarities and differences between the hard (crisp) C-means (HCM), the fuzzy C-means (FCM), and the possibilistic C-means (PCM)? Be as specific as possible, but don't just write down the update equations.

8.8. a. Use the final partition matrix from Problem 8.3 as fuzzy labels for two classes. Compute the memberships for $y_1 = \begin{pmatrix} 2.5 \\ 2 \end{pmatrix}$ and $y_2 = \begin{pmatrix} 0.5 \\ 0.5 \end{pmatrix}$ using the fuzzy 2-NN and fuzzy 3-NN algorithms.

b. Now suppose that X is a small piece of a larger training set, and the fuzzy labels for X are given as follows:

	x_1	x_2	x_3	x_4	x_5	x_6	x_7	x_8
$U = u_1$	0.8	0.6	0.0	0.0	0.0	0.6	0.0	0.0.
u_2	0.2	0.4	1.0	1.0	1.0	0.4	1.0	1.0

Compute the crisp and fuzzy class labels for $y = \begin{pmatrix} 2.5 \\ 2.5 \end{pmatrix}$ using the crisp 5-NN and fuzzy 5-NN.

8.9. a. Give a variation to Eq. 8.9 for a fuzzy k-NN and for a possibilistic k-NN.

b. Can you think of other ways to assign soft labels to training points?

CHAPTER 9

Fuzzy Measures and Fuzzy Integrals

Many problems in decision making can be cast as in the framework of fusion of multiple sources of information. For example, doctors may request several tests to arrive at a diagnosis. These tests, including patient history, all supply partial evidence for possibly more than one conclusion. An expert diagnostician combines the results of the tests with "worth" of the individual assays, as well as groups of them, to support or refute particular conclusions. Some tests taken individually may only provide limited confidence in a decision, but taken as a group greatly increase that support. One of the advantages of fuzzy set theory is the wide range of computational mechanisms to implement such fusion of information. In this chapter, we develop one of these powerful frameworks, the fuzzy integral. It is used to combine partial (objective) support for a hypothesis from the standpoint of individual sources of information together with (possibly subjective) weights of various subsets of these sources of information.

All classifiers are either implicit or explicit fusion engines. A simple Bayes classifier with Gaussian probability density functions fuses the values from individual features into a class decision. The covariance matrices carry the information about the connection between features for each class. The same is true for the rule-based systems in Chapter 7 and the crisp/fuzzy/possibilistic k-NNs of Chapter 8. Here, we consider the situation of explicit information fusion with one of our favorite frameworks, fuzzy integrals.

9.1 FUZZY MEASURES

The fuzzy integral is based on the concept of fuzzy measures,[1] generalizations of probability measures, which in themselves will be shown to be effective to combine information in certain applications. Consider a finite set $X = \{x_1, x_2, \ldots, x_n\}$ of

[1] Fuzzy measures, and the more specific belief and probability measures, address the ambiguity axis of uncertainty: how likely can the answer to a question be found in various subsets of the sources of information.

Fundamentals of Computational Intelligence: Neural Networks, Fuzzy Systems, and Evolutionary Computation, First Edition. James M. Keller, Derong Liu, and David B. Fogel.
© 2016 by The Institute of Electrical and Electronics Engineers, Inc. Published 2016 by John Wiley & Sons, Inc.

sources of information. Each x_i can be a diagnostic test, a feature (e.g., color, texture, or shape) in a segmentation problem, a particular pattern recognition algorithm, and so on. While only finite sets are considered here, the theory of fuzzy measures and fuzzy integrals can be extended to infinite sets (see Grabisch *et al.* [2000] and Wang and Klir [1993]).

Let 2^X denote the power set of X, that is, the set of all (crisp) subsets of X. A fuzzy measure, g, is a real-valued function $g : 2^X \rightarrow [0, 1]$, satisfying the following properties:

1. $g(\emptyset) = 0$ and $g(X) = 1$
2. $g(A) \leq g(B)$,   if $A \subseteq B$ (9.1)

Note that the normal additivity condition of probability theory, that is, $g(A \cup B) = g(A) + G(B)$ if $A \cap B = \phi$, is replaced by the weaker condition of monotonicity (property 2). All probability measures are fuzzy measures (prove this—Exercise 9.1). So you already know many fuzzy measures. But there are many more fuzzy measures over a given set than there are probability measures. Consider the set $X = \{x_1, x_2, x_3\}$. Table 9.1 lists the subsets of X and a few fuzzy measures over X. Measure g_1 represents total ignorance—there is no proper subset of X with nonzero measure, g_2 signifies total confusion—all nonempty subsets of X have full measure, and g_3 is a measure that gives complete certainty to $\{x_1\}$. We all recognize g_4, a probability measure. Measure g_5 might correspond to the statement that "the whole is greater than the sum of its parts", whereas g_6 is the opposite, that is, "the whole is less than the sum of its parts." In other words, the sources of information in g_5 support each other; in combination they are much stronger than probability would suggest. Conversely, g_6 represents a situation where there is some conflict in the sources, that is, combinations don't give as strong a support as the corresponding probability measure would supply. These examples show precisely what a fuzzy measure actually "measures," at least in this context, that is, our confidence or belief that each given subset of sources of information will provide the answer to the question at hand.

With the infinite possibilities for fuzzy measures even in the low cardinality cases, how do we find useful fuzzy measures (other than probability functions)? You sure

TABLE 9.1   Subsets of Set with Three Elements and a Few Fuzzy Measures

Subset	g_1	g_2	g_3	g_4	g_5	g_6
ϕ	0.0	0.0	0.0	0.0	0.0	0.0
$\{x_1\}$	0.0	1.0	1.0	0.2	0.2	0.2
$\{x_2\}$	0.0	1.0	0.0	0.3	0.2	0.2
$\{x_3\}$	0.0	1.0	0.0	0.5	0.2	0.2
$\{x_1, x_2\}$	0.0	1.0	1.0	0.5	0.6	0.3
$\{x_1, x_3\}$	0.0	1.0	1.0	0.7	0.7	0.4
$\{x_2, x_3\}$	0.0	1.0	0.0	0.8	0.8	0.5
$X = \{x_1, x_2, x_3\}$	1.0	1.0	1.0	1.0	1.0	1.0

can't try them all. (How many numbers need to be specified for a fuzzy measure over a finite set $X = \{x_1, x_2, \ldots, x_n\}$?) There are many approaches to construct interesting measures. One popular tactic is to restrict the freedom a little by adding additional constraints to the class of fuzzy measures. Exercise 9.3 gives one interesting and useful such family of fuzzy measures, possibility and necessity measures. We now develop a very general form of fuzzy measure that has found great utility in theory and application.

For a fuzzy measure g, let $g^i = g(\{x_i\})$. The mapping $x_i \rightarrow g^i$ is called a fuzzy density function. The fuzzy density value g^i is interpreted as the (possibly subjective) importance of the single information source x_i in determining the answer to a particular question, perhaps the similarity of two genes or the label of a particular object. Fuzzy measures are quite general since they require only two simple properties to be satisfied. However, it is often the case that the densities can be extracted from the problem domain or supplied by experts. We do know that for $X = \{x_1, x_2, \ldots, x_n\}$, there are only n numbers needed to represent the densities. One key to using fuzzy measures involves finding ones that can be built out of the densities. One of the most useful classes of fuzzy measures is due to Sugeno [1977]. A fuzzy measure g is called a Sugeno measure (g_λ fuzzy measure) if it additionally satisfies the following property [Sugeno, 1977]:

3. For all $A, B \subseteq X$ with $A \cap B = \emptyset$,

$$g_\lambda(A \cup B) = g_\lambda(A) + g_\lambda(B) + \lambda g_\lambda(A)g_\lambda(B), \quad \text{for some } \lambda > -1 \qquad (9.2)$$

Unless needed, the subscript λ will be omitted for simplicity. If the densities are known, the value of λ for any Sugeno fuzzy measure can be uniquely determined for a finite set X using Eq. 9.2 and the facts $X = \bigcup_{i=1}^{n} \{x_i\}$ and $g_\lambda(X) = 1$, which leads to solving the following equation for λ:

$$(1 + \lambda) = \prod_{i=1}^{n}(1 + \lambda g^i) \qquad (9.3)$$

Exercise 9.3 asks you to verify this for a specific case.

This equation has a unique solution for $\lambda > -1$ [Sugeno, 1977]. The right-hand side of Eq. 9.3 is a polynomial of degree n with respect to λ. It could have as many as n real roots, so Eq. 9.3 could have as many as $n - 1$ real solutions. How can it have only 1? It is perhaps the coolest polynomial on the planet. Its coefficients are the symmetric functions of the densities

$$G(\lambda) = \prod_{i=1}^{n}(1 + \lambda g^i) = 1 + \lambda \sum_{j=1}^{n} g^j + \lambda^2 \sum_{j=1}^{n-1} \left(\sum_{k=j+1}^{n} g^j g^k \right)$$

$$+ \lambda^3 \sum_{j=1}^{n-2} \left(\sum_{k=1}^{n-1} \left(\sum_{l=1}^{n} g^j g^k g^l \right) \right) + \cdots + \lambda^n g^1 g^2 \cdots g^n$$

First note that $\lambda = 0$ is a solution to Eq. 9.3. If $\lambda = 0$, what does that say about the fuzzy measure g and about the sum of the densities (Exercise 9.4)? For values other than 0, Sugeno showed that Eq. 9.3 has a unique solution for $\lambda > -1$. Here we give a simple calculus-based proof adapted from Tahani and Keller [1990].

Proposition 9.1   For information sources $X = \{x_1, x_2, \ldots, x_n\}$ and densities $\{g^i\}$, $0 < g^i < 1$, there is a unique $\lambda > -1$ and $\lambda \neq 0$ satisfying Eq. 9.3.

Proof: It is sufficient to show that the polynomial $G(\lambda) = \prod_{i=1}^{n}(1 + \lambda g^i)$ and the line $\lambda + 1$ intersect only at one point $\lambda > -1$ and $\lambda \neq 0$. Now,

$$G'(\lambda) = \sum_{i=1}^{n} g^i \left(\prod_{\substack{j=1 \\ j \neq 1}}^{n} (1 + \lambda g^j) \right) \text{ and } G''(\lambda) = \sum_{k=1}^{n} \left(\sum_{\substack{i=1 \\ i \neq k}}^{n} \left(g^k g^i \prod_{\substack{j=1 \\ j \neq i \\ j \neq k}}^{n} (1 + \lambda g^j) \right) \right)$$

For $\lambda > -1$, $G(\lambda)$, $G'(\lambda)$, and $G''(\lambda)$ are all greater than zero, that is, $G(\lambda)$ is a positive strictly increasing real-valued function that is concave upon the open interval $(-1, \infty)$. Furthermore, $G(0) = 1$ and $G'(0) = \sum_{i=1}^{n} g^i$. The slope of the line $L(\lambda) = \lambda + 1$ is 1 and $L(0) = 1$. If $\sum_{i=1}^{n} g^i < 1$, then $G(\lambda)$ crosses $L(\lambda)$ at $\lambda = 0$ with a slope $G'(0) < 1$ (obvious to you?). Since $G(\lambda)$ is a polynomial of at least degree 2 strictly increasing on $(-1, \infty)$, it follows that it must cross $L(\lambda)$ exactly one more time for some large enough value of λ. (Draw a picture to give yourself the intuition behind the calculus. G must be above L on $(-1, 0)$ and below L right after 0, but strictly increasing and concave up.)

A similar argument holds for $\sum_{i=1}^{n} g^i < 1$ QED.

So, Sugeno λ-measures form a subclass (although still quite large) of general fuzzy measures that have the property that they can be completely determined by an assignment of worth (density) to each singleton source of information. For our pattern recognition types, you can imagine running some basic classifier on each source separately on training data and using correct classification rates to help set the densities. Or, in the algorithm/sensor fusion domain, each source has been trained and so, some value can be assigned to measure of the singleton source. Note that once the densities are fixed, the measures of all subsets of sources are computable from a recursive application of Eq. 9.2. Even though Sugeno measures form a large subset, they don't allow a person to say something like "neither of these two sources is particularly strong, but together they answer the question." More on this later.

While we don't present Dempster–Shafer (D–S) belief theory [Shafer, 1976], there is a strong connection between Sugeno measures and belief/plausibility measures of D–S. Belief and plausibility measures are particular subclasses of fuzzy measures with additional constraints. In the statistical literature, they are referred to as lower and upper probabilities—belief is a lower bound for the assumed underlying probability, while plausibility generates an upper bound. It turns out that a Sugeno λ-measure is a belief measure if $\lambda > 0$ ($\sum_{i=1}^{n} g^i < 1$) and a plausibility measure if $\lambda < 0$ ($\sum_{i=1}^{n} g^i > 1$) [Banon, 1981].

EXAMPLE 9.1 SUGENO FUZZY MEASURE

To illustrate the calculation of a Sugeno fuzzy measure, suppose $X = \{x_1, x_2, x_3\}$ and $g^1 = 0.2$, $g^2 = 0.3$, and $g^3 = 0.1$. Note that the resulting measure in this case cannot be a probability measure because the densities, that is, the measures of the singleton subsets, do not add up to 1. Then, Eq. 9.3 becomes

$$1 + \lambda = (1 + 0.2\lambda)(1 + 0.3\lambda)(1 + 0.1\lambda)$$

Expanding and collecting terms, λ must be the solution of the quadratic equation $0.006\lambda^2 + 0.11\lambda - 0.4 = 0$. While there are two solutions, only one of them, $\lambda = 3.2$, is greater than -1, as guaranteed by the theory. Hence, the complete fuzzy measure is shown in Table 9.2.

TABLE 9.2   Sugeno Fuzzy Measure for $X = \{x_1, x_2, x_3\}$ with Densities 0.2, 0.3, and 0.1

Subset	Measure Value
ϕ	0.0
$\{x_1\}$	0.2
$\{x_2\}$	0.3
$\{x_3\}$	0.1
$\{x_1, x_2\}$	$0.2 + 0.3 + 3.2(0.2)(0.3) = 0.69$
$\{x_1, x_3\}$	$0.2 + 0.1 + 3.2(0.2)(0.1) = 0.36$
$\{x_2, x_3\}$	$0.3 + 0.1 + 3.2(0.3)(0.1) = 0.5$
$X = \{x_1, x_2, x_3\}$	$0.69 + 0.1 + 3.2(0.69)(0.1) = 1.01 \approx 1.0$
	$0.36 + 0.3 + 3.2(0.36)(0.3) = 1.006 \approx 1.0$
	$0.5 + 0.2 + 3.2(0.5)(0.2) = 1.02 \approx 1.0$
	(we rounded-off the intermediate results)

Many applications exist where fuzzy measures are used to make decisions or to build similarity values. Proponents of D–S theory subscribe to this approach. Instead of a set of sources of information, X is viewed as the set of all possible answers to a given question, like the suspects in a crime. Each piece of evidence serves to create a measure over the subsets of X (the perp was left-handed, or had blond hair, or was a particular person seen leaving the crime area, etc.). A rule of aggregation (Dempster's rule) allows these individual belief measures to be combined [Shafer, 1976]. The end result is a measure of confidence that the perpetrator lies in each of the many subsets of X. Of course, the police and DA would like all of the belief focused on a singleton subset of X, while the defense wants the final belief distribution to be much less specific, that is, to create a situation that is not "beyond a reasonable doubt."

Closer to home for this book, fuzzy measures can be used to create a fuzzy similarity value between objects described by phrases or sentences from structured vocabularies. The following example shows an application of fuzzy measures to build a similarity between sets of annotations coming from an ontology, specifically, the Gene Ontology [Popescu *et al.*, 2006]. See Xu *et al.* [2008] for a detailed exposition of the creation and use of fuzzy measures in biological ontologies.

9.2 FUZZY INTEGRALS

Let X be a set and let h : X → [0, 1] be a function that provides the support of a given hypothesis from the standpoint of each source of information, called a partial support function. Suppose g : 2^X → [0, 1] is a fuzzy measure. Then, the Sugeno fuzzy integral is defined by

$$\int h(x) \circ g = \sup_{E \subseteq X} \left[\min(\min_{x \in E} h(x), g(E)) \right] = \sup_{\alpha \in [0,1]} [\min(\alpha, g(A_\alpha)] \qquad (9.4)$$

where $A_\alpha = \{x | h(x) \geq \alpha\}$ [Sugeno, 1977].

For the finite case (the case that this chapter really addresses), reorder X = $\{x_1, x_2, ..., x_n\}$ to get X = $\{x_{(1)}, x_{(2)}, ..., x_{(n)}\}$ so that $h(x_{(1)}) \geq \cdots \geq h(x_{(n)})$. Then, the Sugeno fuzzy integral reduces to

$$S_g(h) = \max_{i=1}^{n} \left[\min(h(x_{(i)}), g(A_i)) \right] \qquad (9.5)$$

where $A_i = \{x_{(1)}, ..., x_{(i)}\}$ (see Exercise 9.6).

The reader is referred to Grabisch *et al.* [2000] and Wang and Klir [1993] for an extensive theoretical background on fuzzy measures and fuzzy integrals.

The Sugeno integral of a partial support function is the best pessimistic agreement between the objective evidence (h) and the (possibly subjective) worth of that evidence (g). This is shown graphically for a set of 10 sources of information in Figure 9.1.

In the case of 10 information sources, the measure g would need to be calculated on 1022 subsets. However, if g is a Sugeno λ-measure, these computations can be streamlined using a recursive application of Eq. 9.2. Specifically,

$$g(A_1) = g(\{x_{(1)}\}) = g^{(1)}$$
$$g(A_i) = g^{(i)} + g(A_{i-1}) + \lambda g^{(i)} g(A_{i-1}), \quad \text{for} \quad i > 1 \qquad (9.6)$$

If the number of information sources is not too large, then any measure can be precomputed and indexed so that as the function values are sorted, corresponding measure values are extracted from the list.

The original definition given by Sugeno for the fuzzy integral is not a proper extension of the Lebesgue integral, that is, the integral from Calculus, in the sense that the Lebesgue integral is not recovered when the measure is additive. To avoid this

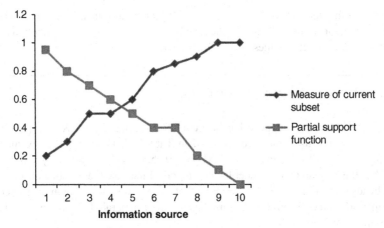

FIGURE 9.1   Graphical interpretation of Sugeno integration of a partial support function with respect to a measure. The set X is sorted so that the function is nondecreasing (curve marked by squares) and the measure of the sets A_i increases. In the continuous case, the value of the integral would be the intersection of these two curves. For finite sets, the value is the square or diamond below this intersection, and closest to it.

drawback, Murofushi and Sugeno [1991] proposed the Choquet fuzzy integral, referring to a functional defined by Choquet in a different context. Let h be the partial support function on X with values in [0,1] and g be a fuzzy measure. The Choquet integral is

$$\int_X h(x) \circ g = \int_0^1 g(A_\alpha)d\alpha, \quad \text{where} \quad A_\alpha = \{x \mid h(x) \geq \alpha\} \tag{9.7}$$

If X is a discrete set, the Choquet integral can be computed as follows:

$$C_g(h) = \sum_{i=1}^{n} \left[h(x_{(i)}) - h(x_{(i+1)}) \right] \cdot g(A_i) \tag{9.8}$$

where X is sorted so that $h(x_{(1)}) \geq \cdots \geq h(x_{(n)})$, $h(x_{(n+1)}) = 0$, and $A_i = \{x_{(1)}, \ldots, x_{(i)}\}$.

It is also informative to write the discrete Choquet integral as a (nonlinear) weighted sum of these values in which the weights depend on their order. Define

$$\delta_i(g) = g(A_i) - g(A_{i-1}), \quad \text{for} \quad i = 1, 2, \ldots, n \tag{9.9}$$

where we take $g(A_0)$ to be 0. Then,

$$C_g(h) = \sum_{i=1}^{n} \delta_i(g) \cdot h(x_{(i)}) \tag{9.10}$$

Note that, in the general case, the sum in Eq. 9.10 is a nonlinear combination of the values of h because the ordering of the original arguments $x_1, \ldots, x_n$ depends upon the relative sizes of the values of the function h.

EXAMPLE 9.2   FUZZY INTEGRALS

As a simple illustration, suppose that the set of information sources is $X = \{x_1, x_2, x_3\}$ with the Sugeno fuzzy measure specified in Table 9.2. For the partial evaluation function $h : X \to [0, 1]$, given by $h(x_1) = 0.9$, $h(x_2) = 0.7$, and $h(x_3) = 0.2$, we calculate the Sugeno and Choquet fuzzy integrals. First note that the function values are already sorted in descending order so that there is no need to reorder the set of information sources. From Eq. 9.5, the Sugeno fuzzy integral of h with respect to the fuzzy measure is given by

$$S_g(h) = (0.9 \wedge 0.2) \vee (0.7 \wedge 0.69) \vee (0.1 \wedge 1.0) = 0.69$$

Similarly, Eq. 9.8 produces a Choquet fuzzy integral of

$$C_g(h) = (0.9 - 0.7)(0.2) + (0.7 - 0.2)(0.69) + (0.2 - 0.0)(1.0) = 0.59$$

Both Sugeno and Choquet fuzzy integrals act as generalized expectation operators, that is, their values lie between the minimum and maximum values of $h(x_1), \ldots,$ $h(x_n)$. This property and several other elementary properties of fuzzy integrals are studied in Exercises 9.9–9.15.

Following the idea in Popescu *et al.* [2006], for each $\alpha \in [0, 1]$, Wilbik and Keller [2013] defined a similarity index between two sets of linguistic protoform summaries (Q As are P) from λ-fuzzy measures defined over the two sets of summaries. The average of the two measure is a measure and the "obvious" aggregation across α is the Sugeno integral of the function $h(x) = 1 - x$ with respect to that measure.

EXAMPLE 9.3   FUZZY INTEGRALS AS NONLINEAR FILTERS

Let X be a finite set, for example, X may represent the pixels in a neighborhood of a given point, called W_x here. The idea here is to replace the center pixel in a window by the integral of the values in the window. Selection of different fuzzy measures yields different types of filters. In Keller *et al.* [2000], several examples of Choquet fuzzy integral filters are given, with references to filters represented as Sugeno integrals. Assume that all neighborhoods are of size n. If the measure g is additive with all densities equal to 1/n, then the filter is the simple local average. Suppose $n = 2k + 1$. If the measure, g_x is defined to be

$$g_x(A) = \begin{cases} 1, & \text{if} \quad |A| \geq k \\ 0, & \text{else} \end{cases} \tag{9.11}$$

then the filter is the median filter. This is easy to see using Eq. 9.10 because δ from Eq. 9.9 will be nonzero for only one value of the index, which is the index required to "pick off" the median of the input values. In fact, replacing k with any i between 1 and $2k + 1$ in the above definition yields the ith order statistic (including the maximum for $i = 1$ and the minimum for $i = 2k + 1$).

In this image processing example, let W_x be a neighborhood of a point x. Choquet integral filters can also represent linear combination of order statistic (LOS) filters defined by

$$LOS_{W_x}(h)(\mathbf{x}) = \sum_{\mathbf{x}_k \in W_x} w_k h(\mathbf{x}_{(k)}) \tag{9.12}$$

where the weights, $\{w_1, \ldots, w_n\}$ satisfy the requirements that $w_i \in [0,1]$ and $\sum_{k=1}^{n} w_k = 1$, and the function values are sorted in descending order. This operator can be seen as a fuzzy integral operator by defining the measure g according to

$$g(A) = \begin{cases} 0, & \text{if } A = \phi \\ \sum_{j=1}^{i} w_j, & \text{if } |A| = i \end{cases} \quad \text{(see Exercise 9.16)} \tag{9.13}$$

They have also been referred to as generalized order filters by Grabisch and are useful for implementing robust estimators, such as the alpha-trimmed mean. These filters can also be referred to as ordered weight average (OWA) filters if the weights have a linguistic interpretation, such as "at least two neighbors are bright" or "many neighbors are highly textured" [Yager, 1988, 1993, 1996, 2004]. The point is that a wide variety of linear and nonlinear filters can be modeled by the same operator with just a change in the measure that defines the integral. Hence, if you had training data, that is, if you could specify input windows and desired responses, just using these simple measures, you could hunt over a large subclass of Choquet integrals for optimal or near-optimal filters that match the desired outputs. How can you do this search? By now, you probably have some ideas, but stay tuned.

9.3 TRAINING THE FUZZY INTEGRALS

The key to using fuzzy integrals to fuse multiple sources of information is to construct fuzzy measures that specify the worth of all subsets of sources of information. For many applications, Sugeno measures are employed. As noted above for these fuzzy measures, only densities (the worth of each singleton information source) need to be specified. There are numerous methods to automatically learn either densities or entire measures if training data are available [Tahani and Keller, 1990; Keller *et al.*, 1994; Grabisch *et al.*, 2000; Keller *et al.*, 2000; Mendez-Vazquez *et al.*, 2008]. These techniques dramatically increase the applicability of fuzzy integrals for general information fusion.

Consider a two-class pattern recognition problem ω_0 and ω_1. In the framework of fuzzy integration, the set $X = \{x_1, \ldots, x_n\}$ represents the names of features, algorithms, sensors, or other sources of information (including human intelligence) that are brought to bear on the question of deciding the class label of an object O. Note that in this case x_i is not the vector of values as is the common notation for other classifiers (like the k-NN). Instead, the partial support function plays that role. To be precise, we should write

$$h(O) = \begin{pmatrix} h(O; x_1) \\ h(O; x_2) \\ \vdots \\ h(O; x_n) \end{pmatrix}$$

to denote that h maps the object into a set of numbers in $[0,1]$ from the standpoint of each source of information. That's pretty tedious, so as long as there is no confusion, we assume that we know we evaluate each object, drop the designation "O," and just write $h(x_1), \ldots, h(x_n)$. Under the most straightforward interpretation, we want the fuzzy integral of objects in ω_0 to have a value 0 and those in ω_1 to map to 1. Since it is harder to extend this framework to multiple classes, alternately, we could search for two measures g_0 and g_1 such that

$$\int h \circ g_0 > \int h \circ g_1, \quad \text{if } O \in \omega_0 \quad \text{and} \quad \text{conversely if } O \in \omega_1 \tag{9.14}$$

This clearly extends to multiple classes, although it complicates training (learning the measures). You can even consider the situation where different sources of information are used to assess individual class confidence. This generalization requires separate h functions and measures for each class. The nice thing about this general formulation is that the number of sources of information need not be the same for each class; sources of information that provide confidence for class ω_i can be different from those supplying evidence for ω_j. The class evaluation is effectively decoupled, that is, the resulting integrals behave like possibilities as they will not necessarily sum to 1. While there is some contention to this statement, we believe that Keller et al., [1986, 1987] and Tahani and Keller [1990] represent the first papers that frame the pattern recognition problem as one of fuzzy integration.

However you formulate the problem, the key is to learn the appropriate measure(s). If you choose Sugeno λ-measures, possibility measures, measures like those in Example 9.3, or other so-called "decomposable" measures, then the problem reduces to finding n density values instead of learning $2^n - 2$ numbers (why is it $2^n - 2$ instead of 2^n?). For these measures, the learning task is linear instead of being exponential, which makes a big difference as n grows. Additionally, it is usually easier to determine a worth for each singleton source of information than to specify worth for all possible combinations of sources. The common approach to learn densities from a set of training data where class labels are known is to train a one source (usually a feature) classification algorithm and to calculate densities based on how well the training data (or a validation set of data) is separated by the one source algorithm. For example, Keller *et al.*, [1986, 1988] simply used the relative amount of overlap between the histograms of a feature for the various classes (on training data) to generate densities, mapping the overlap inversely into the interval [0,1]. They built λ-measures and performed segmentation with the Sugeno integral. Another early implementation of this approach for segmentation is given in Example 9.4. It uses possibility measures instead of λ-measures.

EXAMPLE 9.4   FUZZY INTEGRALS FOR SEGMENTATION

Since fuzzy integrals can fuse local information at any resolution, it is natural to use it for image segmentation [Keller *et al.*, 1986, 1988], which is just a pattern recognition problem "in the small." Here again, the trick to employing fuzzy integrals successfully is to determine the measures for the various regions. Since the integration process produces soft class decisions, this approach to segmentation not only groups pixels together but also assigns region labels to them.

As an example, Yan [Yan, 1993, Yan and Keller, 1991] utilized possibility integrals, that is, Sugeno fuzzy integrals with respect to possibility measures, to segment color images of natural scenes. Possibility measures were used because of their simplicity in assigning the measure to arbitrary sets. In particular, if $g^i = g(\{x_i\})$ represents the densities for a possibility measure for one of the classes, then

$$g(A) = \max_{x_i \in A} \{g^i\} \tag{9.15}$$

Yan used a recursive variation of random search to hunt for a set of densities to separate three classes from two features (intensity and position) extracted at each pixel. Using one image for training (hand segmented), he generated a huge number of training data patterns. Densities for each class were chosen randomly (and exhaustively given a quantization of the interval [0,1]). The h functions for each pixel for each class were generated from the normalized histograms of the training data for the intensity feature, and from heuristic membership functions for position. The best set of densities, at the coarsest quantization level, was found from the classification results on the training data. Small intervals (at the next lower order of magnitude) were chosen around each of the winning densities, and the process was repeated. This was continued until no better segmentation results on the training image were obtained with increased quantization. Figure 9.2a shows an intensity training image, while Figure 9.2b depicts the output of the segmentation of the training image at the completion of the training phase. This set of densities was then used to segment several additional test images of the same basic type, that is, outdoor scenes with roads, trees, and sky. Figure 9.3 displays one such test.

FIGURE 9.2 Training image. (a) Intensity training image. (b) Fuzzy integral segmentation.

FIGURE 9.3 Test image. (a) Intensity testing image. (b) Fuzzy integral segmentation.

Since this is a book on computational intelligence, it should be jumping out at you that learning densities from training data for either a Sugeno or Choquet integral-based pattern recognition problem is exactly suited to evolutionary computation techniques. The obvious structure has chromosomes having n genes or particles having n slots, one for each density required. The fitness function is class assignment of the evaluation of the fuzzy integral on the training data. The generalized problem can be framed as one of coevolution. Example 9.4 can be thought of as an overly simplistic attempt at guided random search. Evolutionary computation algorithms are efficient ways to hunt for good sets of densities for fuzzy integral pattern recognition since the chromosomes have a nice and compact gene structure.

In an effort to link the problem of learning densities to concepts from neural networks, Keller and Osborn [1996] created a training algorithm that used a "reward–punishment" approach similar to that used in neural networks to train the fuzzy densities for each class in a pattern recognition problem. Initially, the densities for each class started out at the same value, for example, $1/n$. For a given labeled object instance, the integrals were calculated for each classification hypothesis. If the largest integral value did not correspond to the correct classification, training must be done. The offending fuzzy integrals were punished by decreasing the densities that supported their integral values, while the correct class had its supporting densities increased. This tended to raise the integral value of the correct class integral and lower the value of those that were misclassifying the input. This process was continued until all objects were correctly classified. They proved a theorem that characterized the calculation of the new λ after adjustment of the densities, showing that the problem and it's solution were well defined.

Why not just use gradient descent on the density values themselves? The answer is that you can, but you have to be careful. You need a criterion function that is differentiable and then you have to be able to take the partial derivatives to formulate the algorithm. The obvious criterion function is one of least squares involving the Choquet integral. For the two-class, one-measure problem, let $O_0 = \{O_{01}, \ldots, O_{0n_0}\}$ be the training data from class ω_0 and $O_1 = \{O_{11}, \ldots, O_{1n_1}\}$ be the corresponding training set for ω_1. Let α_0 and α_1 be the target values for the integral for the respective classes (could be $\alpha_0 = 0$ and $\alpha_1 = 1$). Define

$$E^2 = \sum_{i=1}^{n_0} \left(C_g(h(O_{0i})) - \alpha_0\right)^2 + \sum_{i=1}^{n_1} \left(C_g(h(O_{1i})) - \alpha_1\right)^2 \qquad (9.16)$$

Rewriting Eq. 9.8 to highlight the dependence on the object under consideration, we have

$$C_g(h(O)) = \sum_{i=1}^{n} [h(O; x_{(i)}) - h(O; x_{(i+1)})] \cdot g(A_i) \qquad (9.17)$$

If the measure is a λ-measure, we can take the partial derivatives $\partial E^2 / \partial g^k$. Evaluating these partial derivatives requires the ability to compute $\partial g(A_i) / \partial g^k$. Note that from Eq. 9.6, $g(A_i)$ is defined recursively by $g(A_1) = g(\{x_{(1)}\}) = g^{(1)}$ and $g(A_i) = g^{(i)} + g(A_{i-1}) + \lambda g^{(i)} g(A_{i-1})$ for $i > 1$.

If λ was a constant for all time, this would be a straightforward calculation. However, λ is a function of the densities (Eq. 9.3). Hence, to obtain $\partial g(A_i) / \partial g^k$, it is also necessary to compute $\partial \lambda / \partial g^k$. This has to be done with implicit differentiation. In Mendez-Vazquez *et al.* [2008], it is shown that for $\lambda \neq 0$,

$$\frac{\partial \lambda}{\partial g^k} = \frac{\lambda^2 + \lambda}{(1 + \lambda g^k)\left[1 - (1 + \lambda) \cdot \sum_{i=1}^{n} \left(g^i / (1 + \lambda g^i)\right)\right]} \qquad (9.18)$$

Can you derive this? This looks messy, but once formulated, gradient descent is easily applied to fuzzy density estimation for Sugeno measures. Mendez-Vazquez *et al.* reformulated the criterion to be one of discriminative training using a measure for each class and requiring only that, given an training object, the Choquet integral for the correct class be greater than that for the other class (actually, this was even generalized to the multi class, multimeasure situation). This led to better overall results in general in a real landmine detection scenario. One of the main advantages is that you don't have to guess about what values to pick for α_0 and α_1. It turned out for the landmine detection data set, $\alpha_0 = 0$ and $\alpha_1 = 1$ were not that good as choices.

There have been attempts to incorporate neural network models to learn measures for fuzzy integration, but it is hard to build in the monotonicity constraints into that framework. The following example shows one successful approach that uses self-organizing feature maps (SOFM) to help generate densities in a word recognition application.

EXAMPLE 9.5   HANDWRITTEN CHARACTER AND WORD RECOGNITION

This example of learning measures for pattern recognition comes from the work of Gader and Chiang and has to do with estimating confidence in word recognition hypotheses based on several partial estimates of character confidence [Gader *et al.*, 1996, 1997]. Handwriting recognition systems are useful for automated document processing systems, such as mail sorting, and also in handheld computing systems. Here we discuss a novel use of fuzzy integrals on Kohonen self-organizing feature maps [Kohonen, 1982; Kohonen *et al.*, 1992] as an aid in handwritten character classification [Chiang and Gader, 1997].

Consider the handwritten word shown in Figure 9.4.

Suppose that a lexicon for matching contains the strings "Richmond" and "Edmund." We can try to match the word in Figure 9.4 to these strings as shown in Figure 9.5.

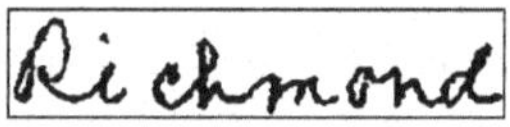

FIGURE 9.4   An example of a handwritten word.

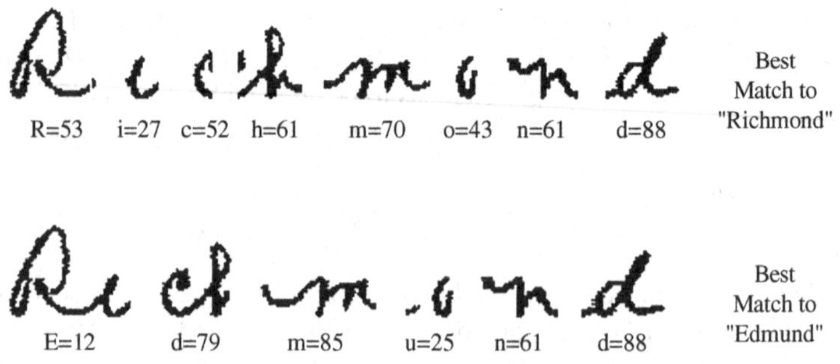

R=53   i=27   c=52   h=61   m=70   o=43   n=61   d=88   Best Match to "Richmond"

E=12   d=79   m=85   u=25   n=61   d=88   Best Match to "Edmund"

FIGURE 9.5   Ways of matching the word image in Figure 9.4 to the strings "Richmond" and "Edmund."

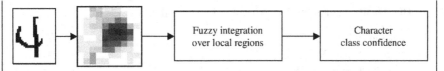

FIGURE 9.6   Example to show how fuzzy integration can be used on the SOFM.

In the standard approach to processing each input string, the word image is split into pieces, called segments. Each segment is matched to a character in the corresponding lexicon string. Each match is assigned a score. The question is how to assign an overall score to the match between a word image and a string given the matches between the segments of the image and the characters in the string. A particular problem that can arise is that a piece of one or more characters can look something like the "c" attached to a piece of the "h" in the figure above. These pieces are outliers and their effect can be diminished using the robustness properties of the Choquet integral because it inherently satisfies the principle of least commitment.

When properly trained, a SOFM provides a spatially organized two-dimensional array representing an input character as shown in Figure 9.6. Each element of the array is called a node. A region of the SOFM is a connected collection of nodes. When a digital image of a handwritten character is presented as input to the SOFM, each node takes on a value between 0 and 1. These values are referred to as activation values. Ideally, characters with similar shapes produce high activation values in similar regions and low activation values in similar regions as indicated in the figure. Higher activation is indicated by darker colors.

The fuzzy integral filter methodology worked as follows: For each character class "a," "b," "c," and so on, regions were identified for which that character class produced high outputs in the SOFM. Each region was considered as a set of information sources. A measure was defined on each region by defining densities according to how often a node in the SOFM was high for a given class. When an unknown character was presented to the SOFM, a fuzzy integral was calculated on each region using the activation values as the h-function and the measures defined off-line. The maximum fuzzy integral output over all the regions associated with each class was taken as the confidence that the input character was from that class.

A more precise description is the following: For each class ω, let $A_\omega = \{X \mid X$ is a region for class $\omega\}$. A given class may have several regions of high activity in the map. For each region $X \in A_\omega$, define a λ-fuzzy measure $g_{X,\omega}$. Let

N_ω^i = number of times class ω wins at node i,

N^i = number of times node i wins, and

B_ω^i = number of nodes at which class ω wins at least once and that there are four neighbors of node i.

The density values for X are defined by the following:

For class $\omega = 1, 2, \ldots, 26$ Do
   For each region $X \in A_\omega$
      For each node $i \in X$

$$g^i_{X,\omega} = \frac{N^i_\omega}{N^i \cdot B^i_\omega}$$

Intuitively, the density of a node reflects its degree of importance in terms of relative winning frequency. This classification algorithm helped to increase word recognition performance from 80 to 86% on real-data sets from the US Post Office. Details can be found in Chiang and Gader [1997].

You might decide to learn the entire measure from training data instead of focusing on the densities. Certainly, you can consider the least-squares criterion function above (or the equivalent one for the Sugeno integral) as a fitness function in an evolutionary computation algorithm. As long as n is not too big, $2^n - 2$ genes in a chromosome (or particle) is no longer unreasonable from a computation standpoint. The challenge with an EC approach to learn entire measures comes with crossover and/or mutation. Specifically, if values for the measure of some subsets are swapped or if the values are mutated, say by the addition of a random number, constraint 2 of the fuzzy measure ($g(A) \leq g(B)$ if $A \subseteq B$) may be violated. This is not impossible to resolve, but care must be taken to ensure that the constraint is satisfied after reproduction. For example, if $g(A)$ is changed, then all supersets of A can be bumped up to at least $g(A)$ to satisfy the constraint. A similar kind of care was needed in crossover to maintain constraints in the standard form of solving the traveling salesman problem (see Chapter 11).

Grabisch and Nicolas [1994] transformed learning measures for the Choquet fuzzy integral with least-squares criteria into a quadratic programming (QP) problem. Specifically, consider learning the measure for a Choquet integral mapping function for a desired set of training data $T = \{(O_j, \alpha_j) | j = 1, \ldots, m\}$:

$$C_g(h(O_j)) = \sum_{i=1}^{n} [h(O_j; x_{(i)}) - h(O_j; x_{(i+1)})] \cdot g(A_i) = \alpha_j, \quad \text{for} \quad j = 1, \ldots, m \quad (9.19)$$

Then, the goal is to find a measure g that minimizes

$$E^2 = \sum_{j=1}^{m} (C_g(h(O_j)) - \alpha_j)^2 \quad (9.20)$$

To formulate the optimization problem, define $g_{i_1 i_2 \cdots i_k} = g(\{x_{i_1}, \ldots x_{i_k}\})$. Using lexicographic ordering, we combine the $2^n - 2$ unknown measure values into a vector:

$$g = (g_1, g_2, \ldots, g_{12}, g_{13}, \ldots, g_{123 \cdots n-1}, \cdots g_{23 \cdots n})^t \quad (9.21)$$

For any subset A of X, and object O_j, let

$$\Gamma_{j,A} = \begin{cases} h(O_j; x_{(i)}) - h(O_j; x_{(i+1)}), & \text{if } A = A_i \text{ for some i} \\ 0, & \text{else} \end{cases} \quad (9.22)$$

For each j, define the vector

$$\Gamma_j = (0, \ldots, 0, h(O_j; x_{(1)}) - h(O_j; x_{(2)}), \ldots, h(O_j; x_{(n-1)}) - h(O_j; x_{(n)}), 0, \ldots, 0)^t$$
(9.23)

Note that $\Gamma_j \in R^{2^n-2}$ with only n−1 possible nonzero values corresponding to those indices that occur in the definition of the Choquet integral. Hence, $C_g(O_j) = (\Gamma_j^t \cdot g) + h(O_j; x_{(n)})$ and so

$$
\begin{aligned}
E^2 &= \sum_{j=1}^{m} \left(C_g(h(O_j)) - \alpha_j \right)^2 \\
&= \sum_{j=1}^{m} \left(\Gamma_j^t \cdot g + [h(O_j; x_{(n)}) - \alpha_j] \right)^t \cdot \left(\Gamma_j^t \cdot g + [h(O_j; x_{(n)}) - \alpha_j] \right) \\
&= \sum_{j=1}^{m} \left(g^t \Gamma_j \Gamma_j^t g + 2[h(O_j; x_{(n)}) - \alpha_j] \Gamma_j^t g + [h(O_j; x_{(n)}) - \alpha_j]^2 \right) \\
&= g^t \left(\sum_{j=1}^{m} \Gamma_j \Gamma_j^t \right) g + 2 \left(\sum_{j=1}^{m} [h(O_j; x_{(n)}) - \alpha_j] \Gamma_j^t \right) g + \sum_{j=1}^{m} [h(O_j; x_{(n)}) - \alpha_j]^2
\end{aligned}
$$

which is equivalent to

$$E^2 = g^t D g + \Gamma^t g + \alpha^2 \tag{9.24}$$

where

$$D = \left(\sum_{j=1}^{m} \Gamma_j \Gamma_j^t \right), \Gamma = \left(\sum_{j=1}^{m} 2(h(O_j; x_{(n)}) - \alpha_j)\Gamma_j \right), \text{ and } \alpha^2 = \sum_{j=1}^{m} [h(O_j; x_{(n)}) - \alpha_j]^2.$$

Now, consider the constraints, a total of $n(2^{n-1} - 1)$, on the measure, g. We require

$$g_1 - g_{12} \le 0$$
$$g_1 - g_{13} \le 0$$
$$\vdots$$
$$g_1 - g_{123\cdots n} \le 0,$$
$$\vdots$$
$$g_{123\cdots n-1} - g_{123\cdots n} \le 0$$
$$\vdots$$
$$g_{23\cdots n} - g_{123\cdots n} \le 0$$

You could list all of the inequalities that arise from the montonicity condition on the measure, but there is a smaller subset that is sufficient by thinking of the subsets of X as a partially ordered lattice and then you only need to enforce the constraints from one level to the next. For example, $g_1 - g_{123} \leq 0$ is already guaranteed by $g_1 - g_{12} \leq 0$ and $g_{12} - g_{123} \leq 0$. This can be written in matrix form as

$$
A \cdot g + b =
\begin{pmatrix}
1 & 0 & \cdots & 0 & -1 & 0 & \cdots & 0 & 0 & \cdots & 0 \\
1 & 0 & \cdots & 0 & 0 & -1 & \cdots & 0 & 0 & \cdots & 0 \\
\vdots & \vdots & \cdots & \vdots & \vdots & \vdots & \cdots & \vdots & \vdots & \cdots & \vdots \\
0 & 0 & \cdots & 0 & 0 & 0 & \cdots & 1 & 0 & \cdots & 0 \\
\vdots & \vdots & \cdots & \vdots & \vdots & \vdots & \cdots & \vdots & \vdots & \cdots & \vdots \\
0 & 0 & \cdots & 0 & 0 & 0 & \cdots & 0 & 0 & \cdots & 1
\end{pmatrix}.
$$

$$
\begin{pmatrix}
g_1 \\
\vdots \\
g_{12} \\
g_{13} \\
\vdots \\
g_{123\cdots n-2} \\
\vdots \\
g_{123\cdots n-1} \\
g_{134\cdots n} \\
\vdots \\
g_{234\cdots n}
\end{pmatrix}
+
\begin{pmatrix}
0 \\
\vdots \\
0 \\
0 \\
\vdots \\
0 \\
\vdots \\
-1 \\
-1 \\
\vdots \\
-1
\end{pmatrix}
\leq 0 \tag{9.25}
$$

where A is of size $\left(n\left(2^{n-1} - 1\right)\right) \times \left(2^n - 2\right)$, b is of size $\left(n\left(2^{n-1} - 1\right)\right) \times 1$, and b is the vector of all 0's except for the last n entries that are of value -1. Note that the -1 entries correspond to the fact that $g_{123\cdots n} = 1$ and so, the last set of inequalities that involve the measure of the whole space have that measure explicitly given, and that we only need to specify the inequalities for the sets directly "below" the whole space in a lattice representation. For example, the row of A that is shown with a 1 and the rest zeros is there to encode the inequality $g_{123\cdots n-1} - g_{123\cdots n} \leq 0$. You have to ponder Eq. 9.25 for a while to understand the dimensions involved. Hence, the search for the measure g reduces to a quadratic program of the form

$$
\text{Minimize} \frac{1}{2} g^t \cdot D \cdot g + \Gamma^t \cdot g \tag{9.26}
$$
$$
\text{subject to } A \cdot g + b \leq 0 \text{ and } 0 \leq g \leq 1
$$

Note, $D = \left(\sum_{j=1}^m \Gamma_j \Gamma_j^t\right)$ need only be scaled by a factor of 2 and our inequality, $A \cdot g + b \leq 0$, can be adjusted via multiplying by -1, that is, $(-A) \cdot g - b \geq 0$. This formulation has been used to find a single measure and has been extended to find multiple measures in pattern recognition problems.

EXAMPLE 9.6   RECOVERING A KNOWN MEASURE VIA EQ. 9.26

This example is courtesy of Professor Derek Anderson, Mississippi State University, who needed to construct it for another reason. Using the above quadratic programming approach, we learn a fuzzy measure that fits a set of data, T, for the case of $n=3$ and $m=8$. The α_j values are generated below using an OWA operator with the weights $\mathbf{w}^t = (.6 \quad .3 \quad .1)$. For example, in T, $\mathbf{O}_1 = \begin{pmatrix} 0.68 \\ 0.53 \\ 0.81 \end{pmatrix}$, and so

$$\alpha_1 = (0.6) \cdot 0.81 + (0.3) \cdot 0.68 + (0.1) \cdot 0.53 = 0.743$$

Specifically, we solve for **g** below using the following QP (re)formulation.

$$\text{Minimize } \frac{1}{2}\mathbf{g}^t \cdot D \cdot \mathbf{g} + \boldsymbol{\Gamma}^t \cdot \mathbf{g}$$
$$\text{subject to } \mathbf{0} \le \mathbf{g} \le \mathbf{1} \text{ and } A \cdot \mathbf{g} \le b$$

The corresponding set of matrices and vectors are as follows:

$$O = \begin{pmatrix} 0.68 & 0.53 & 0.81 \\ 0.74 & 0.99 & 0.86 \\ 0.45 & 0.07 & 0.08 \\ 0.08 & 0.44 & 0.39 \\ 0.22 & 0.10 & 0.25 \\ 0.91 & 0.96 & 0.80 \\ 0.15 & 0 & 0.43 \\ 0.82 & 0.77 & 0.91 \end{pmatrix}, \quad \alpha = \begin{pmatrix} 0.743 \\ 0.926 \\ 0.301 \\ 0.389 \\ 0.226 \\ 0.929 \\ 0.303 \\ 0.869 \end{pmatrix}, \quad g = \begin{pmatrix} g_1 \\ g_2 \\ g_3 \\ g_{12} \\ g_{13} \\ g_{23} \end{pmatrix},$$

$$Z = \begin{pmatrix} 0 & 0 & 0.13 & 0 & 0.15 & 0 \\ 0 & 0.13 & 0 & 0 & 0 & 0.12 \\ 0.37 & 0 & 0 & 0 & 0.01 & 0 \\ 0 & 0.05 & 0 & 0 & 0 & 0.31 \\ 0 & 0 & 0.03 & 0 & 0.12 & 0 \\ 0 & 0.05 & 0 & 0.11 & 0 & 0 \\ 0 & 0 & 0.28 & 0 & 0.15 & 0 \\ 0 & 0 & 0.09 & 0 & 0.05 & 0 \end{pmatrix}$$

where Z is the matrix of Γ_j entries and

$$
D = \begin{pmatrix}
0.2738 & 0 & 0 & 0 & 0.0074 & 0 \\
0 & 0.0438 & 0 & 0.110 & 0 & 0.0622 \\
0 & 0 & 0.2086 & 0 & 0.1392 & 0 \\
0 & 0.0110 & 0 & 0.0242 & 0 & 0 \\
0.0074 & 0 & 0.1392 & 0 & 0.1240 & 0 \\
0 & 0.0622 & 0 & 0 & 0 & 0.2210
\end{pmatrix},
$$

$$
\Gamma = \begin{pmatrix}
-0.1709 \\
-0.0922 \\
-0.2504 \\
-0.0284 \\
-0.1996 \\
-0.2362
\end{pmatrix}
$$

The constraints (note that there is no g_{123} here, it's got to be 1 anyway) are given by

$$
A = \begin{pmatrix}
1 & 0 & 0 & -1 & 0 & 0 \\
1 & 0 & 0 & 0 & -1 & 0 \\
0 & 1 & 0 & -1 & 0 & 0 \\
0 & 1 & 0 & 0 & 0 & -1 \\
0 & 0 & 1 & 0 & -1 & 0 \\
0 & 0 & 1 & 0 & 0 & -1 \\
0 & 0 & 0 & 1 & 0 & 0 \\
0 & 0 & 0 & 0 & 1 & 0 \\
0 & 0 & 0 & 0 & 0 & 1
\end{pmatrix}, \quad b = \begin{pmatrix}
0 \\
0 \\
0 \\
0 \\
0 \\
0 \\
1 \\
1 \\
1
\end{pmatrix}
$$

After optimization, that is, after running the quadratic program, the resulting fuzzy measure is

$$
\tilde{g} = \begin{pmatrix}
0.6 \\
0.6 \\
0.6 \\
0.9 \\
0.9 \\
0.9
\end{pmatrix}
$$

This measure is the desired OWA measure, that is, sets of equal size cardinality have equal measure value and the OWA weights are found by the following:

$$\mathbf{w}(1) = g_1 = g_2 = g_3 = 0.6$$
$$\mathbf{w}(2) = g_{12} - \mathbf{w}(1) = g_{13} - \mathbf{w}(1) = g_{23} - \mathbf{w}(1) = 0.3$$
$$\mathbf{w}(3) = g_{123} - \mathbf{w}(2) = 1 - .9 = 0.1$$

9.4 SUMMARY AND FINAL THOUGHTS

As extensions to probability measures, fuzzy measures provide a plethora of ways to build credibility of subsets of objects or evidence in answering a question. They continue to be directly useful in decision-making applications. In the context of pattern recognition or fusion of information where the set over which the measure is defined is a set of sensors/algorithms/features/intelligence, and where for each object or instance under consideration is an evaluation of confidence with respect to each source of information, fuzzy integration is a great mechanism to fuse this objective evaluation with the (possibly subjective) worth of the sources. We introduced both the Sugeno and Choquet integrals as generalized expectation operators, considered some of their properties, and described a few of the learning algorithms for real problems. There is a rich and evolving theory and new applications continue to arise.

EXERCISES

9.1. Show that all probability measures are fuzzy measures, that is, if a set function $g : 2^X \rightarrow [0,1]$ satisfies $g(\emptyset) = 0$ and $g(X) = 1$ and $g(A \cup B) = g(A) + g(B)$, if $A \cap B = \phi$, then it must also satisfy $g(A) \leq g(B)$ if $A \subseteq B$.

9.2. Let $X = \{x_1, \cdots x_n\}$ and $F : X \rightarrow [0,1]$ be a fuzzy subset of X.
Define

$$g(\emptyset) = 0$$
$$g(A) = \max\{1 - F(x)\}, \quad \text{for all } x \in A$$

Is g a fuzzy measure? Justify your answer. Consider conditions on the fuzzy set F.

9.3. Let $F : X \rightarrow [0,1]$ be a normal fuzzy subset of X and let $Pos : 2^X \rightarrow [0,1]$ be a function defined by

$$Pos(A) = \begin{cases} 0, & \text{if } A = \phi \\ \max_{x \in A}(F(x)), & \text{if } A \neq \phi \end{cases}$$

A. Show that Pos is a fuzzy measure (called a possibility measure).

B. Show that $Pos(A \cup B) = Pos(A) \vee Pos(B)$ for all subsets A and B of X.

C. Define a function $Nec(A) = 1 - Pos(A^c)$. Show that Nec is also a fuzzy measure, called a Necessity measure.

D. What is the relationship between $Pos(A)$ and $Nec(A)$?

E. Why do you think they are called "possibility" and "necessity"?

9.4. Show that Eq. 9.2 holds for $X = \{x_1, x_2, x_3\}$ and densities $\{g^1, g^2, g^3\}$.

9.5. If $\lambda = 0$ in Eq. 9.2, show that g is a probability measure and that $\sum_{i=1}^{n} g^i = 1$.

9.6. Let $g^1 = 0.4$; $g^2 = 0.3$; $g^3 = 0.2$

Calculate the complete Sugeno λ-measure on $X = \{x_1, x_2, x_3\}$.

Subset A	g(A)
$\emptyset$	
$\{x_1\}$	
$\{x_2\}$	
$\{x_3\}$	
$\{x_1, x_2\}$	
$\{x_1, x_3\}$	
$\{x_2, x_3\}$	
$\{x_1, x_2, x_3\}$	

9.7. Show that Eq. 9.4 holds for finite sets. *Hint:* What are the level sets for the sorted fuzzy set $\{h(x_{(1)}), \ldots, h(x_{(n)})\}$?

9.8. Given the measure

Subset A	g(A)
$\emptyset$	0
$\{x_1\}$	0.5
$\{x_2\}$	0.1
$\{x_3\}$	0.3
$\{x_1, x_2\}$	0.7
$\{x_1, x_3\}$	0.9
$\{x_2, x_3\}$	0.5
X	1.0

Calculate the Sugeno and Choquet fuzzy integrals for the following functions:

i. $h(x_1) = 0.3$; $h(x_2) = 0.5$; $h(x_3) = 0.9$

ii. $h(x_1) = 0.5$; $h(x_2) = 0.5$; $h(x_3) = 0.5$

9.9. Given a finite set $X = \{x_1, x_2, \ldots, x_n\}$, a measure $g : 2^X \rightarrow [0, 1]$, and a function $h : X \rightarrow [0, 1]$, show that both the Sugeno and Choquet integrals of h with respect to g are generalized expected values, that is, they both lie between the minimum and maximum of the values $h(x_1), \ldots, h(x_n)$.

9.10. Show that if $h(x) = a$ for all x, then $S_g(h) = a$, and $C_g(h) = a$.

9.11. Show that if $g_1(A) \leq g_2(A)$ for all $A \subset X$, then $S_{g_1}(h) \leq S_{g_2}(h)$ (same for the Choquet integrals).

9.12. Show that if $h_1(x) \leq h_2(x)$ for all $x \in X$, then $S_g(h_1) \leq S_g(h_2)$ (same for the Choquet integrals).

9.13. Is the Sugeno/Choquet integral linear, that is, is $S_g(a \cdot h + b) = a \cdot S_g(h) + b$, where a and b are real numbers in [0,1]? (Same equation substituting C_g for S_g). You can assume that the linear combinations remain in the interval [0,1], although that is not strictly required. Supply a justification or a counterexample.

9.14. What can you conclude if $S_g(h) = 0$? Is it true that the function h must be identically 0?

9.15. Let h be the characteristic function for a subset A of X, that is, $h(x) = \begin{cases} 1, & \text{if } x \in A \\ 0, & \text{if } x \notin A \end{cases}$. What is the integral (Sugeno or Choquet) of this h with respect to a given fuzzy measure g?

9.16. Verify that g in Eq. 9.13 is a fuzzy measure and that $C_g(h)$ (Eq. 9.12) is the corresponding LOS operator.

9.17. Generate Eq. 9.18, that is, compute and simplify $\partial \lambda / \partial g^k$.

9.18. Using the notational convention in Eq. 9.21 with three sources of information, intuitively describe the fusion operation performed by the Choquet integral with the following measures:

   i. $g = [0, 0, 0, 1, 1, 1, 1]$

   ii. $g = [1, 1, 1, 1, 1, 1, 1]$

   iii. $g = [1/3, 1/3, 1/3, 2/3, 2/3, 2/3, 1]$

   iv. $g = [0, 0, 0, 0, 0, 0, 1]$

9.19. Write the complete matrix A and vectors **g** and **b** in Eq. 9.25 for a set of information sources X that has four elements.

9.20. Fill in all the details in Example 9.6, that is, hand-compute α, Z, D, A, and **b**.

9.21. Use the Sugeno measure in Table 9.2 and redo Example 9.6 with the same input training data O. You will have to write the QP program (or use the Matlab or other package version). Do you recover the measure in Table 9.2?

9.22. Give an interpretation of the Sugeno and Choquet fuzzy integrals; compare and contrast them: you might want to illustrate with examples.

PART III

EVOLUTIONARY COMPUTATION

CHAPTER 10

Evolutionary Computation

10.1 BASIC IDEAS AND FUNDAMENTALS

Evidence suggests that life has been evolving on Earth for over 3 billion years [Schopf, 2006]. Three billion years is a long time. With a few exceptions, we each live for less than a century and thus it's difficult to truly comprehend a segment of time that's as long as 3 billion years. Fortunately, the fossil record gives us a glimpse into times long since past. Bones, fossilized footprints, and an occasional bug in amber can provide clues about what life was like in ancient times. Even a casual examination of the world around us now, and the worlds we can envision from history, leads to an obvious observation: Nature's organisms are, and have often been, wonderfully adapted to their environments.

In the grand scale of time, these beautiful adaptations are ephemeral. They last for mere moments relative to the history of the universe. Scientists estimate the average life span for various species to be between 1 and 16 million years [Buzas and Culver, 1984; Liow et al., 2008]. So, to use a fuzzy term, that's about 10^7 years.[1] If we compare a time frame on the order of 10^7 years with the universe's 1.3×10^{10} years, it's fair to say that on average every organism on the planet lasts for just a blink of the universe's eye.

But for that brief blink of time, time and again, nature has found a way to solve some very challenging engineering problems. You may not have considered life to be a series of engineering problems, but let's consider some examples.

1. How is it that a fly can walk upside down on a ceiling? The answer, which is common to over 300 observed insect species, is that fly feet have multiple fat footpads to provide surface area, and tiny hairs on the feet excrete a sticky substance of sugars and oils. They are literally adhered to the ceiling, glued sufficiently to walk upside down, but also weakly enough that they can become airborne just before a flyswatter arrives [Binns, 2006].

---

[1] Some species likely endure much less than 1 million years, at best leaving just traces in the fossil record, and thus making it very difficult to determine exactly how long they were here.

Fundamentals of Computational Intelligence: Neural Networks, Fuzzy Systems, and Evolutionary Computation, First Edition. James M. Keller, Derong Liu, and David B. Fogel.

2. Speaking of sticky situations, how is it that spiders do not stick to their own webs? The answer is again one of delicate engineering. Spiders' legs are protected by a nonstick chemical coating and are covered with hairs that reduce exposure to the sticky part of the web. Spiders also move around adeptly on their webs, crawling in ways that minimize adhesive forces [Briceño and Eberhard, 2012].

3. What about geckos? Like flies, they seemingly defy gravity as they climb walls and hang upside down (even by just two legs). Do they use glue, like flies do? In a sense, they do, but the glue they use isn't a substance, it's a force. Gecko feet have millions of tiny hairs called *setae*, and these interact with the surface by way of van der Waals forces to attract them to the surface [Autumn et al., 2002]. In essence, geckos use these forces to "hang in there."[2]

These ingenious inventions of nature are examples of evolution by variation and natural selection. Evolution is a simple concept, but it was resisted for three-quarters of a century after first being proposed by Charles Darwin and Alfred Russel Wallace in 1859 [Mayr, 1988]. Ultimately, a synthesis of biological thought took place in the 1930s and 1940s, which serves as the foundation of the modern understanding of evolution.

Despite the simplicity of the concept, evolution is a complex process, itself a compilation of numerous nonlinear interacting processes. Thus, any concise description of evolution will omit many important details. However, the essence of evolution—variation and selection—can be illustrated conveniently and cogently by way of a series of mapping functions that connect genetics with behavior.

Living organisms act as a duality of their genotype (the underlying genetic coding) and their phenotype (the manner of response contained in the behavior, physiology, and morphology of the organism). This genotype–phenotype pairing may be viewed as a pairing across two state spaces, **G** and **P**, that are *informational (genetic)* and *behavioral (phenotypic)*, respectively. Figure 10.1 shows how these spaces are connected [Lewontin, 1974; Atmar, 1992, 1994]. Specially, four functions map elements in **G** and **P** to each other.

These four mappings can be described as follows.

$$f_1 : \mathbf{I} \times \mathbf{G} \to \mathbf{P}$$
$$f_2 : \mathbf{P} \to \mathbf{P}$$
$$f_3 : \mathbf{P} \to \mathbf{G}$$
$$f_4 : \mathbf{G} \to \mathbf{G}$$

The function f_1 is called *epigenesis*. This maps an element $g_1 \in G$ and indexed set of symbols $(i_1, \ldots, i_k) \in \mathbf{I}$, where **I** is the set of all such environmental sequences into

---

[2] Researchers have found that the gecko's ability to adhere to surfaces depends in part on temperature and humidity, and others have suggested that lipids found in gecko footprints may also play a role in their adhesion, so a complete understanding of "gecko technology" may not yet be at hand [Niewiarowski et al., 2008; Hsu et al., 2011].

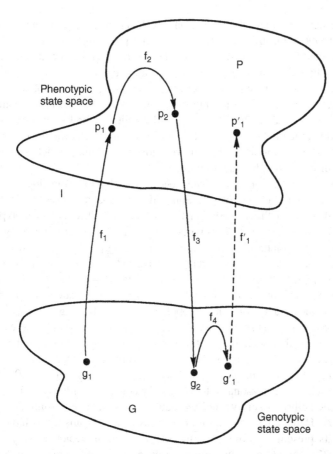

FIGURE 10.1 The evolution of a population within a single generation. Evolution can be viewed as occurring as a succession of four mapping functions—*epigenesis, selection, genotypic survival*, and *mutation*—relating the genotypic information state space and the phenotypic behavioral state space.

the phenotypic space **P** as a particular collection of phenotypes p_1, whose development is modified by its environment. It's important to recognize that the same set of genetics can correspond to alternative observed behaviors when exposed in different environments. The function f_2, *selection*, maps phenotypes p_1 into p_2. As natural selection operates only on the phenotypic expressions of the genotype [Mayr, 1960; Hartl and Clark, 1989; Reed and Bryant, 2004], the underlying coding g_1 is not involved in function f_2.

Before describing functions f_3 and f_4, it may be helpful to emphasize this property of selection. Natural selection acts on the behaviors generated by organisms and thus only indirectly on their underlying genes. Within the past few decades, there has been a great deal of focus on evolution from the viewpoint of genes, how they change in frequency over time, and what selective advantages may accrue to individual genes. It has been written sometimes that evolution *is* a change in gene frequencies (see Smith and Smith [2006]), but this view has the twofold problem of (i) omitting or under-stating the connection of the environment to genes in order to generate behaviors, and (ii) omitting the effects of selection on behaviors altogether. Some have argued that evolution does not even require a change in gene frequencies [Sober, 2000], but, importantly, function is what is pertinent to natural selection. When a cheetah chases a gazelle, it is the speed, agility, eyesight, and stamina of both that are being tested directly, and not their genes *per se*, except indirectly and in a particular environmental setting. This is a very important facet of evolutionary modeling, which will be covered again later in this and subsequent chapters.

Returning to the mappings between genes and behaviors, function f_3, genotypic survival, describes the effects of selection and migration processes on G. For those phenotypes that survive the selection function f_2, there is a collection of genotypes g_2 that correspond to those surviving phenotypes. (The collection g_2 differs from g_1 because selection and migration have removed individuals from g_1.) These are now the focus of function f_4, *mutation*, which maps the representative codings $g_2 \in$ G to the point $g_1' \in$ G. This function represents the "rules" of genetic variation—individual gene mutations, recombination, inversion, and so forth—that occur in the creation of offspring from parents. With the creation of a new population of genotypes g_1', one generation is complete. Evolutionary adaptation occurs over successive iterations of these mappings.

With the view that selection is the primary evolutionary force that prevails in shaping the behavioral (phenotypic) characters of organisms [Mayr, 1988; Hoffman, 1989; Brunk, 1991; Wilson, 1992; Rundle et al. 2003; Steinger et al., 2003; Lyytinen et al., 2004], it becomes reasonable to address the concept of "fitness," where fitness is defined as the ability to survive and reproduce in a specific environment [Hartl and Clark, 1989]. Here fitness refers to the aptness of an organism's behaviors in the context of its present environment.[3] In essence then, as depicted many years ago [Wright, 1932], evolution occurs on a landscape (topography) that is the result of a function describing the fitness or suitability of being adapted to the environment. Figure 10.2 illustrates this concept.

An *adaptive landscape* will look very familiar to an engineer, mathematician, or computer scientist, for it is immediately suggestive of the concept of function optimization. Rather than having a purpose of searching for a maximum, minimum, or perhaps a saddle point on the landscape, evolution moves without purpose. But its movements appear purposeful in that the results of evolution reflect opportunities to improve fitness (or obversely to reduce predictive error[4]). Selection eliminates

---

[3] There are many definitions of fitness, each tailored for a specific application of the term. For example, Fisher [1930] described the fitness of a gene as the per capita rate of increase (of a genotype, based on the average effect of gene substitution). Gene-based definitions of fitness are numerous in the so-called *selfish genetics* [Dawkins, 1976]. The utility of these definitions for doing anything other than counting genes is beyond the scope of this text. Interested readers are referred to Fogel [2006] and Fogel and Fogel [2011] for further information.

[4] Living organisms can be viewed as being in a constant state of predicting what is coming next in their environment. Better predictions equate to less surprise, and being surprised in nature is quite often a really bad outcome (e.g., consider the surprise a fly gets when it is met by a frog's tongue).

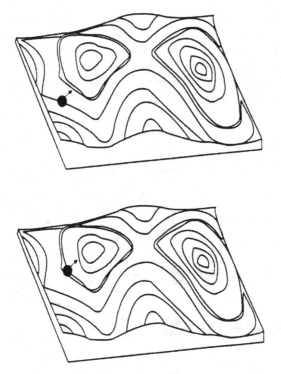

FIGURE 10.2 Evolution of an inverted adaptive topography. A landscape is abstracted to represent the fitness of alternative phenotypes and, as a consequence, alternative genotypes. Rather than viewing individuals or populations as maximizing fitness and thereby climbing peaks on the landscape, a more intuitive perspective may be obtained by inverting the topography [Atmar, 1979; Templeton, 1982; Raven and Johnson, 1986]. Populations proceed down the slopes of the topography toward valleys of minimal predictive error.

organisms that are less fit relative to others. Variation provides an unending source of searching for new possibilities. Coupled together, variation and selection provide a mechanism for searching over a landscape, resulting in outcomes that often appear highly engineered or "designed" for survival.

Note that unlike many engineering optimization problems, which may be of the form:

$$\text{Find } x \mid f(x) \text{ is minimized}$$

where x is a scalar value and f(x) is a real-valued function of x, problems of optimization in evolution involve numerous variables acting simultaneously. An extremely fast animal, such as cheetah, can accelerate from a standing start to 60 mph (96 kph) within 3 s, but their chases rarely last for more than 100 yards of distance or 1 min of time. Their highly energy-intensive bursts of speed come at the expense of endurance. There are always trade-offs like these across behavioral traits.

Not only is there an interaction between behavioral traits, recall that these traits arise as a function of the interaction with an environment and with some genetic basis. The relationship between genes and behaviors is not one-to-one. Instead, the relationship is one-to-many in some cases and many-to-one in others, simultaneously. Figure 10.3 provides a graphic that illustrates these effects, which are described by the terms *pleiotropy* (\plī-ˈä-trə-pē, one gene has many effects) and *polygeny* (one effect

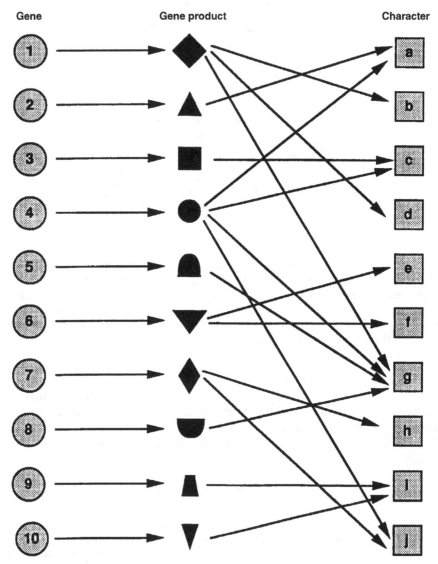

FIGURE 10.3 The effects of pleiotropy and polygeny. Pleiotropy is the effect where a single gene affects multiple phenotypic characters. Polygeny is the effect where a single phenotypic character is affected by multiple genes. (*Source:* After Mayr [1963].)

FIGURE 10.4   This is a drone fly (*Eristalis tenax*). It mimics a honey bee. If honey bees and other insects with similar warning patterns did not exist, there would be no selective pressure for the fly to have evolved this pattern. The fly does not have a stinger. (*Source:* From http://www. dpughphoto.com/pitt_ county_arboretum.html.)

coming from many genes). As one example, about 40% of cats with white fur and blue eyes are also deaf. There's no reason to expect deafness to correlate with fur and eye color, but the complexity of the cat's genetic makeup links these effects [Hartl and Jones, 2005]. (What's more, white cats that have one blue eye and one yellow eye are more likely to be deaf only on the blue-eyed side.) Even single genetic point mutations can have manifold effects on resulting behaviors, and thus on an organism's fitness.

Moreover, fitness, which may be difficult to measure instantaneously, is a function of the environment that includes all the other organisms present. Consider the example shown in Figure 10.4. It looks like a bee, but it's not. It's a fly. It gets the benefit of looking like a bee, which may be a big benefit if there are other bees around. In the absence of bees and other insects with similar warning patterns, this particular pattern would be of much less or even no particular value.[5] Thus, fitness is a temporal function of the world in which the organism lives. Adaptive landscapes look simple in three dimensions, but they are in fact highly complex, span numerous dimensions, and vary in time.

Having introduced the concepts of variation and selection on an adaptive topography, as well as noting the ingenious evolutionary inventions found in nature, it's natural to explore the possibility of simplifying evolutionary processes in software (*evolutionary algorithms*) that can be used for engineering purposes. These purposes

---

[5] Warning patterns are common across diverse species and are described with the term *aposematism* (from the Latin apo = away, sematic = sign).

include optimization, design, and learning. It's also of interest then to study the mathematical properties of these algorithms.

Such efforts have been ongoing for at least 60 years, and for the past two decades under a moniker of *evolutionary computation*. The simplest evolutionary algorithm can be viewed as a search procedure that generates potential solutions to a problem, tests each for suitability, and then generates new solutions. It's important to understand how this process differs from exhaustive search or blind random search.

10.2 EVOLUTIONARY ALGORITHMS: GENERATE AND TEST

Consider a sample space $\mathbf{S}$, from which individual potential solutions $s_1, \ldots, s_k$ can be selected ($|\mathbf{S}| = k$). In principle, the cardinality of $\mathbf{S}$ can be infinite, but there is always some limit of precision on a computer, so no loss of generality occurs here when treating a finite number k. Suppose there is a real-valued function f(s) for which it's desired to find a maximum. An exhaustive search of $\mathbf{S}$ can be described in pseudocode by the following procedure:

```
i = 1;
best = i;
bestscore = f(s(i));
repeat
   i = i + 1;
   if f(s(i)) > bestscore then
   begin
      best = i;
      bestscore = f(s(i));
   end
until (i == k);
```

This will indicate the best solution s(best), which has the corresponding greatest score f(s(best)). While such a procedure is certain to find the best solution, the time required to find it may be prohibitive. As will be discussed in the next chapter, many practical problems pose search spaces that are trans-computationally large, meaning that the size of the space is greater than 10^{100}. There are only about 10^{18} s in the history of the universe, so in these cases you could examine 10^{80} solutions every second, and still not be done in over 13 billion years![6] (And after another 13 billion years, no one would be around to applaud you for finding the solution.)

---

[6] The value 10^{80} happens to correspond to estimates of the number of protons in the observable universe. So, imagine doing calculations on par with the number of protons in the observable universe every second for 13 billion years and still not completing your task.

Instead of searching exhaustively, consider searching at random, blindly choosing a series of possible solutions $s_1, \ldots, s_n$ from S, where the sequence of length $n \leq k$ is chosen uniformly without replacement. Mathematically, this means that for the first sample s_1, each solution in S has a $1/k$ probability of being selected. After s_1 is selected, every remaining solution in S has a $1/(k-1)$ probability of being selected as s_2, and so forth. In pseudocode, the procedure looks like this:

```
i = 1;
selected = U(1, k); //
chosen uniformly at random over the integers 1, . . . , k
best = selected;
bestscore = f(s(selected));
repeat
   i = i + 1;
   remove s(selected) from S;
   selected = U(1, k - i + 1); //chosen uniformly over the
      remaining solutions
   if f(s(selected)) > bestscore then
   begin
      best = selected;
      bestscore = f(s(selected));
   end
until (i == n);
```

The procedure can be performed such that it completes in a reasonable amount of time by selecting n appropriately. Unfortunately, this procedure often performs poorly on real-world problems (although it has some interesting mathematical properties on average across all possible problems, which will be presented in greater depth in Chapter 11).

Note that both the exhaustive procedure and the blind random sampling choose each next solution without regard to what has been chosen previously. As suggested by the term *blind*, blindly searching for things is often a big handicap. Traditional search procedures described in mathematics are not blind. They explore for a maximum (or minimum) by utilizing information from the function being searched. Often this involves use of the gradient or higher order statistics that allow a search algorithm to move rapidly in a direction that is presumed to be beneficial (i.e., leading to higher or lower values depending on whether the task is to maximize or minimize). These procedures can converge quickly on a maximum or minimum, but run the risks of stalling at a saddle point—where the gradient is zero—or becoming trapped in maxima or minima that are only optimal locally. Figure 10.5 illustrates an example.

Evolutionary algorithms operate in two ways that are fundamentally different from traditional gradient methods. First, rather than executing a point-to-point search, they incorporate a *population* of solutions, each individual solution competing for survival. Second, instead of utilizing gradient information from the response surface

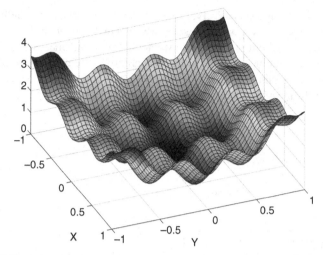

FIGURE 10.5 A function that provides opportunities for a gradient-based search algorithm to stall at local maxima, minima, or saddle points. The function is $f(x, y) = x^2 + 2y^2 - 0.3\cos(3\pi x) - 0.4\cos(4\pi y) + 0.7$.

being searched, they utilize random variation to explore for new solutions. In pseudocode, the basic procedure looks like this:

```
g = 0; // g is for generation number
choose initial population; //
often this is selected at random from S
repeat
     evaluate population; //assign a score to each individual
     select parents; // based on the scores, choose a subset
       of individuals
     generate offspring; //based on the parents, create
       offspring
     g = g + 1;
until (g == max)
```

Illustrating this procedure with an example, suppose we want to find the minimum of the function $f(x) = x^2$. The function is convex and continuous, thus gradient methods would be more appropriate (the Newton–Gauss method would be most appropriate). But the function will serve for instructional purposes. The value g represents the number of generations completed. It starts at zero. The initial population is chosen. This might be done at random from a broad range of **S**. It requires choosing a population size, for purposes that will be made clear shortly, which we will describe with the term $(\mu + \lambda)$. Each of the $(\mu + \lambda)$ solutions is evaluated in light of $f(x)$. The μ individuals with the best scores are selected to become parents. These parents then become the basis for generating λ offspring and the generation counter g is incremented by 1.[7] The process continues until a maximum number (max) of generations is completed.

The relationship between parent and offspring defines *inheritance*. If there is no inheritance, then the procedure is essentially like a blind random search conducted multiple samples at a time. Suppose instead that the value of an offspring is related to its parent as follows:

$$x' = x + N(0, \sigma)$$

where x' is the offspring of x and $N(0, \sigma)$ is a Gaussian random variable with zero mean and standard deviation σ. If $\sigma \to \infty$, then the search is again like a blind random search. If the $\sigma \to 0$, then there is complete inheritance from parent to offspring and there's no search at all because each offspring merely replicates its parent. But with $0 < \sigma < \infty$, the procedure becomes a stochastic parallel search with many interesting mathematical properties. These will be discussed further in Chapters 11 and 13.

But first, let's examine an example of a simple evolutionary algorithm at work. Let's consider the function $f(x, y) = x^2 + y^2$ and use an evolutionary approach to approximate the answer that minimizes the function. We'll start with a population of 100 candidate solutions that we choose uniformly at random from the range $(x, y) \in (-10, 10)^2$. Figure 10.6 shows our initial population. In this very basic approach, we'll have each parent generate one offspring by mutating the (x,y) coordinates. Mutation is accomplished by adding a Gaussian random variable with zero mean and a fixed standard deviation of 0.01 to the x- and y-coordinates of the parent. After the creation of 100 offspring, we choose 100 best solutions from among the parents and offspring to be parents for the next generation. Figure 10.7 shows the improvement in the score of the best surviving solution at each generation. Figure 10.8 shows the distribution of the 100 surviving solutions at generation 100. After generation 100, you'll note that the solutions are distributed from -0.005 to 0.005 in each dimension, confined in a box that's 0.01 units on each side. A box that is 0.01 on a side has an area that is 2.5×10^{-7} of the area of the initial box we started with, which was 20 units on a side. So, one way of judging the improvement that the evolutionary algorithm offers in this case is that that narrows the area of the best answer by a factor of over 1 million in 100 generations.[8]

---

[7] The nomenclature of μ parents and λ offspring comes from decades of use in one subset of evolutionary computation called *evolution strategies*. It is now commonplace to use this notation in evolutionary algorithms broadly.

[8] There are more efficient ways to construct an evolutionary algorithm for this problem. There are many more efficient algorithms that are constructed exactly for this problem, such as Newton–Gauss. The example here is intended only to illustrate a simple case.

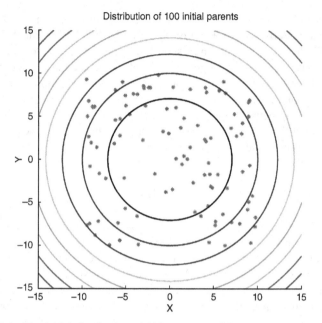

FIGURE 10.6   The initial distribution of 100 parent candidate solutions to the problem of minimizing $f(x, y) = x^2 + y^2$. Solutions are chosen uniformly at random in a box from -10 to 10 in x and y. The contours of the function depict lower error as they approach the origin.

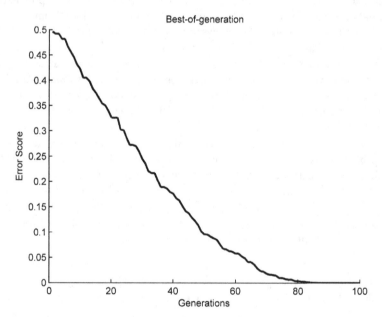

FIGURE 10.7   The rate of optimization of the best solution as a function of the number of generations. The rate slows after about the 50th generation as the population approaches the origin. The "error score" is $f(x, y) = x^2 + y^2$.

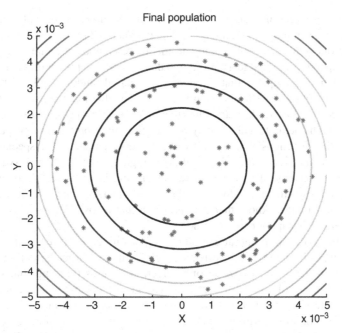

FIGURE 10.8   The distribution of surviving solutions at generation 100. All solutions are in the range from −0.005 to 0.005 in x and y. This area is 2.5×10^{-7} of the initial area, which is one way of assessing the improvement that the evolutionary approach was able to offer over 100 generations.

10.3   REPRESENTATION, SEARCH, AND SELECTION OPERATORS

Evolutionary algorithms are often viewed in terms of the data structures that are used to represent solutions, the search operators that are applied to those data structures, and the selection operators that determine which solutions in a population will influence the creation of the next generation.

With some mathematical contortionism, it's possible to encode solutions into representations that are completely unintuitive and difficult to work with. For example, in the simple case above in which the goal was to minimize the real-valued function $f(x, y) = x^2 + y^2$, the most natural representation for a solution (x, y) is a pair of real values. It's the most natural because it allows you to think intuitively about the problem with an image in mind that corresponds well to the shape of the function.

Alternatively, it's possible to encode any real value into, say, some number of binary digits to a given degree of precision. For example, we could encode the integers −7 to 7 with four digits, where the first digit is 1 if the number is positive and 0 otherwise, and the remaining three digits are the binary representation of the integer. Thus, [0 1 1 1] would be the number −7. With very long bit strings, we could essentially do what computers do in representing real-valued numbers with lots of precision, but it wouldn't be very intuitive.

Choosing a representation for a problem must be done in light of what search operators (i.e., variation operators) are going to be applied to that representation. It's infeasible, or certainly not optimal, to consider representation separately from search. Variation operators applied to a current population modify a probability mass/density function over the sample **S** that provides for a biased random walk in that space, searching for solutions of better quality. It's not possible to decouple the effects of search operators from the data structures that they are operating on. Some of the more common search operators include *mutation* functions that apply to a single parent, and *recombination* functions that apply to two or more parents. One special case of recombination is *crossover*, which takes contiguous segments and swaps them between two parents. The choice and effects of different search operators in alternative cases will be explored further in Chapters 11 and 13; however, for now it is important to know that it is provable that across all problems no single search operator is superior to any other.

The effects of search operators cannot be partitioned from representation, nor can they be partitioned from the effects of selection operations. Selection can vary in effect from *strong* to *weak*. Strong selection ensures that only the very best solutions in a population will serve as parents for the next generation. Weak selection offers nonzero probability that weaker solutions may also serve as parents. One common form of weak selection is called *proportional selection*, in which parents are chosen with probabilities that are proportional to their relative fitness (under the assumption of strictly positive fitness values). Strong and weak selection may also be used to describe the percentage of a population that is retained as parents. For example, in the case illustrated earlier, if $\mu \gg \lambda$, then this would be weak selection, whereas if $\mu \ll \lambda$, then this would be strong selection. Certain heuristics are sometimes applied to ensure that the best solution in a population is not lost under weak selection. These heuristics are generally described as *elitist*.

Search and selection are intertwined in effects. A very narrow search can be made effectively broader by using weak selection. This would allow subpar solutions to become parents, thus expanding the application of the narrow search operation. Similarly, a broad search can be made effectively narrower to some extent by using strong selection. Some mathematical relationships between search and selection are described in Chapter 12.

It may be helpful to view evolutionary algorithms in the form of an update equation:

$$\mathbf{r}[t + 1] = k(v(\mathbf{r}[t]))$$

where $k(\cdot)$ is the selection operation, $v(\cdot)$ is the variation operation(s), $\mathbf{r}[t]$ is the population at time t under representation $\mathbf{r}$, and there is an implicit generation of $\mathbf{r}[0]$, for example, starting by sampling uniformly at random across all possible solutions. The use of bold notation for $\mathbf{r}$ emphasizes that there is a population of multiple solutions forming a vector. Formally, r is a transformation on $s \in \mathbf{S}$, such that r(s) is a solution s being represented as r(s), or r for convenience.

This update equation applies to all search algorithms, including exhaustive search, blind random search, hill-climbing, and evolutionary algorithms. But when applied to describe the behavior of evolutionary algorithms, it illustrates that the algorithm designer has certain choices to make about initialization, representation, variation, and selection. Variation is random, and often so is selection. Thus, the equation is a stochastic update equation that describes the probability of sampling from different locations in **S** at each generation t. To a certain extent, design choices for any one of these operators can be compensated by choices made in implementing the others. For a given problem, suitable choices will yield greater probabilities for more quickly locating solutions of interest. Across all problems, there are no choices that will be uniformly superior [Wolpert and Macready, 1997].

10.4 MAJOR RESEARCH AND APPLICATION AREAS

10.4.1 Optimization

The most common application of evolutionary algorithms comes in optimization. The problems addressed span numerical and combinatorial optimization, with representation and operators tailored to the specific problems. Many real-world problems also pose constraints, which must be addressed to generate feasible solutions. They also may pose multiple criteria, and thus require either a tailored function to aggregate the degrees of achievement across the criteria into a single overall value or a mechanism for trading off more success in one criterion against less success in others.

There are literally thousands of published applications in the literature of evolutionary algorithms. These applications range in diversity, including biomedical pattern classification, industrial scheduling, financial forecasting, video game character control, and many others. Evolutionary algorithms can often be hybridized easily with other approaches and can also accommodate domain-specific knowledge that may be available. Specific attention to evolutionary algorithms for optimization will be given in Chapter 11 and with extensions to games in Chapter 12.

10.4.2 Design

Design is often connected intrinsically to optimization. Suppose there is an electronic circuit that must be designed to fulfill a particular function. If the degree to which it performs the function can be written objectively in mathematical terms, evolving a circuit to meet the specified needs is a problem of optimization. Sometimes, however, it is the case that no clear knowledge of an objective function is available to measure the worth of potential solutions to a problem. In this case, real-world *in situ* experimentation can provide a means for evolving solutions. For example, suppose the task was to evolve a hearing aid for a certain person. Only the person in question can truly provide guidance on whether or not any audio transformation being applied by a hearing aid is assisting in providing greater clarity of sound. Having a human in the loop providing guidance for selection is termed *interactive evolutionary computation*. Some further examples will also be provided in Chapter 13.

10.4.3 Learning and Games

Learning often refers to the case of adjusting strategies for accomplishing a task based on feedback about the quality of performance being generated. Classic examples involve learning mapping functions to classify patterns, for instance using neural networks to classify cases of malignancy in mammograms [Fogel et al., 1998]. In this case, evolutionary algorithms can be used to adjust the weights between nodes in a neural network as well as the topology of the network simultaneously. When available data in a learning application are static, the task is essentially another form of direct optimization: adjusting available parameters to maximize or minimize some objective function (e.g., minimizing the squared error between the output of a neural network and the desired target value for each pattern).

This situation becomes intricate when the task requires learning in the face of dynamic data, such as controlling a nonlinear system. For example, the famous cart–pole system (see Section 12.5.1) presents a dynamic learning problem of having to prevent a pole connected to a moving cart from falling over. The problem can be made very challenging by adding additional poles [Wieland, 1991a, 1991b].

Learning strategies in control systems can be made even more demanding when the system being controlled is itself purposeful. This is the essence of *gaming*, in which two or more players allocate resources in order to achieve desired objectives. These objectives may be antithetical, such as in checkers or chess. When one player wins, the other necessarily loses. In other cases, however, the objectives may seem to offer the possibility for mutual cooperation, as in a game called the iterated prisoner's dilemma. Evolutionary algorithms can be used to model each player in games such as these. This concept is explored further in Chapter 12.

10.4.4 Theory

As with all mathematical algorithms, it is of interest to analyze and evaluate evolutionary algorithms in terms of their efficiency, rates of convergence, quality of evolved solution, computational requirements, and so forth. Mathematical proofs regarding the properties of algorithms are often more compelling than mere empirical evidence. But it may be very difficult or even impossible to formulate purely theoretical analyses of evolutionary algorithms on problems of interest. Indeed, often the problems of greatest interest are the ones for which mathematical tools are lacking. Through a combined effort of analysis and experiment, it's possible to gain insight into some properties of evolutionary algorithms that appear important for designing efficient searches on specific problems. It's also possible to gain insight to reject some conventional wisdom that has been proven either incorrect or unhelpful, thus saving wasted time and effort. Basic theoretical considerations, along with empirical supplementation, are offered in Chapter 12, with a perspective of helping the practitioner gain best his or her own intuition about how evolutionary algorithms work and how best to apply them in problem solving.

10.5 SUMMARY

Evolution provides inspiration for solving complex problems. The problems that living organisms address every day are routinely much more complicated than the problems found in textbooks. The fly that clings upside down to the ceiling must not only avoid being eaten but also find food and water (often the water is found in food, like fruit), and importantly it must maintain viability as it develops from a larva to a pupa to a fly, and then it must make other viable flies, otherwise there will be no more flies. The tremendous complexity of life is enormous compared to a problem like classifying which song you are listening to from a short snippet (as with the Smartphone "app" Shazam®) or identifying which species of iris a flower belongs to based on the length and width of sepals and petals [Fisher, 1936].

Sometimes our own problems can seem trivial by comparison to what nature is solving. However, solving these problems can help make our own lives more enjoyable. Problems can sometimes be addressed effectively by standard mathematical techniques, such as gradient-based optimization; however, it's often the case that nonlinear aspects of problems and their solutions (for instance, in the case of neural networks or fuzzy systems) render traditional problem-solving approaches inadequate. In these cases, evolutionary algorithms—simplified models of nature's process of evolution by random variation and selection—may offer a robust alternative.

Evolutionary algorithms have been explored for over six decades, dating back to some of the earliest applications of "modern" computing in the 1950s [Fogel, 1988]. These algorithms have been used to discover solutions to problems that pose multiple local optima, in noisy conditions, that vary over time, and in light of objectives that are not associated with typical least-squares minimization or linear constraint satisfaction. The field of evolutionary computation has a rich history of application and theoretical development. The following chapters are aimed to assist you as a problem solver better understand and apply these techniques.

EXERCISES

10.1. Living organisms are a duality of their *genotype* and their *phenotype*, expressed in a given environment. What do these terms refer to? Are there any examples of nonliving systems that have a similar duality of a genotype and a phenotype?

10.2. What is meant by the terms *pleiotropy* and *polygeny*? If one of your genes is altered, say, by a simple mutation, could this change more than one of your behavioral traits? Why or why not?

10.3. Evolution can be viewed as an optimization process, but it is important not to confuse optimization with perfection. Some describe evolution as a "satisficing" process, in which solutions have to be "just good enough." Why does evolution fail to generate perfect solutions to the problems organisms face?

10.4. Despite the inability for evolution to create perfect solutions, it can often engineer creations that are pretty stunning, if you can see them. One strategy for survival mentioned in the chapter deals with warning signs (aposematism), such as those found on honey bees or their mimics. Another strategy is to not call attention to yourself, to hide in the background. This is called *crypsis*. Have a look at the lizard found at http://en.wikipedia.org/wiki/Crypsis in the upper-right photograph. (Have a look, that is, if you can find it at all.) Explore other cases of crypsis and suggest which examples are most impressive to you.

10.5. Evolutionary algorithms differ from traditional search algorithms in several ways. They rely on a population of solutions instead of just one to form a basis for searching via random variation and selection. Explain intuitively how variation, selection, and representation are connected.

10.6. In the example provided in this chapter, an evolutionary algorithm was described to find a minimum of $f(x, y) = x^2 + y^2$ by adjusting values of the solutions based on Gaussian variation. Consider some other forms of variation, such as by adding a uniform random variable or a Cauchy random variable? What is your intuition about the advantages and disadvantages of these random variations relative to a Gaussian variation. Recall that a Gaussian distribution has the shape of a bell curve. A uniform distribution is constant across a defined minimum to maximum range. A Cauchy distribution appears a lot like a Gaussian distribution but has fatter tails and infinite variance. (You can create a Cauchy random variable by taking the ratio of two standard Gaussian random variables.)

10.7. A French biologist and the Nobel Prize winner, Francois Jacob (1920–2013) wrote in 1977 that evolution works more like a tinkerer than an engineer [Jacob, 1977]. (Jacob's paper is easily accessible online.) Describe what Jacob meant by this remark. How does the evolutionary process of problem solving differ from that of a human engineer solving a problem?

CHAPTER 11

Evolutionary Optimization

Since early in the history of evolutionary computation, simulated evolution has been used as a search mechanism to find optimal solutions to problems of interest. In the 1950s, some used evolutionary methods to design computer programs that would control a robot [Friedman, 1956] or craft a simple computer program [Friedberg *et al.*, 1958]. In the 1960s, others used evolutionary methods to design physical devices [Rechenberg, 1965] or create predictive models of time series data [Fogel *et al.*, 1966]. Still others sought evolutionary methods for generating strategies in games [Reed *et al.*, 1967; Fogel and Burgin, 1969]. The examples here are meant only to be illustrative and not exhaustive. There is a rich and interesting history to the field of evolutionary computation, as described in *Evolutionary Computation: The Fossil Record* [Fogel, 1998]; however, in virtually all of these cases, evolution was used as a foundation for searching a space of possible solutions to a problem using methods of random variation and selection.

In speaking about evolutionary algorithms used for optimization, or in other words *evolutionary optimization*, it is important to recognize that a problem to be solved must be well defined. This means that any possible solution to the problem must be comparable to another possible solution. It is not enough to identify merely the most desired outcome as "the solution." Often in the real world, we are not presented with sufficient resources or there are other constraints that prevent us from obtaining that "golden future." For a problem to be well defined, at least in the sense that is meant here, any possible solution to the problem must be assessable relative to all others.

Most often, the comparisons between two or more candidate solutions are based on quantitative measures of how well a proposed solution meets the needs of the problem. For example, you may have a problem of finding a minimum path from some point to another point separated by obstacles. Any proposed path that avoids the obstacles and gets from the first point to the other point has a certain distance. That distance provides a quantitative measure of the quality of the path. In this case, as in golf, smaller numbers are better. Other cases might be designing a schedule for a factory that maximizes profitability, or finding a new potential drug that binds to a specific protein with minimum free energy.

Fundamentals of Computational Intelligence: Neural Networks, Fuzzy Systems, and Evolutionary Computation, First Edition. James M. Keller, Derong Liu, and David B. Fogel.
© 2016 by The Institute of Electrical and Electronics Engineers, Inc. Published 2016 by John Wiley & Sons, Inc.

Evolutionary algorithms are quite flexible, however, and it is not always the case that purely quantitative measures of a solution's "fitness" are required. For example, suppose that you may want to find a blend of coffee that you find most enjoyable from a mixture of five different types of coffee beans. You could try random combinations of blends and seek to evolve the overall blend that you like best. You could rate each blend on a numeric scale, say from 0 to 10 with 10 being the best coffee you've ever tasted. Then you could make a quantitative comparison between blends. (There's a problem here though because the best coffee that you've ever tasted might be surpassed by a new blend that is the new best coffee that you've ever tasted, and by our scoring system, each would receive a score of 10, unless you retaste and rescore coffees at each generation.) But instead you could simply rank the blends in order of preference, or in the most primitive case simply assert that you like one blend more than another. (See Herdy [1997] for an example of evolving coffee blends.)

The use of qualitative or even fuzzy descriptors of measures of fitness are most often found in what's called *interactive evolutionary computation*, in which a human provides a judgment about the quality of proposed solutions. Putting a human in the loop of assessing the merits of alternative solutions is typically slow relative to evolutionary algorithms that can compute a quantitative measure of a solution's quality. Nevertheless, there are certain applications that require a human's assessment, such as judging which type of music or art is preferred [Takagi, 2001], or the quality of a hearing aid or tinnitus masking device [Fogel, 2008].

This chapter focuses on quantitative evolutionary optimization, in which there is a numeric description—a function—that operates on a potential solution and returns either a single real number or multiple numbers that describe the value of the solution. Within evolutionary optimization, as with all engineering, there are essentially two forms of optimization problem.

One form is numeric. For example, find the point (x, y) such that $f(x, y) = x^2 + y^2$ is minimized, which we saw in Chapter 10. Here, the solution space is $\Re^2$ and $f(x, y)$ can be used as a measure of solution quality (lower is better because it's a minimization problem).

The other form is combinatoric. For example, given a collection of tools, each with a certain weight, find the combination that has the greatest combined weight that will fit in a bag that can hold only 25% of the weight of all the tools combined. This is called a *knapsack problem*. In this case, you aren't searching for a point in $\Re^n$, but rather for a combination of items that can be listed. In the case here, the order of the listing doesn't matter; the tools can go in the bag in any order. In other problems, the order of presentation of the items makes a big difference. For example, think about optimizing the arrival of supplies at a construction site.

This chapter provides an introduction to both numeric and combinatorial evolutionary optimization. It also describes some of the mathematical properties of representation and selection operators, and of evolutionary algorithms broadly. Some important extensions of the basic application of evolutionary algorithms for optimization are also covered, including handling constraints and allowing the evolutionary algorithm to learn how to optimize its own search parameters in a process called *self-adaptation*.

11.1 GLOBAL NUMERICAL OPTIMIZATION

11.1.1 A Canonical Example in One Dimension

In the previous chapter, we saw an example that used an evolutionary algorithm to address a simple problem in two dimensions. Let's briefly consider something even simpler first: the case of searching for a point $x \in \Re$, such that $f(x) = x^2$ is minimized. Since the function $f(x)$ is quadratic, finding the minimum by calculus methods is straightforward.[1] But suppose we didn't know that the function $f(x)$ was actually $f(x) = x^2$. Suppose it was just a "black box" that responded with a number any time we put a number in the box. We put the number 4 in the box and the box says 16. We put the number −4 in the box and again the box says 16. We need to find a number that minimizes what the box generates. This is called *black box optimization*.

An evolutionary approach to finding the minimum number could be as follows. Suppose we form a population of candidate solutions, $x_1, \ldots, x_\mu$, where there are μ "parents." We select these parents at random from a portion of real numbers, say uniformly between a lower limit of −100 and an upper limit of +100. In pseudocode, this would be as follows:

```
i = 0;
repeat
    i = i + 1;
    x[i]= U(-100,100);
until (i == μ);
```

where $U(-100, 100)$ is a uniformly distributed random variable over the interval $(-100, 100)$.

Each of these parents is varied randomly to create more solutions, $x_{\mu+1}, \ldots, x_{\mu+\lambda}$, where there are λ "offspring."[2] We could conduct this random variation in many ways, but one typical method is to add a random number from a standard Gaussian distribution (i.e., a mean of zero and standard deviation of 1) to a parent to create an offspring.

For the sake of simplifying notation, assume that in this case $\lambda = \mu$ and thus each parent creates one offspring (although there is no limitation in evolutionary algorithms about the number of offspring that can be created from a parent). Then, in pseudocode, the process of creating offspring from parents would be as follows:

---

[1] Take the derivative $f'(x) = 2x$ and set it equal to zero and solve for x ($x = 0$). Then take the second derivative $f''(x) = 2$ and note that it is positive, thus the point $x = 0$ is a minimum of $f(x)$.

[2] The notation of μ parents and λ offspring is standard nomenclature in evolutionary algorithms and developed within the offshoot of evolutionary computation known as evolution strategies that emerged in Germany in the 1960s. It's important not to be confused by the use of μ, which has a long history in statistics of describing the mean of a population or a random variable.

```
i = 0;
repeat
    i = i + 1;
    x[μ + 1] = x[i] + N(0,1);
until (i == μ);
```

where $N(0, 1)$ denotes a standard Gaussian random variable (also known as a "standard normal").

At this point, we have 2μ random ideas about what to put in the black box. We now have to test each idea and see what the box says. In pseudocode, this would be as follows:

```
i = 0;
repeat
    i = i + 1;
    score_x[i] = f(x[i]);
until (i == 2μ);
```

We then rank order the 2μ solutions in terms of their scores from lowest to highest. The μ best-ranking solutions then become the new parents for the next generation. In pseudocode, the process is as follows:

```
InitializePopulation;
repeat
    CreateOffspring;
    ScoreEveryone;
    SelectNewParents;
until (done);
```

The question of when to halt this procedure is often a matter of how much time is available to compute a solution or what level of quality is required. Here, we might continue the process of variation and selection for some number of generations g, or until the value of the best solution is lower than a threshold, say, 10^{-6}. This threshold would work in our problem if we had a hunch that zero was the minimum (which it is for $f(x) = x^2$).[3] If you construct a simple evolutionary algorithm for this problem, you'll find that it quickly locates solutions that have a score of less than 10^{-6}, but not as quickly as some other search methods, such as bisection or gradient search, and certainly not as quickly as calculus if we knew the function inside the black box ahead of time.

11.1.2   A Canonical Example in Two or More Dimensions

Having reviewed the canonical example in a single dimension, it is easy to extend it to the canonical case of two (or more dimensions), as we did in the example in

---

[3] Alternatively, a stopping rule could be to halt when improvement from one generation to the next is below a threshold, or improvement over several generations is below a threshold.

Chapter 10. Parents are chosen from the space $\mathfrak{R}^n$, where n is the number of dimensions. Offspring can be created by randomly varying each dimension of the parent, and also by combining or averaging across parents. Then, all the solutions are assessed and the best are retained to be parents of the next generation.

With regard to creating offspring from parents, traditionally, methods that use a single parent to create a single offspring are described under the heading of *mutation*, whereas methods that seek to combine multiple parents to create offspring are described with the term *recombination*. There are various forms of recombination that have their origins in inspiration from nature.

One method of recombination is called *crossover*. This operates on the following two solutions:

$$x_{11}, x_{12}, \ldots, x_{1n}$$

$$x_{21}, x_{22}, \ldots, x_{2n}$$

where x_{12} denotes the second parameter of the first solution and n is the number of parameters (dimensions). A crossover point is selected, usually at random, and two new solutions are created by splicing the first part of the first solution with the second part of the second solution, and vice versa. For example, suppose the crossover point was 3, then the two offspring would be

$$x_{11}, x_{12}, x_{23}, \ldots, x_{2n}$$

$$x_{21}, x_{22}, x_{13}, \ldots, x_{1n}$$

This "one-point" crossover operator has the sometimes undesirable property of forcing segments that are near each end of the solution vector to remain together. Thus, a multipoint[4] crossover operator can be employed, which treats the solution vectors more like rings in which sections can be exchanged, rather than strings in which a transition is made from one to the other. The limiting form of this, called *uniform crossover* [Fraser and Burnell, 1970; Syswerda, 1991], selects one component from either parent at random without regard to maintaining continuous segments and exchanges them.

Another form of recombination is *blending*. This averages parameters of parent solutions when creating offspring. For example, the two parents

$$x_{11}, x_{12}, \ldots, x_{1n}$$

$$x_{21}, x_{22}, \ldots, x_{2n}$$

could create

$$(x_{11} + x_{21})/2, (x_{12} + x_{22})/2, \ldots, (x_{1n} + x_{2n})/2$$

In general, there is no need to use a simple arithmetic mean; a weighted arithmetic mean or even a geometric mean may be useful in certain circumstances.

---

[4] This is called "n-point" in evolutionary algorithm literature, but here n refers to the number of crossing points not the dimension of the problem.

Recombination, in terms of both crossover and blending, can be extended to more than two parents. There are no restrictions that evolutionary operators follow the mechanics of the operations as found in nature.

It's almost always the case that evolutionary algorithms are employed when an optimization problem has multiple dimensions rather than just one dimension. In pseudocode, the process can be described as

```
InitializePopulation;
repeat

    CreateOffspring;       //mutate and/or recombine
    ScoreEveryone;
    SelectNewParents;

until (done);
```

11.1.3 Evolution versus Gradient Methods

If the problem at hand presents a smooth, convex, continuous landscape (e.g., $f(x, y) = x^2 + y^2$), then gradient or related methods of optimization will be faster in locating the single optimum point. On the other hand, if the problem presents a landscape with multiple local optima (e.g., $f(\mathbf{x}) = An + \sum_{i=1,\dots,n} [x_i^2 - A \cos(2\pi x_i)]$, where A is a constant, see Figure 11.1), then gradient methods will likely fail to find the global

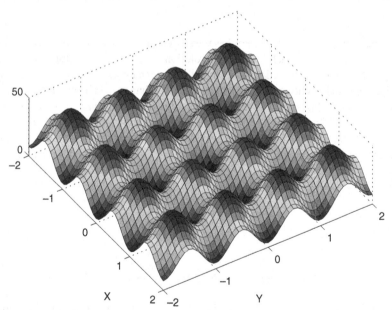

FIGURE 11.1   A multimodal surface $f(\mathbf{x}) = 20 + \sum_{i=1,\dots,10} [x_i^2 - 10 \cos(2\pi x_i)]$, known as Rastrigin's function with n = 2 and A = 10 (see text). This function is sometimes used as a test case for evolutionary algorithms and other techniques to assess the challenge of overcoming local optima and saddle points.

optimum solution because each gradient is associated only with the local optima created by the effects of the cosine function.[5] This may present an opportunity for evolutionary optimization to be used effectively. If the landscape is discontinuous and/or not smooth, then gradient-based approaches may be inapplicable—it may not be possible to compute gradients—and thus other "generate and test" methods such as evolutionary algorithms again may present an opportunity for addressing the problem successfully.

11.2 COMBINATORIAL OPTIMIZATION

To illustrate the use of evolutionary algorithms for combinatorial optimization, let's consider the canonical case of addressing the traveling salesman problem. The problem is as follows. There are n cities. The salesman starts at one of these cities and must visit each other city once and only once and then return home. The salesman wants to do this in the shortest distance. The problem then is to determine the best ordering of the cities to visit. This is a difficult problem because the total number of possible solutions increases as a factorial function of the number of cities. More precisely, for n cities, the total number of possible solutions is $(n - 1)!/2$. For a small problem such as $n = 10$, there are 181,440 different paths to choose from. For $n = 100$, the number of different paths is on the order of 10^{150}. Figure 11.2 shows the relationship between the number of cities and the number of possible tours. By way of comparison, as mentioned earlier, there are about 10^{18} s in the approximately 13 billion year history of the universe. So, when there are so many possibilities to choose from, enumerating them all and determining the best one is infeasible.

The traveling salesman problem is NP-hard. There are no known methods for generating solutions in a time that scales as a polynomial function of the number of cities, n. There are heuristics that are available for this canonical form of the traveling salesman problem, and some of these can be effective even for large n given certain cases of the general problem but for the time being let's focus on an evolutionary approach to the problem.

First we must create a data structure to represent a solution. One such structure is a list of cities to be visited in order. For example, given the seven cities, a potential solution is

$$[1, 3, 2, 5, 7, 6, 4]$$

in which the salesman is presumed to start at city 1 and return to city 1 after visiting city 4. Other data structures are also possible. For example, since the salesman's circuit is a series of links between cities (more formally *edges* on a *graph*), the representation could be in the form of a series of links, such as

$$[(1, 3), (3, 2), (2, 5), (5, 7), (7, 6), (6, 4), (4, 1)]$$

---

[5] This function is known as Rastrigin's function and was introduced in evolutionary algorithm research many decades ago as a test case to determine how effective evolutionary methods could be on functions that present many local optima. At the time of this writing, there is a nice illustration of Rastrigin's function at en.wikipedia.org/wiki/Rastrigin_function.

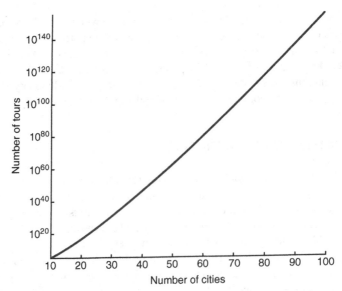

FIGURE 11.2 The number of possible tours in a traveling salesman problem as a function of the number of cities. There are $(n-1)!/2$ possible tours for n cities. The formula is derived noting that whenever you start, there are $(n-1)$ possible next cities, then $(n-2)$, then $(n-3)$, and so forth; however, it doesn't matter which direction you traverse the tour, thus the $(n-1)!$ options can be divided by 2. Problems with 70 cities are already close to presenting 10^{100} possible options.

Some people may find this more intuitive; however, since the ordered list of cities is quite straightforward, so for our purposes here, we'll focus on that representation.

Next, we must be able to score any potential solution. Here, the score associated with any solution, f, is the length of the total path through the cities:

$$f = \sum_{i=1}^{n-1} d(c_i, c_{i+1}) + d(c_n, c_1)$$

where $d(a, b)$ is the distance between city a and city b, and c_i is the ith city. Lower scores are better.

Then we must determine a method for creating offspring solutions from parent solutions. Given the representation of an ordered list, the following are some of the potential methods:

1. *Select and replace:* Choose a city at random along the list and replace it at a random place along the list.
2. *Invert:* Choose two cities along the list at random and invert the segment between those cities.
3. *Protect and randomize:* Choose a segment of the list to be passed from the parent to the offspring intact, and then randomize the remaining cities in the list.

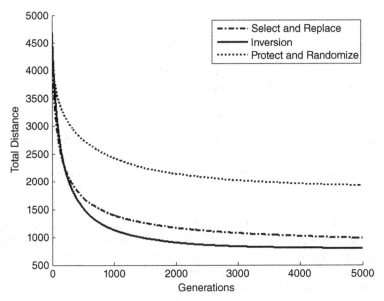

FIGURE 11.3   The best tour length for each average of 50 trials of a (50 + 50) evolutionary algorithm using permutation encoding on a 100-city traveling salesman problem as a function of the number of generations. The legend shows the results of different mutation operators: dotted = protect and randomize, dashed = select and replace, solid = invert a segment. At 5000 generations, the mean best tour lengths based on these mutation operators were 1932.6, 988.4, and 807.0 units, respectively.

Figure 11.3 shows a comparison of results for these three operators. Each curve represents the average score of the best solution in the population at the specified generation. The results suggest that the two-point inversion operator generates better solutions faster than the other two mutation operators on this problem. (We'll discuss how to use statistics to make conclusions about these sorts of comparisons in later examples.)

Let's consider as well a recombination operator we could try on the traveling salesman problem. One possibility is an operator called partially mapped crossover (PMX).[6] The operator works on two parents by choosing a segment of the first parent to move directly to the offspring. It then moves the feasible parts from the second parent to the offspring. Finally, it assigns the remaining values to the offspring based on the indexing in the first parent. Here's an example.

Suppose we have two parents

$$[3, 5, 1, 2, 7, 6, 8, 4]$$

---

[6] PMX is also called partially matched crossover but, as introduced in Goldberg and Lingle [1985], it was described as partially mapped crossover.

and

$$[1, 8, 5, 4, 3, 6, 2, 7]$$

and suppose that the segment [1, 2, 7] is selected to be saved from parent 1. Then, at this step the offspring could be written as

$$[+, +, 1, 2, 7, +, +, +]$$

where the + symbol in a placeholder for an undetermined component of the solution. Next, the feasible elements of parent 2 are copied to the offspring. These are the values that do not already appear in the offspring. So, the offspring becomes

$$[+, 8, 1, 2, 7, 6, +, +]$$

Finally, we look at the first position, which was a 1 in parent 2 but we can't copy that because 1 already appears in the offspring. So, we look at where 1 appears in parent 1 and see that the corresponding value for parent 2 in that position is 5. So, 5 goes in the first place in the offspring. Similarly, 4 goes in the seventh place and 3 goes in the eighth place, and we end up with

$$[5, 8, 1, 2, 7, 6, 4, 3]$$

Figure 11.4 shows the average results on 50 trials of applying the PMX operator on the same 100-city traveling salesman problem instead of the other mutation operators. It also shows the results of combining PMX with inversion, with the probability of having an inversion set at 0.5 per offspring. The results show that the PMX operator

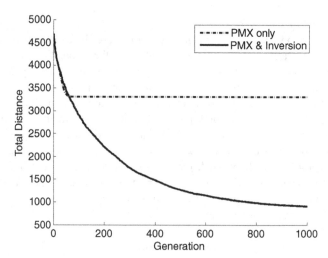

FIGURE 11.4   The average of 50 trials with the same 100-city traveling salesman problem as studied previously comparing the best tour as a function of the number of generations for PMX and PMX plus a 0.5 probability of applying inversion. On average, PMX alone stalls at a relatively poor solution. By including inversion, the search for improved solutions can continue.

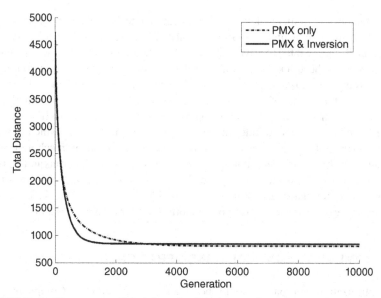

FIGURE 11.5 The average of 20 trials on the 100-city traveling salesman problem comparing PMX + inversion (0.5 probability/offspring) with inversion alone. The combination of PMX and inversion offers more rapid improvement on average for about the first 3000 generations. After that, the results suggest relying on inversion alone.

alone stalls at a mean tour length that's just less than 3000 units. It does this because selection has eliminated the variation in the population and the PMX operator cannot generate anything new. It can only recombine existing solutions. Once those solutions look the same, PMX (or any standard crossover operator) becomes incapable of searching further. The results show that adding in the possibility of mutating offspring via inversion allows the evolutionary search to proceed toward improved solutions.

It's natural to want to compare the results of the PMX + inversion approach with inversion alone to see if PMX is offering any benefit to the search. Figure 11.5 shows the results of that comparison through 10,000 generations. Up to generation 3000, the combination of PMX and inversion generated faster improvement than inversion alone, on average. After the 3000th generation, inversion alone did better on average at fine-tuning toward better solutions.

At this point, if you are thinking that perhaps it would make sense to use different variation operators at different times during an evolutionary search, rather than rely only on a fixed set of operators in a certain framework (e.g., PMX + inversion at 0.5 probability/offspring), that's great! We'll cover more on the self-adaptation of variation operators later in this chapter.

Returning to the broader concept of combinatorial optimization problems, there is a wide variety of these problems that can be addressed using evolutionary algorithms. Some of these are extensions of canonical problems, such as the traveling salesman problem. For example, consider the problem of optimizing the routing of delivery trucks for a major logistics company. Each truck must be loaded optimally in order to provide each driver the opportunity to deliver the products on time. Not all customers are equally important; some deserve higher priority. Not all drivers are capable of handling the same equipment: Some are rated to drive larger vehicles. Determining the best way to allocate the materials to the trucks, assigning the drivers to their vehicles, and routing the vehicles to their destination present a complicated real-world problem.

There are other forms of combinatorial problems that are related to optimal subset selection problems. These are common in mathematical regression, in which you must explain observed data using independent variables but must decide which subset of variables provide the most explanatory power. They arise in many machine learning applications in the form of determining the appropriate number of neurons and topology for a neural network, or the number and shape of membership functions in a fuzzy control system. Similarly, data structures of variable size such as finite-state automata and symbolic expression trees also pose opportunities for combinatorial optimization.

11.3 SOME MATHEMATICAL CONSIDERATIONS

Any optimization approach should be amenable to analysis. Computers have increased in speed so dramatically that much of the analysis that used to be performed analytically (i.e., by mathematical theorem and proof) can be done effectively by empirical observation. Put another way, it's often possible to arrive at a good understanding of the mathematical properties of an approach through statistical estimation based on repeated sampling of a given procedure on a particular problem of interest. This approach is limited, however, to that particular set of observations and it can be difficult to generalize those results to other problems.

Early in the development of evolutionary computing techniques, there were broad beliefs that certain choices would offer superior performance generally. For example, one such belief was that binary representation would offer an intrinsic advantage over other representations regardless of the problem and that a selection method called *proportional selection* would provide an optimal way to create offspring in the search for improved optimization performance [Holland, 1975]. These and other specifics will be addressed in more detail in this section; however, it's important to note that mathematical analysis has shown that there truly is no single best approach to computational (i.e., computer-based) problem solving generally [Wolpert and Mac-ready, 1997; see also Corne and Knowles [2003] and DuenezGuzman and Vose [2013]. Extending this "no free lunch" principle, it's important to note that there is always a challenge of generalizing the results of an evolutionary algorithm (or any algorithm) on a particular problem to other problems, even though they may appear to be closely related. The onus is on the researchers to effectively make the case if they want to claim a generalized result via empirical observations.

11.3.1 Convergence

Prior to the fast modern-day PC computers, extensive efforts were made to determine the mathematical properties of evolutionary algorithms both broadly and on specific function optimization problems of interest. These efforts continue in order to more precisely describe the behavior of evolutionary algorithms in certain cases; however, for a basic understanding of general properties of evolutionary algorithms, it is important to review fundamental issues regarding their convergence properties.

11.3.1.1 *Convergence with Probability 1*   Evolutionary algorithms can be constructed in the framework of a Markov chain, which provides the mathematical tools to show that these algorithms can converge to an optimum solution with probability 1. Relevant publications can be found in Eiben *et al.* [1991]; Fogel [1992b]; Davis and Principe [1993]; and Rudolph [1994], and one accessible proof is found in Fogel [2006]. Rather than repeat the proof here, the essence of it is to consider different configurations of an evolving population to be states in a Markov chain.

If you don't know what a Markov chain is, it's a stochastic process, which is a fancy way of saying that it's a time-indexed sequence of random variables. The Markov chain is defined over a set of states (e.g., "awake" and "asleep"). The probability of transitioning from one state to another is time invariant and depends only upon the current state. For example, here's a Markov chain for "Awake and Asleep."

$$
\begin{array}{c c}
 & \text{Awake}[t+1] \quad \text{Asleep}[t+1] \\
\begin{array}{c} \text{Awake}[t] \\ \text{Asleep}[t] \end{array} &
\begin{bmatrix} 0.4 & 0.6 \\ 0.8 & 0.2 \end{bmatrix}
\end{array}
$$

In this case, if you are in the state Awake at time t, then you have a 0.4 probability of staying in the state Awake and a 0.6 probability of transitioning to the state Asleep. If you are already in the state Asleep, then you have a 0.8 probability of transitioning to the state Awake, and a 0.2 probability of staying in the state Asleep.

Since the transition probabilities depend on the states in the system and don't depend on any particular time, the probability of being in a state of two time steps in the future can be found by multiplying the probability matrix (described by the letter P) by itself (squaring the matrix). Iterating from one point in time to another, say, k steps in the future, in a Markov chain is accomplished by raising P to the kth power. In our case, the basic transition matrix is written as

$$
P = \begin{bmatrix} 0.4 & 0.6 \\ 0.8 & 0.2 \end{bmatrix}
$$

and thus the two-step transition matrix is

$$
P^2 = \begin{bmatrix} 0.64 & 0.36 \\ 0.48 & 0.52 \end{bmatrix}
$$

So, if you are in the state Awake at a time t, then there is a 0.64 probability that you will be in the state Awake at time $t + 2$. It's natural to think about what state you would be in if time went to infinity. In our case,

$$P^{\infty} = \begin{bmatrix} 0.5714 & 0.4286 \\ 0.5714 & 0.4286 \end{bmatrix}$$

That is, no matter which state you are in at time t, there is (approximately) a 0.5714 probability that you will be in the state Awake as time goes to infinity, and a 0.4286 probability that you will be in the state Asleep as time goes to infinity.

We can apply this same principle to study the long-term behavior of some evolutionary algorithms in which the time history of how the population arrived at its present state is not pertinent to the population's future trajectory, and also the probabilities for transitions for one configuration to another are fixed (stationary). If an evolutionary algorithm is constructed with a form of selection called *elitist selection*, in which the absolute best solution(s) in the population is always retained into the next generation, and if the algorithm is constructed such that a mutation operation can reach any state with nonzero probability (e.g., applying a Gaussian random mutation on all parameters), then the transition matrix for the Markov chain can be written as

$$P = \begin{bmatrix} 1 & 0 \\ R & Q \end{bmatrix}$$

where P is the transition matrix, 1 is a 1×1 identity matrix describing the absorbing state that has the global best solution, R is a strictly positive (all entries > 0) $t \times 1$ submatrix, Q is a $t \times t$ transition submatrix, 0 is a $1 \times t$ matrix of zeros, and t is a positive integer based on the size of the state space. Essentially, if the population already contains the best possible solution, then it is in the state "1" and will stay there forever. If the population doesn't contain the best solution, then the submatrix R describes the probabilities of transitioning to "1" in the next step, and submatrix Q describes the probabilities of transitioning elsewhere.

This is a special case of a more general transition matrix

$$P = \begin{bmatrix} I_a & 0 \\ R & Q \end{bmatrix}$$

where I_a is an $a \times a$ identity matrix.

In the limit, as k tends to infinity,

$$\lim_{k \to \infty} P^k = \begin{bmatrix} I_a & 0 \\ (I_t - Q)^{-1}R & Q \end{bmatrix}$$

The components "0" indicate that given infinite time, there is zero probability that the chain will be in a state that is not an absorbing state, and the absorbing state(s) was defined as a state(s) that contains a global optimum due to elitist selection. So, this means that there is convergence with probability 1 to a global optimum. Special properties of the matrix entries R and Q provide for this result, implying a stronger form of convergence called *complete convergence* [Hsu and Robbins, 1947].[7]

This mathematical result is of limited utility because no one has infinite time to wait to discover a globally optimal solution; however, it may be useful to recognize that without elitist selection and without a mutation operation that can reach all possible states, this proof of convergence with probability 1 does not hold. In particular, if crossover is substituted for mutation, then the result does not guarantee convergence to a global optimum, but rather only to a homogenous state in which all solutions are identical and therefore no new solutions are possible.

11.3.1.2 *Premature Convergence*   Evolutionary algorithms that rely predominantly on crossover or other recombination methods to generate offspring can sometimes suffer from what is termed *premature convergence*, which occurs when a population becomes homogeneous at a solution that is not the global optimum (or is less than desirable). The term is often used incorrectly to describe the effect of converging to a local optimum, but the origin of the term applies directly to the case in which no further progress is likely because the population lacks diversity, which effectively stalls an evolutionary algorithm that is heavily reliant on crossover or recombination. (We saw an example of premature convergence when exploring the use of the PMX operator on the traveling salesman problem. The PMX operator became ineffective when the population became homogeneous.)

In some studied cases of evolutionary algorithms that rely heavily on crossover, the likelihood of converging prematurely to a given solution has been shown to be related to the quality of the solution (i.e., there was more likelihood of stalling at a point if that point was of higher quality) [Spears and De Jong, 1997]. The most common methods for overcoming premature convergence include restarting from a new randomized population, using heuristics to move the population to a new collection of points (e.g., by hill climbing), or redesigning the approach. Early literature in evolutionary algorithms often recommended very high rates of crossover and very low rates of mutation [Holland, 1975; Goldberg, 1989], which made premature convergence more likely. Observing repeated premature convergence in an evolutionary algorithm suggests, at least, reconsidering the variation operators that are being used and giving consideration to modifying the probabilities of applying those operators, or creating new variation operators that are better tailored to the problem.

11.3.2 Representation

Designing an evolutionary optimization algorithm requires choices of representation, selection, and variation operators. With regard to representation, as mentioned early in this chapter, for many years in the early formulations of evolutionary algorithms, there

---

[7] Rudolph [1994] showed that some nonelitist evolutionary algorithms can also converge on strictly convex functions.

was a general belief that it would be beneficial to represent solutions using binary strings.

For example, when facing an optimization problem in $\mathfrak{R}^n$, instead of treating solutions directly as a vector $[x_1, \ldots, x_n]$ where $\mathbf{x} \in \mathfrak{R}^n$, the solution would be transformed into a series of bits $[x_1, \ldots, x_k]$, where k defined the length of the bit string. The greater the degree of desired precision, the larger the value of k would need to be. The belief was that longer strings generated more opportunities for an evolutionary algorithm to explore the subspace of possible solutions via substrings, and that this would provide a greater "information flow" [Holland, 1975].[8]

Many problems do not lend themselves easily to a description in binary strings. For example, consider representing a traveling salesman problem as a series of 1's and 0's. This is anything but straightforward. Suppose there are five cities. An intuitive representation would be an ordered list of cities, such as

$$[1\,2\,3\,4\,5]$$

Encoding these in binary could be done as

$$[001010011100101]$$

with each three-bit segment corresponding to the number of a city. But this representation is not easily varied by mutation or recombination. For example, mutating the fifth bit from 1 to 0 yields

$$[001000011100101]$$

and then the second city to visit is city "zero," which doesn't exist. Similarly, crossing two such bit strings would almost certainly generate offspring that would not correspond to legal tours of the available cities.[9]

---

[8] As a historical footnote, this belief originated in the subfield of evolutionary computing known as "genetic algorithms." The core idea of this approach was to view solutions in terms of *building blocks* that can be assembled via crossover. This is of course possible in some problem constructions, but not in others. When real-valued encodings were first tried on problems in $\mathfrak{R}^n$ using this approach, it violated the "building block hypothesis" of genetic algorithms; however, placing more emphasis on mutation and less on crossover was often successful with this real-valued representation and the results were published. This led to an approach called a "real-valued genetic algorithm." That is a misnomer because the core genetic algorithm concept of building block construction within parameters does not take place in the real-valued representation on $\mathfrak{R}^n$. In this case, it's essentially equivalent to what emerged in other branches of evolutionary computation, such as *evolution strategies* [Rechenberg, 1973], *evolutionary programming* [Fogel, 1990], and see Bremermann *et al.* [1966] and others. Differentiating between evolutionary approaches (e.g., genetic algorithms versus evolution strategies) is of dubious scientific value in modern evolutionary computing.

[9] In an ingenious procedure, Grefenstette *et al.* (1985) proposed a binary encoding for the traveling salesman problem that was amenable to recombination based on the order in which cities were removed from a list. But the procedure did not ultimately provide better optimization performance than comparative methods.

Fortunately, the notion that binary representations are universally better than other representations is false. In fact, there is no "best" representation across all problems, and under some conditions there is a provable mathematical equivalence of representations of different cardinality [Fogel and Ghozeil, 1997; see also Radcliffe [1992] and Battle and Vose [1993]]. Thus, the choice of a representation for a particular problem is often a matter of which provides the greatest intuition to the practitioner as the problem solver.

Some important aspects to consider when selecting a representation include the following:

1. The representation should optimally provide immediate information about the solution itself. For example, in the traveling salesman problem, the list of cities is suggestive of the solution.

2. The representation should be amenable to variation operators that are well understood for their mathematical properties and can exhibit a gradation of change. This means that variation operators should be available to make both small changes and big changes to any given parent(s), and that the likelihood of these different-sized changes can be controlled. For example, when searching for $x \in \Re^n$ such that $f(x)$ is minimized, using a Gaussian variation operator on x allows generating offspring that are close to or far from x, and this can be controlled by changing the standard deviation in each dimension (see Section 11.5).

3. Unless the objective is to explore the utility of a novel representation, utilizing representations that have been studied and for which results have been published may allow more systematic and meaningful comparisons.

With experience, you can gain better intuition about the effects of different representations, and how they are coupled with variation operators in order to search a solution space (landscape) for successively better answers to a problem of interest.

11.3.3 Selection

Selection describes either the process of eliminating solutions from an existing population or making proportionally more offspring from certain parents. Some common forms of selection include plus/comma, proportional, tournament, and linear ranking. Each of these has different effects on the likelihood of particular individuals to survive as parents or create offspring, and thus each has conditions that favor or disfavor its utility.

11.3.3.1 Plus/Comma Selection   The notation $(\mu + \lambda)$ and (μ, λ) are now commonplace in evolutionary algorithms and refers to the two cases in which (i) μ parents create λ offspring and the best μ individuals are selected from among the $\mu + \lambda$ to be parents of the next generation, or (ii) μ parents create λ offspring and the best μ individuals are selected only from among the λ offspring to be parents for the next generation. Thus, in "plus" selection, all parents and offspring compete to be parents for the next generation, whereas in "comma" selection, the parents die each generation and a surplus of offspring must be created. Some variations of these approaches include (i) the case of $(\mu + 1)$, which is sometimes referred to as "steady-state" or "continuous" selection, and (ii) allowing parents to survive for some maximum number of generations g before being removed in the comma selection process.

11.3.3.2 Proportional Selection As the name infers, proportional selection picks parents for reproduction in proportion to their relative fitness. (The procedure is sometimes also called *roulette wheel selection.*) Thus, the procedure is constrained to maximization problems on strictly positive fitness scores. If it were desired to find the minimum of $f(x, y) = x^2 + y^2$ using proportional selection, the problem would need to be turned first into a maximization problem, such as find the maximum of $1/f(x, y)$. The probability of selecting an individual in the population for reproduction is determined as

$$p_i = f(i)/\sum_{j=1}^{\mu} f(j)$$

where p_i is the probability of selecting the ith individual, there are μ existing individuals, and $f(i)$ is the fitness of the ith individual. Rather than working directly on the fitness values, proportional selection can work on the relative ranking of solutions (thus making it applicable to minimization problems). Proportional selection is applied to individuals generally by selecting one individual for mutation, or two (or more) individuals for recombination until the population size for the next generation has been filled.

11.3.3.3 Tournament Selection There have been different forms of tournament selection in the history of evolutionary algorithms, but the more common one selects a subset of size q (often $q = 2$) from the existing population and selects the best of those q individuals to be a member of the next generation. The process is repeated until the population is filled. The process can be conducted with or without replacement, that is, individuals that are selected out of a q-tournament can be given an opportunity (or not) to be selected again in another q-tournament. As with proportional selection, tournament selection allows the possibility that solutions that are less than best can propagate into a future generation.

11.3.3.4 Linear Ranking Selection Linear ranking selection maps individuals to selection probabilities according to a prescribed formula based on the rank of the solution (see the earlier remark on proportional selection based on ranking in Section 11.3.3.2). There are many variations of this approach, but one early approach [Baker, 1985] assigned a probability to the ith ranked individual as

$$p_i = (\eta^+ - (\eta^+ - \eta^-)[i - 1]/[\mu - 1])/\mu$$

where p_i is the probability of selecting the ith individual, μ is the number of individuals in the population (sometimes described with λ, see [Bäck, 1994]), and

η^+ and η^- are user-controlled constants constrained by $1 \le \eta^+ \le 2$ and $\eta^- = 2 - \eta^+$. For example, if $\mu = 100$ and $\eta^+ = 1.5$, then $\eta^- = 0.5$, and the probability of selection of the ith-ranked solution in the population is

$$p_1 = (1.5 - (1)(0/99))/100 = 0.0015$$

$$p_2 = (1.5 - (1)(1/99))/100 = 0.00148989\ldots$$

$$p_{100} = (1.5 - (1)(100/99))/100 = 0.0048989\ldots$$

Thus, better solutions are favored over lesser solutions.

11.3.3.5 *Example* Let's examine the effects of different selection operators when combined with a simple evolutionary algorithm to search for the minimum of Rastrigin's function in two dimensions. The function is $f(x,y) = 20 + x^2 - 10\cos(2\pi x) + y^2 - 10\cos(2\pi y)$. We'll initialize a population of 30 candidate solutions uniformly at random from $[-10, 10]$ in each dimension. Variation will be accomplished by a Gaussian mutation, zero mean, with standard deviation equal to 0.25 applied to each dimension. We'll create 60 offspring from 30 parents and compare the results of a (30, 60) selection with those from roulette wheel selection and tournament selection with $q = 2$. Figure 11.6

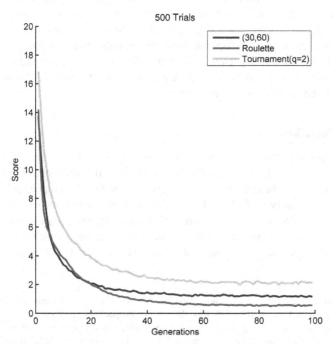

FIGURE 11.6 A comparison of (30, 60) selection, roulette wheel selection, and tournament selection with $q = 2$ on Rastrigin's function in two dimensions. The results, averaged over 500 trials, favor the use of roulette wheel selection in this case.

shows the average results of 500 trials with each method through the first 100 generations. Tournament selection with q = 2 evidenced the slowest rate of improvement, while roulette wheel selection found better solutions on average than did the (30, 60) selection method. It's important to also know the variation of the data in order to assess whether or not these differences are meaningful.

11.3.3.6 *Considerations* There are various extensions to these basic forms of selection. For example, *elitist* selection can be applied, which automatically ensures that the best solution in a population is retained in the next generation. Nonlinear ranking methods can be used [Michalewicz, 1992]. Generally, however, there is a continuum of selection methods from those that are "soft" to those that are "hard." The harder the selection, the faster the better solutions can overtake the population.

This is often described in terms of *takeover time*, which is the expected number of generations required to fill a population with copies of what is initially a single best individual when applying only selection (no variation). As shown by Bäck [1994], for typical settings of the parameters in the procedures above and on given test cases, the order of selection strength from weakest to strongest is proportional selection, linear ranking selection, tournament selection, and then plus/comma selection.

If you are using an evolutionary algorithm that relies heavily on recombination, then diversification is required for the population to search for new solutions (see premature convergence (Section 11.3.1.2)). In this case, it may be advisable to try a weaker form of selection so as to maintain more diversification; however, this may serve as a drag on progress toward an optimal solution. Stronger selection may result in faster optimization, but that optimization may be only toward a local optimum. You must weigh the trade-offs between different rates of progress and the suitability of alternative variation operators that require diversification to greater or lesser degree as those trade-offs pertain to the problem at hand.[10]

11.3.4 Variation

Variation operators provide the means for searching the solution space for improved solutions, or potentially for weaker solutions that could lead to improved solutions. (The latter comes in the case of having a weak selection method that allows less-than-best solutions to survive and be the basis of further exploration.) There are many traditional variation operators, some of which have already been discussed, such as binary mutation, Gaussian mutation, or one-point, n-point, or uniform crossovers. Although some people go into the problem with a set of variation operators in mind

---

[10] As a historical note, for many years, proportion selection was viewed as the optimal selection method based on analysis in Holland [1975]. It was claimed that proportional selection generated "minimum expected losses" when sampling from competing random distributions with different means and variances (analogous to sampling from different subspaces). This was later shown to be false by counterexample [Rudolph, 1997] and by mathematical derivation [Macready and Wolpert, 1998] (i.e., the analysis in Holland [1975] was flawed). Each person must decide which form of selection to employ based on the problem. There is no generally optimal form of selection.

and then seek to adjust the representation and selection operators, it's clear that the choice of variation operators goes "hand in glove" with the choice of representation.

11.3.4.1 *Real-Valued Variation*

When searching $\Re^n$, it is typical to use a Gaussian mutation operator. Recall that a Gaussian distribution is defined by two parameters: the mean and standard deviation (or variance). When applying a Gaussian mutation, the mean is set typically to zero, providing for an unbiased search in the neighborhood of a parent. The size of the neighborhood is defined by the standard deviation, σ; the smaller the standard deviation, the smaller the search neighborhood, although for any $\sigma > 0$, there is a nonzero probability of a Gaussian random variable returning any value between $-\infty$ and ∞. But the probability of a large move away from 0 may be very small. For example, with $\sigma = 1$, the probability of a zero-mean Gaussian random variable returning a value greater than 1 or less than -1 is approximately 0.32. For $\sigma = 0.5$, this probability is approximately 0.05. Aspects of controlling the setting of σ will be described in Section 11.5; however, note that the progress that a search based on a Gaussian random mutation operator will make is highly dependent on the value(s) of σ (in each dimension).

At times, it may be desirable to have a larger probability of creating a greater distance between an offspring and its parent. Rather than increasing the value of σ in a Gaussian mutation, an alternative is to employ a Cauchy-distributed random mutation. A Cauchy random variable is constructed by taking the ratio of two independent identically distributed Gaussian random variables. (For example, using a random number generator, create two independent Gaussian-distributed numbers and divide one by the other.)

The Cauchy distribution has the interesting mathematical properties of being symmetric and yet having no explicit mean or standard deviation (i.e., the expected value of the random variable is undefined and so are all higher moments of the distribution). The Cauchy distribution has "fatter tails" than a corresponding Gaussian distribution, thus the probability of mutating a real-valued parameter to a greater extent is markedly greater. This can be helpful for escaping from locally optimal solutions. Section 11.5 addresses approaches to trading-off between Gaussian and Cauchy mutation operators in real-valued evolutionary optimization.

11.3.4.2 *Multiparent Recombination Operators*

Earlier discussion highlighted the use of one-point, n-point, and uniform crossovers, as well as *blending* recombination that averages components of multiple individuals. Note that there is no limit to relying on two parents. Nature provides inspiration for evolutionary algorithms but it is up to the designer to find what inspires him or her. In some cases, it may be beneficial to recombine elements or blend parameters of three or more solutions. When applying blending recombination, averaging components across multiple solutions, note that the result is an offspring that essentially estimates the mean of the population. The greater the number of individuals that contribute to that blending, the more reliably the offspring will estimate the population mean (centroid). When a population is contained in a locally optimal region of the search space, a blending operator can accelerate convergence toward the local optimum. This comes at the expense of sacrificing searching outside of the local region.

11.3.4.3 *Variations on Variable-Length Structures* Certain representations employ data structures of variable length. For example, consider the case of evolving a neural network that can adapt not only its weights and bias terms but also the number of nodes it uses and the feedback loops that it employs (if any). The data structure used might well be of variable length. As another example, consider the case of evolving a collection of fuzzy membership functions that are used in a fuzzy controller. The number of fuzzy functions, as well as their shape and location, could be subject to evolutionary adaptation. When considering a representation that poses a variable-length data structure, you can consider an array of variation operators with the idea of ensuring the possibility of a thorough search of the solution space. Three other common examples involve finite-state machines, symbolic expressions, and difference equations.

Finite-State Machines Finite-state machines have long been used in evolutionary algorithms for predicting sequences of symbols (time series prediction). Fogel *et al.* [1966] employed Mealy/Moore finite-state automata for this purpose. A finite-state machine was defined by its number of states, its starting state, the input/output function for each state, and the input/state transition function for each state. Suppose that the available symbols were {0, 1}, then one finite-state machine could have the following characteristics (Figure 11.7):

(a) Two states.
(b) State 1 is the start state.
(c) In state 1, input of 0 yields 0, input of 1 yields 1; in state 2, input of 0 yields 1, input of 1 yields 0,
(d) In state 1, input of 0 transitions to state 2, input of 1 remains in state 1; in state 2, input of 0 remains in state 2, input of 1 transitions to state 1.

There are five modes of mutation that follow naturally from the description of the finite-state machine:

(a) Add a state.
(b) Delete a state.
(c) Change the start state.

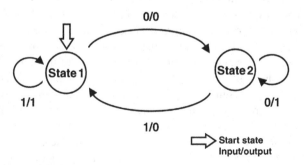

FIGURE 11.7 An example of a finite-state machine as described in the text.

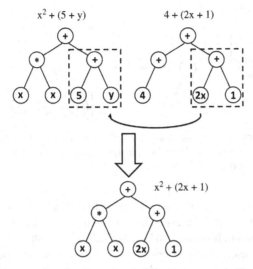

FIGURE 11.8 An example of subtree crossover on two symbolic expressions. In the example, it's desired to have a function that represents $f(x) = x^2 + 2x + 1$. This is accomplished by swapping the right branches of the trees.

(d) Change an input–output relationship.

(e) Change an input–state transition relationship.

By employing nonzero probabilities of applying each of these mutation operators, a parent finite-state machine can create a very different offspring (in terms of its input–output behavior) or one that is very similar. Note that the effects on the sequences that a finite-state machine may generate (i.e., the "behavior" of the machine) vary across these mutation operators. The effect of deleting a state may be much greater than merely adjusting the output a state generates for some particular input.[11]

Symbolic Expressions   Similarly, consider evolving a solution to a problem using a symbolic expression (s–expression).[12] The data structure is of variable length and there are opportunities to modify any terminal node and to perform subtree recombination. Suppose the problem was to find a polynomial expression that returned the value of $x^2 + 2x + 1$ for an input of x. Figure 11.8 shows an example of a subtree recombination

---

[11] In practice, adding and deleting states are limited to a maximum number of states and a minimum of one state, respectively. Also, when deleting states, any transitions that pointed to that deleted state must be redirected. When adding states, it may be desirable to also affect state transitions to ensure that the new state has the possibility of being expressed and thus changing the behavior of the finite-state machine. States that cannot be expressed are sometimes described as *introns* in analogy to sections of DNA that do not translate directly to proteins.

[12] This is often called *genetic programming* (following Koza [1992]), but there's no scientific benefit derived from giving separate names to applying variation operators to different data structures. Similarly, there's no need to describe evolutionary algorithms applied to neural networks, fuzzy systems, or other data structures with some new monikers.

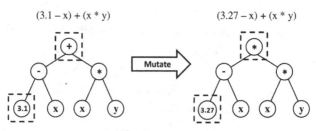

$$(3.1 - x) + (x * y) \qquad\qquad (3.27 - x) + (x * y)$$

FIGURE 11.9 Mutation can be accomplished on tree-based expressions by changing operators or numeric terms, as indicated in the figure. In addition, variables can be modified, and branches can be grown or eliminated.

that would take two suboptimal expressions and combine them to create the correct formula. Figure 11.9 shows an example of mutating specific values associated with nodes in the expression. It would also be feasible to add/construct new branches in the expression tree or delete existing ones. As with the prior example pertaining to finite-state machines, each of these operations may have a different degree of behavioral change on the expression, and thus the probabilities of applying the operators can be adjusted so as to provide for a more (or less) gradual search of the solution space.

Difference Equations   Another example comes in a form of mathematical modeling known as *system identification*. In this case, input–output examples are given for a particular system and the objective is to construct a mathematical model that represents the input–output relationship. This is a typical neural networks application; however, particularly on time series problems, it is more common to address system identification using autoregressive (AR) moving-average (MA) models (or ARMA), which are particular forms of difference equations. These are of the form

$$x[t + 1] = a_0 x[t] + a_1 x[t - 1] + \cdots + a_j x[t - j] + e[t] + b_1 e[t - 1] + \cdots + b_k e[t - k]$$

where $x[t]$ is the observed variable at time t (this is the AR part), $e[t]$ is the random noise occurring at time t (this is the MA part), and $a_0, \ldots, a_j, b_1, \ldots, b_k$ are coefficients. A standard approach employs gradient methods to estimate the coefficients for a model that has a predefined number of lag terms for the AR and MA parts of the model. This is problematic because the best choice of j and k are unknown *a priori*, and thus multiple searches must be conducted over different choices of j and k to provide confidence in a final result. Moreover, some nonlinear time series involve ratios of AR and MA processes, and gradient methods may lead to entrapment at suboptimal solutions (just as backpropagation may do the same when optimizing weights and bias terms of a neural network).

   An evolutionary approach to this problem encodes both the number of lag terms and the coefficients as a single solution, such as

$$[2, 1.5, 0.7, 0]$$

where the first integer is the number of lag terms in the AR process, the next two entries are the coefficients of those lag terms, and then the next integer is the number of lag terms in the MA process. As that is zero, there are no more entries. This specific representation corresponds to the model

$$x[t + 1] = 1.5x[t] + 0.7x[t - 1] + e[t]$$

The following are the variation operators that follow naturally from this representation:

(a) Vary the number of AR terms
(b) Vary the number of MA terms
(c) Vary the coefficients of any/all terms

In addition, it may be feasible to apply recombination operators to combine two or more models. Note that in cases where the degrees of freedom of a model are allowed to increase as part of the search process, a trade-off should be employed that incorporates the goodness-of-fit of the model to the data and also the number of degrees of freedom. An example is described in Section 12.2.2.

11.3.4.4 Considerations One common approach to evolutionary optimization involves employing a high rate of crossover and low rate of mutation. A possible justification for this approach is a view of evolution in nature proceeding by assembling building blocks of genetic code between parents. This view of evolution is controversial (as it was 20 years ago, see Atmar [1994]), but the effectiveness of believing that a particular problem is amenable to solution by assembling building blocks can be tested empirically.

As an example, let's explore this with a function that poses multiple local optima. The function is

$$f(x, y) = \exp(-[(x - 3)^2 + (y - 3)^2]/5) + 0.8 \exp(-[x^2 + (y + 3)^2]/5)$$
$$+ 0.2[\cos(x\pi/2) + \cos(y\pi/2))] + 0.5$$

which is displayed in Figure 11.10. The maximum of the function in this range is 1.6903, which occurs at (3, 3), and let's say we want to find that maximum. In order to assess the possibility for recombining building blocks, we'll choose a binary representation that was familiar within one early school of evolutionary algorithms. The representation takes the real values for x and y and encodes them in eight bits ranging from $[00000000] = -5$ to $[11111111] = 5$. With eight bits of precision, there are 2^8 possible values for x and y, and a precision of 0.039. If we wanted more precision, we could increase the number of bits used to represent the x and y values. For this level of precision, the maximum value of the function is 1.6902.

We can compare the effects of one-point, two-point, and uniform crossovers on this function. Recall that one-point and two-point crossovers select and recombine segments of a solution that might serve as building blocks for new solutions. Uniform crossover doesn't do that. It chooses each component (each bit) from each parent with equal probability. We can also compare the effects of regular binary coding and what's called Gray coding, which uses binary but encodes numbers so that two successive numbers differ by only one bit.

Figures 11.11 and 11.12 show a comparison over 200 trials, starting in a range of −5 to 5 in x and y, with 100 parents making 100 offspring under roulette wheel selection for

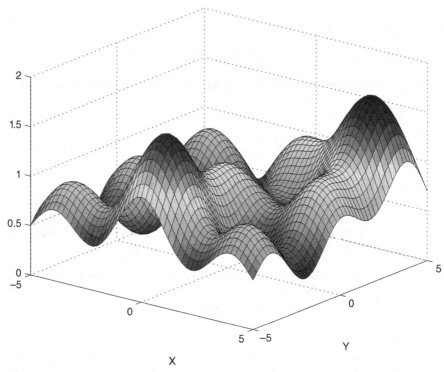

FIGURE 11.10 The function $f(x, y) = \exp(-[(x-3)^2 + (y-3)^2]/5) + 0.8 \exp(-[x^2 + (y+3)^2]/5) + 0.2[\cos(x\pi/2) + \cos(y\pi/2)]) + 0.5$. It reaches a maximum height of 1.6903.

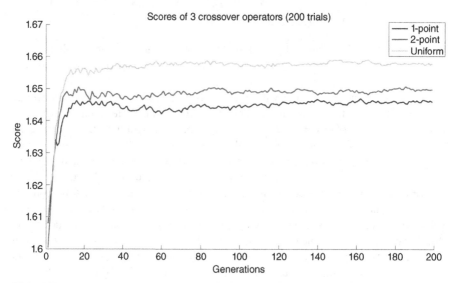

FIGURE 11.11 When averaged over 200 trials using standard binary encoding on the function in Figure 11.10, uniform crossover outperformed two- and one-point crossovers.

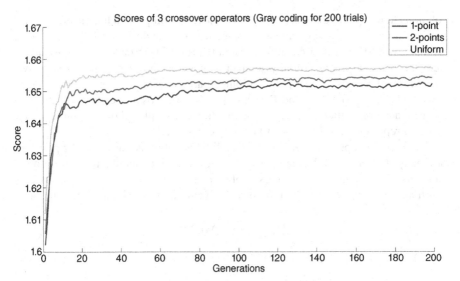

FIGURE 11.12 When averaged over 200 trials using Gray coding on the function in Figure 11.10, uniform crossover again outperformed two- and one-point crossovers, but did not reach as high a score on average as when using standard binary encoding.

each of the three recombination operators using traditional binary encoding and Gray coding. The results indicate that the "building block hypothesis" for bringing together good subsections of solutions does not hold for this problem, as uniform crossover outperforms one- and two-point crossovers under both representations.

One procedure for testing this hypothesis is to compare the effectiveness of an evolutionary algorithm that uses recombination with one that does not on the same problem. Statistical hypothesis testing can then be employed to determine if there is any statistically significant evidence to favor one approach over another. Many test problems in the literature have shown niches where mutation alone has worked better than recombination, and others where it has not [Fogel and Atmar, 1990; Schaffer and Eshelman, 1991; Fogel and Stayton, 1994; Chellapilla, 1997; Luke and Spector, 1998; Spears, 1998; and many others].[13] You might try an evolutionary algorithm that uses a real-valued representation and Gaussian mutation on this function and see how it compares to the results in Figures 11.11 and 11.12.

Another procedure, offered in Jones [1995a], employs what is called *headless chicken crossover*. The procedure takes an existing parent and instead of crossing it over with another existing parent, it crosses the parent with a completely random

---

[13] Despite more than 20 years of empirical and theoretical comparison between different evolutionary optimization methods showing no single approach works best, it's common to observe complacency and a reliance on a particular approach that is "comfortable" to a researcher. This is true across computational intelligence methods, and it's likely also true in science generally. People tend to start with the tools they are most comfortable with, rather than starting with the problem and asking which tools they should employ (see Michalewicz and Fogel [2006]).

solution. At this point, the mechanism of crossover is still present but the effect of the mechanism is purely a large mutation, since it is just random material that is being swapped with the existing parent to create offspring. There have been cases shown (and this was surprising initially to many) in which headless chicken crossover outperformed the usual one-point or other variations of crossover (e.g., Angeline [1997]; see also White and Poulding [2009]). In those cases, greater attention can be placed on mutation operators, or attention can be placed on reformulating the problem representation in order to facilitate crossover.[14]

One variation operator that has received little attention is *inversion*. As proposed originally [Fraser, 1968; Holland, 1975], this operator inverted the index position of elements in a solution string. For example, given a set of values $x_1, \ldots, x_5$ with an additional subscript for the position of those values, an initial solution might be

$$[x_{11}, x_{22}, x_{33}, x_{44}, x_{55}] \qquad (11.1)$$

but after inverting between position 2 and 4, this would be

$$[x_{11}, x_{24}, x_{33}, x_{42}, x_{55}] \qquad (11.2)$$

Thus, if the objective were to, say, minimize the value of $f(\mathbf{x}) = x_1^2 + 2x_2^2 + 5x_3^2 + x_4^2 + x_5^2$ then each of (11.1) and (11.2) above would evaluate to the same $f(\mathbf{x})$. Inversion here did not change the values of $x_1, \ldots, x_5$, it only changed the position of these values in the solution string. The suggested purpose of doing this was to ensure that one-point crossover would have more opportunity to find building blocks between elements of a solution regardless of their initial position in the solution string. Given all the research performed on the effectiveness of variation operators, there has been comparatively little research on the value that could be imparted by utilizing this form of inversion in certain cases.

When designing an evolutionary optimization algorithm keep in mind that the number of variation operations applied in creating an offspring can be greater than 1. It is typical to see an approach such as apply crossover with a probability of 0.9 and mutation with a probability of 0.01, or apply one from a list of variation operators each with equal probability. But a greater exploration of a search space, and also sometimes a more gradual one, can be created by repeating variation operators. For example, when mutating a finite-state machine (Section 11.3.4.3), rather than select only one variation operator, or rather than test for the application of each variation operator once, another parameter could be the number of variations to apply. One way to

---

[14] One of the authors (DF) has seen this many times in more than 30 years of evolutionary algorithm experience: An evolutionary approach does not yield effective results and thus the designer seeks to reformulate the problem to be more amenable *to crossover*. An alternative and potentially more fruitful suggestion is to reformulate the algorithm to be more amenable to yielding *a superior solution*. Crossover is just one of many tools in the virtual Swiss-army knife of the evolutionary algorithmist. Using crossover is not an objective in and of itself.

engineer this is to use a Poisson random variable with a set mean rate, and use a sample from that random variable to determine how many variation operations to apply. This also leads to the concept of having the evolutionary process learn how much variation to apply, which is treated in Section 11.5.

It is easy to become accustomed to a basic plan of applying an evolutionary algorithm for optimization. Some opt for a plan that involves a high rate of crossover between parents and a low rate of mutation. Others opt for no crossover or other form of recombination at all and employ only mutation unless the problem at hand suggests that mutation alone is insufficient or inefficient. With experience, you can demonstrate to yourself that neither of these approaches is optimal generally. It is often beneficial to think imaginatively about how to design variation operators that exploit the characteristics of the objective function being searched in light of the chosen representation and type of selection.

Different variation operators can be effective at different stages of an evolutionary optimization process. That is, it may be that recombination may be most effective early in optimization with mutation serving to fine-tune optimization in a later stage. Or it may be that mutation and recombination actually generate very similar expected rates of progress given the same population and fitness criteria. An example of this is given in Section 11.5.5. The use of static probabilities of applying variation operators has been of limited utility in the experience of one of the authors (DF); in contrast, the concept of having the evolutionary algorithm adjust its own search via reinforcement learning mechanisms has demonstrated utility and is described in more detail in Section 11.5.

11.4 CONSTRAINT HANDLING

Almost all real-world problems are constrained problems. For example, suppose you are designing an optimal schedule for a bus company. They have a specific number of existing buses. That is one constraint. They have a limited budget to purchase additional buses. That is another constraint. They have a limited number of qualified drivers for each different type of bus. That is yet another constraint. Each bus has limited capacity, which is yet another constraint. We could invent more constraints for this problem, including required maintenance, roads that can and cannot be used, the available time for each driver to work each week, and so forth.

Thus, when applying evolutionary algorithms it is important to consider how to treat the constraints of the problem. Some of these constraints are part of the objective, whereas some are part of the parameters of a solution and therefore impose boundary conditions on the search space. Each of these may pose *hard* or *soft* constraints.

A hard constraint is one that, if violated, makes the entire proposed solution worthless. For example, suppose you are designing a new golf club for professional golfers. The United States Golf Association (USGA) requires that all golf clubs be at least 18 in. in length and not more than 48 in. in length. These are hard constraints on your design because anything outside of these limits cannot be used in regulation play.

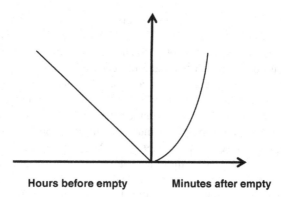

Hours before empty   **Minutes after empty**

FIGURE 11.13 A sketch of a penalty function to describe the cost of a fuel truck arriving too early or too late to a fuel station to refuel the station. The function describes the cost to the company. If the truck arrives too early, the company pays the driver an hourly wage to sit and wait until the station becomes close to empty. If the truck arrives too late, the company loses the opportunity to sell fuel to customers and also loses customers to its competitors.

A soft constraint is one that can be violated, but there is some imposed penalty for violating it, and perhaps the penalty increases with the degree of violation. For example, suppose you are creating a schedule for fuel trucks to refuel gas stations. A constraint would be to design a schedule in which no station is ever empty, which would mean the oil company would have customers waiting at the gas pumps for a refueling truck. But this is not likely to be a hard constraint. If you crafted a schedule that had a gas station out of fuel for 1 min, it might lead to some bad publicity. In fact, it would be more likely to add to bad publicity the longer the station remained empty. So, a penalty function of the form shown in Figure 11.13 could be used to describe this soft constraint.

It is often helpful to craft an objective function that comes in two parts. The first involves the primary criteria of interest and the second involves a penalty function that treats constraint violations. Sometimes these two parts can be simply added together. For example, in information-theoretic model building, there are many criteria that trade off the goodness-of-fit of a mathematical model to available data and the number of parameters (degrees of freedom) that the model employs. For example, Akaike's information criteria (AIC) [Akaike, 1974] can be written as

$$AIC(\mathbf{x}) = -2\ln(L) + 2p$$

where $\mathbf{x}$ is the parameter vector of the model, L is the likelihood function of the model, and p is the number of model parameters. If you don't know what a likelihood function is, you can think of it as a measure of the goodness-of-fit of the model to the available data. Better models generate lower AIC scores. For each new parameter that is added, the likelihood function will be higher (and thus $-2\ln(L)$ will be lower), but it has to be sufficiently better to "pay for" the addition two points that the extra parameter will cost. Here, there is no hard constraint on the number of parameters, but each additional parameter comes at a cost.

When a hard constraint is involved, the objective function can be set to an infinitely bad score if the constraint is violated. For example, suppose that in the above AIC minimization problem there was a constraint not to use more than 10 parameters. Then the objective function could be written as

$$f(\mathbf{x}) = \begin{cases} \text{AIC}(\mathbf{x}), & p \le 10 \\ \infty, & p > 10 \end{cases}$$

The difficulty with this approach is that it sets up areas of the search space that have no information content to direct the evolutionary search to any places of improved performance. Anything in the search space that violates the hard constraint is equally worthless. An alternative is to impose a penalty for violating the hard constraint that increases in effect gradually over successive generations. In this way, the constraint is treated as a soft constraint at first but over time solidifies as a hard constraint. A drawback to this approach is the requirement for tuning a schedule for how to transition the soft constraint to a hard constraint.[15]

Sometimes it is possible to treat constraints in the objective function as constraints in parameters. For example, in the above case, a hard constraint that $p \le 10$ could be handled by ensuring that any mutation that generates $p > 10$ is set aside, or that mutations to values of p that are greater than 10 are reflected into the feasible range. That is, if a parent had $p = 9$ and a mutation were to make p grow by 3, then it would first become 10, and then it would reflect into the feasible range to 9 and then again to 8. This has advantage of always evaluating feasible solutions, but it can have the disadvantage of introducing boundary effects on variation operators that are sometimes difficult to intuit.

Thinking about problem solving is often more important than attempting problem solving. By thinking about the problem at hand, it is often possible to garner insights about that problem and design-specific operators to address the intricacies and constraints posed therein. For example, Michalewicz et al. [1996] studied an n-dimensional problem of maximizing:

$$f(\mathbf{x}) = \left| \left[\sum_{i=1}^{n} \cos^4(x_i) - 2 \prod_{i=1}^{n} \cos^2(x_i) \right] \Big/ \sqrt{\sum_{i=1}^{n} i x_i^2} \right|$$

subject to $\prod_{i=1,\ldots,n} x_i \ge 0.75$, $\sum_{i=1,\ldots,n} x_i \le 0.75\,n$, and $0 \le x_i \le 10$ for $1 \le i \le n$. By recognizing that the maximum point was likely to exist on the boundary condition $\prod_{i=1,\ldots,n} x_i = 0.75$, Michalewicz et al. constructed an evolutionary algorithm that searched only on that boundary. They initialized the algorithm on the boundary

[15] This is analogous to tuning a schedule for reducing the temperature in a simulated annealing procedure [Kirkpatrick et al., 1983].

condition and then constructed a new "geometric crossover" that combines two parents **x** and **y** as follows:

$$[x_1, \ldots, x_n]$$

$$[y_1, \ldots, y_n] \rightarrow [x_1^\alpha y_1^{1-\alpha}, \ldots, x_n^\alpha y_n^{1-\alpha}]$$

where α is selected uniformly at random in the interval [0, 1]. When applied to any two solutions on the boundary condition, this operator returns another solution on the boundary. Finally, they also employed a mutation operator that selected two components of an existing solution and multiplied one by a factor $q > 0$ and the other by $1/q$. By tailoring the operators to the problem in this way, Michalewicz *et al.* [1996] reported finding a new best solution for this problem (better than previous approaches that had relied on traditional forms of evolutionary algorithms).

11.5 SELF-ADAPTATION

When considering a simple evolutionary algorithm, say, one that employs only Gaussian mutation to real-valued components of potential solutions, **x**, it is evident that the setting of the standard deviation of the mutation will affect the degree of progress that can be made toward an optimum. For example, consider Figures 11.14 and 11.15, which show the case in one dimension. The function $f(x) = x^2$ is to be minimized and there is a parent solution at $x = 2$. Using a zero-mean Gaussian

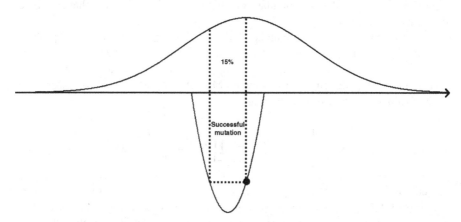

FIGURE 11.14 Suppose that there is a one-dimensional function to be minimized, such as the parabola in the figure, and the current solution is represented by the black dot. Under zero-mean Gaussian mutation, as the step size grows large, the probability of finding a solution that has a better score (corresponding to a lower point on the function) becomes very small. Here the probability is illustrated at about 0.15.

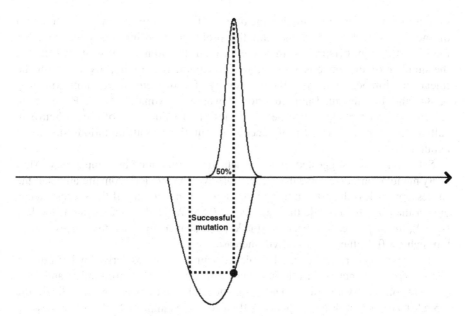

FIGURE 11.15 Suppose we have the same one-dimensional function to be minimized, such as the parabola in the figure, and the current solution is again represented by the black dot, but this time we have a zero-mean Gaussian mutation with a very small standard deviation. As the standard deviation shrinks close to zero, the probability of improving tends to 0.5; however, the amount of improvement at that limit will be close to zero.

distribution for mutation, denoted $N(0, \sigma)$, for increasingly large values of σ, say $\sigma = 10^6$, the likelihood that an offspring will be better than the parent is extremely low (zero in the limit) (see Figure 11.14). In contrast, for infinitesimal values of σ, say $\sigma = 1/$(national debt of the United States), the likelihood that an offspring will be better than the parent approaches 0.5. This increased likelihood of success comes with the drawback of a high likelihood of making very little progress toward the optimum (because the step size is so small) (see Figure 11.15).

The problem shown poses a balance between the step size of the mutation and the expected rate of progress. For very small steps, the likelihood of success can be maximized but the rate of progress is slow. For steps that are too large, the likelihood of success is minimized and the rate of progress is also slow. In between there is an optimum step size for which the likelihood of success is not maximized, but the rate of progress toward the optimum is maximized.

11.5.1 The 1/5 Rule

Rechenberg [1973] studied two minimization problems that are linear and quadratic functions of **x**. Using a $(1 + 1)$ evolutionary algorithm that relied on zero-mean Gaussian mutation, Rechenberg studied the trade-off between the rate of improvement (i.e., how quickly the best solution's score decreases as a function of the number of generations) and the probability of improvement (i.e., the likelihood that an offspring will be better than its parent). It was shown that as the number of dimensions n → ∞, the maximum rate of improvement on the linear function occurred when the probability of improvement was approximately 0.184, and for the quadratic function it was approximately 0.27. Rechenberg selected 0.2 as a compromise between these two values and offered a heuristic called *the 1/5 rule*: The ratio of successful mutations to all mutations should be about 1/5.

Schwefel [1995] suggested a simple method for computing this empirically. After every m mutations, determine the success rate over the previous 10m mutations. If the success rate is less than 0.2, multiply the step size (σ) by 0.85. If the success rate is greater than 0.2, then divide the step size by 0.85. When the success rate is too low (< 0.2), the step sizes will become smaller, and conversely when the success rate is too high (> 0.2), the step sizes will increase.

It is important to recall that the 1/5 rule is a heuristic that was derived only from two simple types of functions, as the number of dimensions tends to infinity, and for a $(1 + 1)$ evolutionary algorithm. There is no general case to make for the utility of the 1/5 rule (see Chellapilla and Fogel [1999] for counterexamples). Still, it illustrates that static parameters for variation operators are very unlikely to lead to the best rates of progress toward optimum solutions.

11.5.2 Meta-Evolution on Real-Valued Parameters

Given the limited utility of the 1/5 rule, more general methods for controlling the degree to which variation operators act coarsely or finely are desired. A common method employed in evolutionary algorithms views the parameters that control the evolutionary search as part of the evolutionary process to be adapted while searching for an optimum. This poses a form of *meta-evolution* in that adaptation takes place on two levels within the evolutionary algorithms: the parameters that are used to evaluate the objective function (*objective parameters*) and the parameters that are used to control variation (*strategy parameters*).

The history of meta-evolutionary approaches extends back to the 1960s [Reed *et al.*, 1967; Rosenberg, 1967; Rechenberg, 1996[16]]. Currently, for real-valued evolutionary optimization, the most common approach follows a procedure offered in Schwefel [1981]. With the problem of minimizing (maximizing) a real-valued function f(**x**), encode a possible solution as (**x**, σ), where **x** is the vector of real-valued objective parameters and σ is the vector of strategy parameters, which designate the positive standard deviation to apply a Gaussian mutation in the given dimension.

---

[16] In personal communication with one of the authors (DF), Ingo Rechenberg remarked that he devised an unpublished self-adaptive approach for evolution strategies in 1967.

Each parent creates an offspring via a two-step process applied across all dimensions, $i = 1, \ldots, n$:

1. $\sigma'_i = \sigma_i \exp\left(\tau(N(0, 1)) + \tau' N_i(0, 1)\right)$
2. $x'_i = x_i + N(0, \sigma'_i)$

where σ'_i is the standard deviation for the offspring in the ith dimension, $N(0, 1)$ is a standard Gaussian random variable sampled once and held at the same value for all i dimensions, $N_i(0, 1)$ is a standard Gaussian random variable sampled anew for each of the i dimensions, τ and τ' are constants that are proportional to $1/\sqrt{2\sqrt{n}}$ and $1/\sqrt{2n}$, respectively, and x'_i is the objective parameter value for the offspring in the ith dimension.

In this process, the standard deviations used in mutating a parent are carried as "genetic" information along with the parent solution. The first step mutates the standard deviations using a lognormal distribution. The second step uses the mutated standard deviations to create the offspring's objective parameters. In this way, standard deviations that lead to improved solutions are retained by selection, while those that lead to solutions of lesser quality perish. In essence, the procedure uses a form of reinforcement learning to update the strategy parameters of the evolutionary search while the search is in progress.

The two-step process above allows the evolutionary algorithm to learn how to shape the Gaussian distribution to better fit the contours of the objective function in the neighborhood of each solution. In this process, the dimensions are treated independently; however, extensions to self-adapting correlated mutations have also been offered in Schwefel [1981] (and for further discussion, see Fogel [2006]). Figure 11.16 provides an

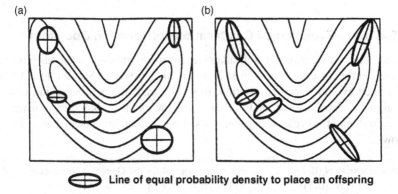

(a)   (b)

⊕ Line of equal probability density to place an offspring

FIGURE 11.16 Self-adapting the mutation distributions on each solution allows new trials to be generated in light of the contours of the response surface. Independent adjustment of the standard deviation in each dimension provides a mechanism for varying the width of the probability contour in alignment with each axis (a). Correlated standard deviations provide a mechanism for generating trials such that the probability contours are not aligned with the coordinate axes (b). (Taken from Fogel [Fogel, 2000, p. 157] and Back et al. [1991].)

intuitive picture of what the self-adaptive process can accomplish. In addition, a common form of modern self-adaptation simplifies step (11.3) to be

$$\sigma_i' = \sigma_i \exp\left(\tau' N_i(0, 1)\right) \tag{11.3}$$

foregoing the use of a "global" Gaussian random variable and relying solely on independent Gaussian random variables for each ith dimension.

11.5.3 Meta-Evolution on Probabilities of Variation Operators

The concept of meta-evolution can be extended to support variation operators that are not in the continuous domain. Self-adaptation can be applied both within a variation operator (controlling how that operator functions) and to determine the likelihood or number of times the variation operator is applied.

For example, when considering a traveling salesman problem, Chellapilla and Fogel [1997b] utilized a permutation representation (an ordered list of cities to visit) and applied self-adaptation to adjust either (i) the length of an inversion operation or (ii) the probability of applying every possible inversion from length 1 to $c/2 - 1$, where there were c cities. Angeline et al. [1996] used self-adaptation to adjust the mutation probabilities of varying finite-state machines.

Thus, self-adaptation is useful potentially not only for applications of evolutionary optimization in $\mathfrak{R}^n$ but also in virtually any real-world application. It can be used to adjust the likelihood of using mutation, crossover, blending recombination, multi-parent recombination, as well as the types of these operators (e.g., one-point, n-point, uniform crossovers). It is extremely unlikely that any static settings for variation operators will be optimal for solving a problem, and the effectiveness of different variation operators changes during the course of the evolution.

11.5.4 Meta-Evolution on Combinations of Variation Operators

An additional extension of self-adaptive meta-evolutionary methods comes in blending different operators. For example, as mentioned in Section 11.3.4.1, in a real-valued evolutionary optimization, there is a trade-off between using Gaussian variation and Cauchy variation. Instead of hand-tuning the probabilities of using these operators, their effects can be adjusted via self-adaptation, implemented potentially as follows:

(1) $\sigma_i' = \sigma_i \exp\left(\tau' N_i(0, 1)\right)$	//step-size control for the Gaussian component
(2) $s_i' = s_i \exp\left(\tau N_i(0, 1)\right)$	//step-size control for the Cauchy component
(3) $\alpha' = \alpha + N(0, \omega)$	//weight to apportion between Gaussian and Cauchy
(4) $x_i' = x_i + |\sin(\alpha')| N(0, \sigma_i') +$	//update with mutation in part by Gaussian and Cauchy
$(1 - |\sin(\alpha')|) C(0, s_i')$	components

where σ_i' and s_i' are the offspring scaling parameters for the Gaussian- and Cauchy-distributed mutations, respectively, τ and τ' are scaling constants, α' is the offspring's adjustment factor for weighting the effects of Gaussian and Cauchy mutations, ω is a stepsize parameter for mutating α, and $N(0, \sigma)$ and $C(0, s)$ represent Gaussian and Cauchy distributions centered at zero with scaling parameters σ and s, respectively. Using the transform $|\sin(\alpha')|$ returns a value between 0 and 1 and thus weights the contribution of the Gaussian and Cauchy components of the overall variation.[17]

11.5.5 Fitness Distributions of Variation Operators

Applying a variation operator to a parent or set of parents generates offspring based on a probability mass or density function. Choosing among variation operators and changing the parameters of those operators alters the likelihood of sampling each point from the solution space. Thus, a probabilistic distribution of offspring fitness scores can be created given a variation operator and the parent or parents that it operates on. In certain simple cases, this *fitness distribution* can be computed mathematically (e.g., leading to Rechenberg's 1/5 rule), but it can be estimated empirically in general cases. Doing so is computationally intensive, as it requires generating a sufficient number of examples of applying an operator to a parent(s) and determining the results statistically; however, the fitness distribution of an operator can yield insight into setting its parameters, how to make them subject to self-adaptation, and whether or not to include them.

For example, Nordin and Banzhaf [1995] quantified the change in fitness that occurred after applying crossover to machine code expressions that represented regression equations. Figure 11.17 shows the histogram of the percentage of fitness change over 35 successive generations. Positive change denotes an improvement. The figure shows that the most common result was either no change in fitness at all (i.e., the offspring's fitness was the same as the parent's fitness) or a 100% decrease in fitness. In this case, the success rate of crossover was very small, leading to the suggestion that in this case (and others [Nordin et al., 1996; Teller, 1996]) crossover was acting as an ineffective macromutation [Banzhaf et al., 1998].

Fogel and Jain [2000] showed that for the case of evolving a neural network to perform the XOR function, the fitness distributions of Gaussian mutation and one-point crossover changed as a function of the progress of the evolution. At certain times, the expected progress was the same regardless of which operator was chosen; at others, mutation or crossover was favored. For additional background on fitness distributions, see Altenberg [1995], Grefenstette [1995], Fogel and Ghozeil [1996], and Fogel [2006].

---

[17] The process suggested here is untested and provides an open area for research, including self-adaptation of ω. See Saravanan and Fogel [1997] and Chellapilla and Fogel [1997a] for other methods of adapting the use of Gaussian and Cauchy variation operators. As an aside, it is easy to imagine applying this form of adaptation to neural networks or fuzzy systems in evolving weights on transfer or membership functions. For example, a neural network can have nodes that are flexible between sigmoid (SIG) and radial-basis functions (RBF) by adapting a parameter α and utilizing it in the manner of αSIG + $(1 - \alpha)$RBF.

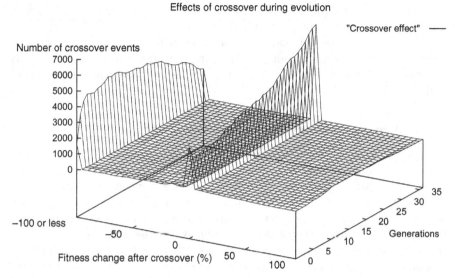

FIGURE 11.17 The fitness change after crossover on problems evolving machine code expressions from Nordin and Banzhaf [1995]. The vast majority of crossovers either resulted in no change in fitness or a large decrease in offspring fitness.

11.6 SUMMARY

Evolutionary algorithms have a long history of being used for optimization [Fogel, 1998]. These procedures have been applied to a wide array of problems in different areas: drug design, finance, video games, robotics, pattern recognition, scheduling, and many more. This chapter has provided an introduction to the application of evolutionary algorithms for optimizing solutions to problems in their canonical forms, including real-valued optimization, combinatorial optimization, and hybrids of these.

The essence of the evolutionary approach to optimization is to utilize a population of contending ideas about how to solve a problem and subject those ideas to random variation and selection in light of objective criteria that provide feedback on the suitability of the solutions. Determining which solutions to maintain as parents follows the concept of natural selection, eliminating the least-fit candidate solutions from the population probabilistically.

Representation, selection, and variation act in concert. There is no best choice to make for these design variables outside of the context of the problem at hand [Wolpert and Macready, 1997; see also English [1996]]. All canonical forms of variation—Gaussian mutation, one-point crossover, binary mutation, recombination by blending—serve not as ends but as beginnings for the evolutionary algorithmist (you!) who must design an effective search of a solution space. Self-adaptive methods can assist in allowing the evolutionary algorithm to learn how to search the solution space via reinforcement of the strategies that have already found better solutions.

It is important to maintain an open mind when designing an evolutionary algorithm. Simple traditional approaches may be sufficient, and if so, they should be used. But imaginative thinking can lead to interesting new approaches (as with the use of "geometric recombination" noted in Section 11.4). For example, consider the problem of optimizing for a solution in $\Re^n$ that minimizes f(**x**), except that the representation is made in polar coordinates instead of Cartesian coordinates (see Ghozeil and Fogel [1996]). Or consider the problem of designing variation operators that can effectively jump from one locally optimal region directly to another in a given application. These sorts of activities may require a fundamental understanding of the properties of the "landscape" posed by the objective function in light of the variation operators [Jones, 1995b]. (Variation operators define the search neighborhood of each parent regardless of the "shape" of the landscape being searched.) Each of these requires specific investigation and standard approaches as found commonly in literature may not be effective.

Evolutionary algorithms offer opportunities for hybridizing with other methods. For example, an evolutionary algorithm can be run for a number of generations and then a gradient method can be used to approach a local optimum more quickly. Evolutionary algorithms also offer significant flexibility in being able to handle problems with constraints (both hard and soft), with multiple criteria. The chapter has not treated the application of evolutionary algorithms to dynamic and noisy environments. Some examples are provided in Chapter 14. It has also not treated many of the mathematical properties of evolutionary algorithms with the depth that would be found in a textbook dedicated solely for that purpose. Students of evolutionary algorithms who would like a more detailed mathematical treatment should refer to Bäck [1996], Vose [1999], Fogel [2006], and Beyer [2010].

EXERCISES

11.1. Write an evolutionary program that addresses the traveling salesman problem as described in Section 11.2, except that instead of evaluating solutions based on minimizing total length, try the following two objectives: (1) minimize total length but have a large penalty for crossing x = 0, and (2) minimize total length but have a large bonus for crossing x = 0. You can use any evolutionary approach you like. Have a look at the solutions that emerge for each variation. Do they each reflect the varied objective of the problem?

11.2. Try another variation of the traveling salesman problem in which you designate a specific city as the starting city and all segments that are traversed from north to south (a higher y-coordinate to a lower y-coordinate) cost 1.5 times the distance of the segment. Do the solutions you generate appear different than those that do not incorporate this penalty?

11.3. Work with Rastrigin's function as described in Section 11.1. Choose an evolutionary algorithm that uses a real-valued representation and zero-mean Gaussian mutations. Implement self-adaptation as described in Section 11.5.2

with the simplified version of updating the standard deviations:

$$\sigma_i' = \sigma_i \exp\left(\tau' N_i(0, 1)\right)$$

using $\tau' = 1/\sqrt{2n}$. Explore the rate of optimization on Rastrigin's function in 2 and 10 dimensions as a function of the population size when trying a $(\mu + 5\mu)$ configuration. For different values of μ, plot the best score attained as a function of the number of generations.

11.4. For the same problem as in exercise 11.3, compare the use of self-adaptation to the 1/5 rule (Section 11.5.1). Does the 1/5 rule provide for faster convergence to better solutions on Rastrigin's function in 2 and 10 dimensions than does simple self-adaptation?

11.5. For the problem illustrated in Section 11.3.4.4, determine the fitness distribution of various operators as follows. Start with a real-valued representation, initialization between $[-5, 5]$ in each dimension, and zero-mean Gaussian mutation with a standard deviation of 0.33 with 100 parents making 100 offspring (as in the text). Run the algorithm for 10 generations and then focus on the best parent. Empirically, determine what fitness values would result from mutating this parent with a zero-mean Gaussian mutation using different fixed standard deviations for x and y. Then compare the best Gaussian mutation that you find with a Cauchy distribution. (You can compute a standard Cauchy distribution by taking N_1/N_2 where N_1 and N_2 are independent $N(0, 1)$ random variables.)

11.6. Now try the same problem, but using a binary encoding as shown in the text. Using one-point crossover and a probability of mutation $= 0.01$/bit, stop the evolution at generation 10 and then determine the fitness distribution of one-point, two-point, and uniform crossovers by selecting the best parent and a random other parent from the population. Does one recombination operator provide a better fitness distribution than another?

11.7. Data for total sunspots are available widely on the Internet. Acquire the data for total sunspots for 2000–2013 and plot the data as a function of year (sunspots on the y-axis, year on the x-axis). Determine the best linear equation to model these data in the form of

$$y = ax + b$$

where a and b are evolved via any evolutionary algorithm that you choose in order to minimize the squared error between each predicted value $(ax + b)$ and the actual value y. Using standard regression software, check how close your values for a and b come to the optimal values. Now try evolving a and b so as to minimize the absolute value of the cube of the error. How much different are a and b from before?

11.8. Develop code structures for finite-state machines and symbolic expressions, as described in Section 11.3.4.3, including variation operators. (You'll need these in Chapter 12.) Test your code by evolving finite-state machines to generate a sequence [100111001] and a symbolic expression to match $x^2 + 2x + 1$. You can use any objective functions that you believe will help the evolution get to the desired outcome.

11.9. For the problem illustrated in Section 11.3.4.4, let's impose a constraint that $x < y - 1$. The domain of (x, y) is still $[-5, 5]$, but now the only feasible candidates for (x, y) must satisfy the constraint. Create a penalty method for addressing this constrained problem and apply an evolutionary algorithm of your choice. How well does it do? Try at least 30 independent trials to assess your results.

11.10. Explain why premature convergence is not necessarily to a point that is locally optima in a continuous space. (That is, the population may converge, but there would be a gradient that would suggest moving to a better point.) When is this more or less likely in evolutionary optimization?

11.11. Indicate the conditions that allow an evolutionary algorithm to converge with probability 1 to a global optimum solution.

11.12. Explain the benefits and drawbacks of self-adaptation in evolutionary algorithms, in terms of adjusting standard deviations in both continuous search and operator probabilities generally.

CHAPTER 12

Evolutionary Learning and Problem Solving

Chapters 10 and 11 have introduced the basic concepts of evolutionary algorithms, particularly as they are applied for optimization. This chapter explores many specific applications of evolutionary algorithms for problem solving. The discussion moves generally from simpler problems to more complex ones, but much of the framework for applying evolution comes in the form of modeling systems so as to predict what they will do next, or perhaps to control what they will do next.

It has been said that "prediction is the keystone of intellect" [Fogel et al., 1966]. When you can predict what is coming next, then you can claim a degree of understanding and take action to have a more desired outcome. From that perspective, intelligence is the property that allows a system to allocate resources to meet goals in a range of environments. Such a system can be a person, a group, a colony of ants, or even a computer running a program. The applications that we'll cover here are aimed generally at assisting us to make more intelligent decisions by providing a better understanding of the world around us.

To do that, we often rely on mathematical models. The basic tenet of mathematical modeling is to provide a description of a system that is neither too complex nor too simple. This is captured in a dictum known as the *maxim of parsimony* or *Occam's Razor*. A mathematical model needs to be sufficiently complicated to make useful predictions about a system. If it is more complicated than necessary, a simpler explanation should be favored. In addition, an explanation that is more complicated than is required can be expected to fail to provide the best predictions about future states of a system owing to the unintended consequences of interactions of the unnecessary parts of the model.

If you're familiar with statistics, parameter estimation, and information theory, much of what follows will also be familiar, even if the application of evolutionary algorithms in these specific examples is novel. If you have not gained experience with statistical methods, the text that follows will serve as an introduction, but you're encouraged to supplement these materials with additional foundational coursework.

Fundamentals of Computational Intelligence: Neural Networks, Fuzzy Systems, and Evolutionary Computation, First Edition. James M. Keller, Derong Liu, and David B. Fogel.

12.1   EVOLVING PARAMETERS OF A REGRESSION EQUATION

12.1.1   A Canonical Example

To provide a simple starting point, let's reconsider the problem of finding a best linear model for n pairs of data (x, y), which we saw in an exercise in Chapter 11 dealing with modeling sunspot data. For linear regression, we have the following equation:

$$y = ax + b + e \tag{12.1}$$

where y is the dependent variable, x is the independent variable, a and b are parameters of the equation, and e represents an unknown noise source, which is assumed typically to be normally distributed, with zero mean, and constant variance; however, for our purposes now, we do not need to make any assumptions about e. The typical approach to finding a and b, given a set of data (x_1, y_1), ..., (x_n, y_n), relies on a least-squares error criterion.

Find a and b such that

$$\sum_{i=1}^{n} (y_i - \hat{y}_i)^2 \tag{12.2}$$

is minimized, where $\hat{y}_i$ is the estimate of y_i that is computed from $ax + b$. This principle of least squares has roots back to the early 1800s in models of astronomy and other physical measurements, having been offered by Legendre.

It is straightforward to compute least-squares coefficients using calculus methods, taking the partial derivatives of the objective function above with respect to each coefficient, setting the result equal to zero, and solving. The set of equations that derive from these operations are called the "normal" equations and take the following form:

$$\sum_{i=1}^{n} y_i = bn + a \sum_{i=1}^{n} x_i \tag{12.3}$$

$$\sum_{i=1}^{n} x_i y_i = b \sum_{i=1}^{n} x_i + a \sum_{i=1}^{n} x_i^2 \tag{12.4}$$

for a constant b and a coefficient a multiplying a single independent variable x. Additional expanded equations are realized with additional independent variables. Note that these formulas for the least-squares values of a and b do not depend on the unknown properties of e. Actually, e doesn't enter into the above formulas at all. If some of the statistical characteristics of e are known then it may be possible to generate confidence intervals around the estimated coefficients; however, for the present discussion, such efforts will be set aside.

The issue at hand concerns the utility of the least-squares estimates. The procedure is so commonplace that it is almost taken for granted as being appropriate. But consider the following situation: Suppose you are listening to active sonar—pings of sound under the water—trying to detect man-made targets (such as a reflection from an underwater explosive). You can make one of two decisions: Either there is or is not a man-made object represented in the data from a particular active sonar pulse. Thus, there are four possible outcomes: (i) correctly classify a man-made object, (ii) correctly declare the absence of a man-made object, (iii) misclassify background noise or other signals as man-made, or (iv) miss a man-made object by declaring the absence of such an object when it is in fact present.

There are two ways to be correct and two ways to be incorrect. It is clear, however, that in any operational setting, correctly identifying an underwater explosive is more important than correctly identifying the absence of that explosive. *Equally correct predictions are not of equal worth.* Similarly, the type I error (a false alarm) may have a very different cost than the type II error (a miss). Here, a miss may be much more costly because of the danger posed by bumping into an underwater explosive.

A least-squares approach attributes zero error to any correct prediction. In the above setting, this is inappropriate. But this is almost always the case in real-world practice. Furthermore, using the squared error penalizes incorrect predictions that are too high or too low, or of type I or type II, based solely on their magnitude. Again, this is rarely appropriate in real-world settings.[1] The least-squares approach to estimating coefficients is mathematically tractable, but can be of limited utility when placed under this sort of scrutiny. It would be useful to have a method for optimizing the coefficients of models that could be responsive to any arbitrary cost function. In this regard, evolutionary algorithms can be particularly useful.

12.1.2 Objectives Other Than Least Mean Squared Error

12.1.2.1 Least Absolute Error   Consider the data shown in Figure 12.1. These 100 points were generated from the equation

$$y = 1.5x + 1 + N(0, 50) \qquad (12.5)$$

where $x \in \{1, \ldots, 100\}$ and $N(0, 50)$ is a Gaussian random variable with zero mean and standard deviation of 50. If given only the data in (x, y) pairs, the least-squares estimates of the slope and intercept are [1.483, 0.778], which generate the line shown in Figure 12.2. The root mean squared residual error is 48.495, or about 0.5 per datum. Suppose that instead of a squared error cost function, the desired evaluation depended on the absolute error between the model's predicted y value and the actual value. A simple evolutionary algorithm can be constructed to optimize the slope and intercept under these new conditions.

Suppose that 100 candidate solutions are allowed in the initial population. Each candidate solution is a two-element vector (i.e., the slope and the intercept), and for convenience, each element is set to zero to start the process. (Alternatively, we could

---

[1] A weighted least-squares approach can be used to give more or less weight to an error depending on the direction of the error and its magnitude, but it is still limited within a squared error framework.

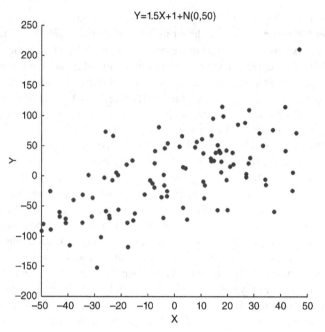

FIGURE 12.1   One hundred points (x, y) generated using the equation y = 1.5x + 1 + N(0, 50).

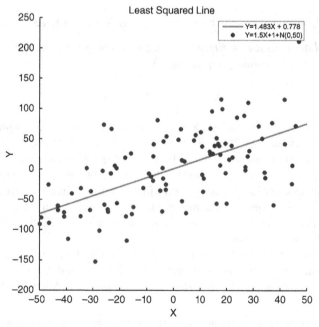

FIGURE 12.2   The least-squares regression line for the data in Figure 12.1.

start by having a random distribution of slopes and intercepts, or any other reasonable choice.) Each of the 100 possibilities in the population generates an offspring by adding a standard Gaussian random variable to each of its elements. If we denote an offspring (a, b) as (a', b'), then this mutation operation is

$$a' = a + N(0, 1) \tag{12.6}$$

$$b' = b + N(0, 1) \tag{12.7}$$

After each parent solution has generated an offspring, all 200 solutions are scored on how well they fit the data in terms of mean absolute error. The 100 best solutions (least sum absolute error) are retained to serve as parents for the next generation, and the process is repeated. Figure 12.3 shows the rate of optimization of the evolutionary process as it converges on its best solution of [1.621, 1.390]. The mean absolute error is 38.299, which is lower than the root mean squared error from the least-squares model above. (It also happens to be lower than the mean absolute error for the least-squares model at 38.432.) There is very little qualitative difference between the two sets of slope–intercept coefficients, but the latter model is optimized for its specific criterion and performs that specific task (minimizing mean absolute error) better than the least-squares model.

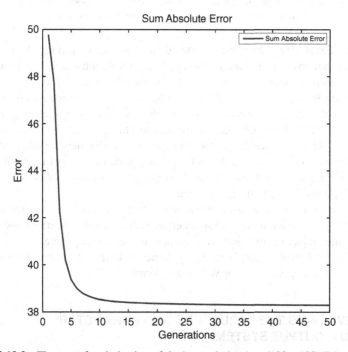

FIGURE 12.3   The rate of optimization of the best solution in a $(100 + 100)$-EA using real-valued representation and fixed standard normal mutations in order to minimize the sum of the absolute errors between the model and the data in Figure 12.1.

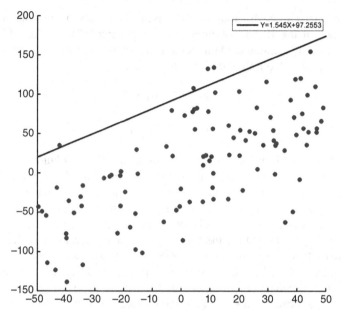

FIGURE 12.4 An evolved line when the objective function is changed from squared error or absolute error across all data to applying the absolute error if the model predicts a value that is too high and squared error if the model predicts a value that is too low.

12.1.2.2 *Extensions to Other Functions*

The situation might be more complicated in that errors where the prediction is greater than the actual value might be scored in terms of absolute error, but errors where the prediction is lower might be scored in terms of squared error. By executing the same evolutionary procedure but substituting this new cost function, we find the evolved result after 50 generations shown in Figure 12.4 and the final best estimates of the slope and intercept are [1.545, 97.2553]. Note how much larger the y-intercept is in this new model. Although it seems unreasonable at first, it reflects the particular chosen payoff function. Here, it costs less to predict on the high side, and the optimal model reflects this bias, reflecting on the high side for virtually all the data.

Error functions can be as complicated as necessary to describe a desired outcome. Section 12.1 has explored only the standard squared and absolute error conditions. In practice, the objective function may comprise many components, be subject to constraints, and even incorporate fuzzy logic in terms of fuzzy weighted error (i.e., describing degrees of error with fuzzy terms).

12.2 EVOLVING THE STRUCTURE AND PARAMETERS OF INPUT–OUTPUT SYSTEMS

Regression equations are often used for predictive modeling. Prediction is the keystone of intellect because intelligent behavior requires acting appropriately in an environment in order to meet a desired goal, or collection of goals.[2] In order to accomplish that feat, you (or another intelligent system) have to be able to anticipate what is going to happen in the future so as to be prepared for it when it happens. It's relatively easy to be a "Monday morning quarterback"[3] and identify what someone should have done, or shouldn't have done, in retrospect. It's far more challenging to predict the future with accuracy and precision. But that's the requirement that natural selection has placed organisms since the dawn of life itself. Creatures that cannot predict what is going to happen next in the real world are often the victims of other creatures who can.

In engineering, it's commonplace to utilize input–output mathematical models to make predictions about the future. The inputs and outputs could be all sorts of things:

1. Measurements of aircraft parameters in flight as inputs and control stick actions as output.
2. Measurements of atmospheric variables as inputs and extent of polar ice caps as output.
3. Historical records of movie ticket sales by genre, actors, release date, and production studio and total revenue as output.

You can surely imagine many other examples. Various mathematical models can be employed to describe these sorts of relationships, each with varying precision and accuracy. To be effective, a model must be able to yield a sufficiently accurate prediction at a suitable time in the future with sufficient precision. It's easy to generate a 100% accurate predictive model: "In the future, something will happen." That is certain, but this model has no precision. It also has no definitive time span (since we do not know with certainly if time will ever come to a stop in the universe). Effective predictive models must incorporate appropriate precision and accuracy over an appropriate time frame.

As mentioned earlier, engineers and other practitioners often use a so-called *maxim of parsimony*, also known as *Occam's Razor*, to help design mathematical models. This maxim, as paraphrased by Albert Einstein, says to keep models as simple as possible and no simpler.[4] If a model is too simple, it will not be sufficient to explain the process that is being modeled. For example, it's not possible to make a mathematical model of a falling object with high fidelity using only a linear function. As we know from physics class, this requires a quadratic function. On the other hand, mathematical models can be made so complex that they can "explain" all of the available historical data, but what they really do in those cases is fit the noise in the

---

[2] This was the basis of Lawrence J. Fogel's doctoral dissertation [Fogel, 1964].

[3] The saying derives from the situation in which National Football League games are played mainly on Sunday. On Monday, it is easy to say that the quarterback of a losing team should have done something different. It is another version of the famous idiom "hindsight is 20–20."

[4] See https://en.wikiquote.org/wiki/Albert_Einstein for a historical view of the development of this attribution.

data—perfectly—and thus these overly complex models tend to predict poorly because future data are by definition uncorrelated to the noise in past data.

These aspects of mathematical modeling are very important generally, and quite pertinent in the application of evolutionary algorithms for predictive modeling. Without *a priori* knowledge of a system to be studied, the range of available models is unlimited. Practical experience over many generations of modeling has generated different classes of models that may help explain certain processes. These are only guides and you should always be open to considering other mathematical models. The following sections provide examples of evolving models within three classes of input–output systems that are commonplace in time series modeling.

12.2.1 Evolving ARMA(X) Models

12.2.1.1 Fundamentals An autoregressive moving average (ARMA) model is one that consists of two parts. The autoregressive part indicates that future values of an observed variable are a function of prior values of the observed variable. For example, the model

$$y[t+1] = a_0 y[t] + a_1 y[t-1] \qquad (12.8)$$

indicates that the predicted value for the observed variable y at time $(t+1)$ is the weighted sum of the value of y at time t and the value of y at time $(t-1)$, weighted by a_0 and a_1, respectively. This model is an AR model. It is a regression equation in itself (hence "autoregressive"). The concept of the moving average comes from additional terms that model noise in the system, such as

$$y[t+1] = a_0 y[t] + a_1 y[t-1] + c_0 e[t] + c_1 e[t-1] \qquad (12.9)$$

where the new terms are $e[t]$ and $e[t-1]$, the values of random noise at time t and $t-1$, respectively, weighted by c_0 and c_1, respectively. Often, the actual noise in a system is unknown and modeled as a function of the difference between a predicted value of $y[t+1]$ and the observed value. Models of the form (12.9) are sometimes written in shorthand notation as

$$A(q)y[t] = C(q)e[t] \qquad (12.10)$$

where $A(q)$ and $C(q)$ are polynomials in the so-called shift operator q^{-1}. For example, if

$$y[t+1] = 2y[t] + 4y[t-1] + 3e[t] - 2e[t-1] \qquad (12.11)$$

then $A(q) = \begin{bmatrix} 2 & 4 \end{bmatrix}$ and $C(q) = \begin{bmatrix} 3 & -2 \end{bmatrix}$. This can be extended in the case where a system is a function of its own past values as well as external inputs (such as an airplane that has inertia and also is affected by control surface changes). In this case, the system is described as

$$A(q)y[t] = B(q)u[t] + C(q)e[t] \qquad (12.12)$$

where $B(q)$ is a polynomial in the shift operation q^{-1} and $u[t]$ is the control input. This is then called an ARMAX model (ARMA + eXternal inputs).

12.2.1.2 *Model Optimization*

If you decide to model a system with an ARMA or ARMAX model, you will have to determine the appropriate number of lag terms in each polynomial $A(q)$ and $C(q)$ (and $B(q)$ if the model is ARMAX) and also the appropriate values of the parameters in the polynomials. You might imagine if data were generated from (12.9) using standard Gaussian noise[5] for $e[t]$. But suppose nobody told you that there were two lag terms in $A(q)$ and two lag terms in $C(q)$. Following a standard approach that every student of system identification learns in school, you'd have to guess at the form of the model. Does it seem best to use one lag in $A(q)$ and three in $C(q)$? How about 10 lags in $A(q)$ and 1 in $C(q)$? There are infinite possibilities.

Once you choose a model form, you could employ traditional gradient-based methods to find the best estimates for the parameters in $A(q)$ and $C(q)$ in terms of minimizing the squared error of model outputs to observed data. These methods are sometimes called recursive prediction error methods (RPEM) because they use the predictive error to update the model parameters in a recursive manner—conceptually this is similar to backpropagation in neural networks, and it also has all the drawbacks of a local search mechanism just as backpropagation does. If the model form is appropriate and the noise is stationary, then the gradient RPEM methods can locate optimal parameter values. If the model form is not appropriate or the noise is nonstationary, then the RPEM methods can yield answers that are only locally optimal and quite possibly insufficient for the task at hand.

Evolutionary algorithms offer different capabilities for exploring model development in these cases. One option is a straightforward extension of what was demonstrated for the linear regression problem in (12.1), in which an evolutionary algorithm can be used to optimize the parameters of a mathematical model and potentially overcome the local optima that might entrap an RPEM method. In this approach, you still have to select the order of the model (the number of lag terms) and the evolutionary algorithm adapts the weights on each term in the model. While that's a possible advantage over RPEM methods, the fact that evolutionary methods do not need to rely on gradient information offers a broader advantage in that a search can be made for appropriate model order at the same time parameters are optimized.

Information Criteria As mentioned earlier, arbitrarily complex models can be constructed to fit any available data. Thus, it is important to be able to assess the complexity of a model in light of the degree of fit it offers to those data. This trade-off of complexity versus goodness-of-fit has been quantified in several different approaches over many years of development in statistical information theory. These approaches have yielded alternative formulas that can be used as objective functions in evolutionary optimization. So it's important to have a basic understanding of the most common of these approaches.

---

[5] Recall, a standard Gaussian variable has a mean of zero and standard deviation of 1.

One of the most common is called Akaike's information criteria (AIC) [Akaike, 1974]. This incorporates the principle of maximum-likelihood estimation (MLE) and a penalty for the number of free parameters that are incorporated in a model. Maximum-likelihood estimation is a method of parameter estimation that involves finding parameters so that the observed data have maximum likelihood. The AIC of a set of parameters, θ, in a predictive model is given by

$$AIC(\theta) = -2\ln(f(y|\theta)) + 2p \qquad (12.13)$$

where p is the number of independently adjusted parameters in θ, y is the observed data, and $f(y|\theta)$ is the likelihood function. (The term θ is used here to be consistent with control theory notation, which is where the AIC was first derived.) The "best" model minimizes the AIC.[6] If you aren't familiar with MLE methods, then (12.13) may look mysterious, but you can view it in two parts. The first part treats the goodness-of-fit of the model (with the parameter vector θ). The better the fit to the observed data, the greater the $\ln(f(y|\theta))$, and thus the smaller the $-2\ln(f(y|\theta))$. Remember that we want to minimize (12.13) so that better fitting models yield lower scores. The second part involves the cost for each parameter, where there are p parameters. Essentially, the goodness-of-fit component is penalized by adding a factor that is twice the number of free parameters. Each parameter that you add has to pay for itself by lowering $-2\ln(f(y|\theta))$ by at least two points. Otherwise, the AIC would favor not including that parameter.

An alternative to the AIC is called the minimum description length (MDL) principle [Rissanen, 1978, 1984]. This approach finds an optimum structure size that yields the shortest description of observed data in an information-theoretic sense. The criterion is

$$MDL(\theta) = -\log_2(f(y|\theta)) + 0.5p\log_2 n \qquad (12.14)$$

where there are n total observations of data and the other symbols retain the same meaning as in (12.13). The difference between (12.13) and (12.14) is most importantly in the form of the penalty function for the number of parameters. In the MDL, each additional parameter must pay for $0.5\log_2 n$ of reduction in $-\log_2(f(y|\theta))$, otherwise it hasn't "paid its own way." Both MDL and AIC have interesting mathematical properties that suggest their use or disuse [Kashyap, 1980; Fogel, 2000]. It's often reasonable to employ both criteria and consider any differences that result in choosing which model is "optimal."

Two other criteria that should be noted are the predicted squared error (PSE) method [Barron, 1984] and an older method employed for optimal subset selection in regression called C_p [Mallows, 1973]. The PSE statistic is

$$PSE = TSE + 2\sigma^2(p/n) \qquad (12.15)$$

---

[6] The derivation of AIC comes from Kullback–Liebler information theory statistics, which is beyond the scope of this book.

where σ^2 is a prior estimate of the true error variance that does not depend on a particular model, p is the number of coefficients in the model (analogous to p in the AIC and MDL criteria), and TSE is the average squared error of the model for n observations. The model that achieves the lowest PSE is considered best. Again, the lower the error of the model, the lower the TSE and thus the lower the PSE, but every additional parameter p costs $2\sigma^2/n$ points.

The C_p statistic [Mallows, 1973] has been more commonly used in statistical model building for optimal subset selection in regression analysis and is calculated as

$$C_p = SS_E(p)/\sigma^2 - n + 2p \qquad (12.16)$$

where $SS_E(p)$ is the sum of the squared errors between the observed and estimated outputs of the model, n is the number of observations, p is again the number of parameters in the model, and σ^2 is an unbiased estimate of the variance of an error term. Lower overall model error makes the first term lower, which is offset by the addition of twice the number of parameters required to obtain that model error. The expected value of $C_p = p$ if the fitted model has negligible bias, but $C_p > p$ otherwise. Generally, small values of C_p that are close to p are desired.

These four criteria, AIC, MDL, PSE, and C_p, are potential ways to trade off a model's goodness-of-fit for the number of degrees of freedom (parameters) that the model employs. Traditional gradient methods require fixing a model form first and then finding parameter values that optimize the objective function (such as lowest mean squared error), and then repeating that for multiple model forms, and then only at the end comparing results in terms of these sorts of information criteria. In contrast, evolutionary algorithms can be used to search over a landscape of varying model forms and parameter values simultaneously.

Example Consider the following example of applying an evolutionary approach to modeling real-world data in light of the AIC [Fogel, 1992b]. A time series of ocean acoustic data was recorded in the arctic with a sampling rate of 22 kHz. The data were observed to be predominantly "quiet ocean" with intermittent, short-duration ice cracks. These cracks (signals) lasted for less than 2.5 ms and were of relatively low signal-to-noise ratio (SNR) (i.e., about 0 dB as measured by taking 10 times the log 10 of the ratio of the estimated variance of the signal to the estimated variance of the noise). Three of these ice cracks were extracted for modeling. The data were scaled to be within a range of [−1, 1]. ARMA models of the time series were considered as a possible representation for the ice cracks. Figure 12.5 shows a time series of the first ice crack that occurs near the end of the depicted waveform.

The evolutionary approach started with a population of 250 candidate models of the form

$$A(q)y[t] = C(q)e[t] \qquad (12.17)$$

These were created initially at random by choosing model orders for A(q) and C(q) uniformly over the integers $\{1,\ldots,8\}$ and initializing the associated parameters

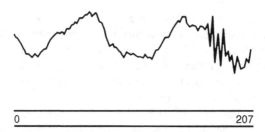

FIGURE 12.5 The waveform (time series) for data from the first ice crack. The crack appears as the larger amplitude impulse near the end of the time series (207 samples). (From Fogel (1992b).)

uniformly over the interval $[-0.5, 0.5]$. To review then, each candidate solution in the population consisted of a pair of integers, each being between 1 and 8 inclusive, which represented the number of lag terms in $A(q)$ and $C(q)$, respectively, along with a parameter vector of variable length with a real-valued entry for each parameter specified in $A(q)$ and $C(q)$. For example, one solution might have been

$$[2, 3, -0.1, 0.2, 0.4, 0.3, -0.02]$$

which could correspond to the model, with two lags in y and three lags in e:

$$y[t+1] = y[t] - 0.1y[t-1] + 0.2y[t-2] + e[t] + 0.4e[t-1] + 0.3e[t-2] - 0.02e[t-3]$$
$$(12.18)$$

where $y[t]$ is the amplitude of the ocean wave at time t and $e[t]$ is the standard Gaussian noise process. The coefficient on $y[t]$ and $e[t]$ is often assumed to be 1.0.

The performance of each candidate was defined in terms of its overall AIC value, that is, the sum of the independent AIC scores when the chosen model was used as a one-step ahead predictor of each individual ice crack time series. One offspring was created from each parent by randomly altering the order of the $A(q)$ and/or $C(q)$ polynomial (0.5 probability of each). The number of lags to be increased or decreased was chosen at random with respect to a Poisson distribution with a rate of 0.1. Each polynomial coefficient was altered by adding a Gaussian random variable with mean zero and variance inversely proportional to the fitness score.[7] In this procedure, an individual could be mutated by changing both its model form (order of $A(q)$ and $C(q)$) and its parameters simultaneously, thus providing a search over model construction and parameter optimization at the same time.

Competition for survival was handled by conditioning on the each contending solution in turn and comparing it with 10 other randomly selected solutions. If the contending solution offered a superior AIC, it received a "win." When performed across all contending solutions, this version of tournament selection focused on the

---

[7] More precisely, this was accomplished using an "affine" function, which is a function that uses a linear transformation and a translation, such as $f(x) = ax + b$.

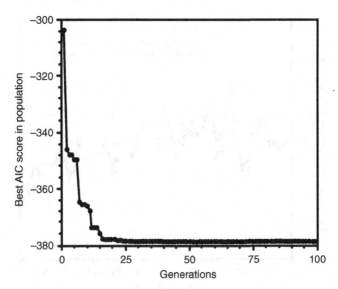

FIGURE 12.6 The rate of optimization of the best solution in the population for fitting the available ice crack data in terms of AIC as a function of the number of generations. (From Fogel (1992b).)

solutions with the most wins to be parents for the next generation. Overall, evolution was iterated for 100 generations.

Figure 12.6 shows the evolutionary optimization of the best model in the population at each generation in terms of AIC score. You'll note that there was a rapid improvement starting from randomized models to those that more appropriately fit the available data and did so with as few parameters as feasible. The best-evolved model after 100 generations had four lag terms in A(q) and one lag term in C(q):

$$A(q) = [\,1 \quad -0.7800 \quad -0.4040 \quad -0.1278 \quad -0.3552\,]$$
$$C(q) = [\,1 \quad 0.5479\,] \tag{12.19}$$

Figure 12.7 shows the predictive fit of these models to the available ice crack data for each of the three modeled events. The mean squared error per prediction over each sampled amplitude was 0.0402. Since there is no reason to believe that ice cracking is actually created by a 4-lag AR, 1-MA process, the result of the evolutionary modeling wasn't to show that evolutionary could find "the right model," but rather to provide a reasonable mathematical approximation to the physical process of ice cracking. Visually, to use a fuzzy term, the degree of fit looks *reasonable*.

For comparison, a recursive prediction error method (RPEM) was applied to the same data. Recall that if you wanted to use an RPEM method generally, you'd have to select the model form by hand and then use the gradient method of RPEM to find values for parameters. If you restricted attention to all models with up to 8 lags in AR and MA, that would be 64 possible models to consider. To facilitate the comparison,

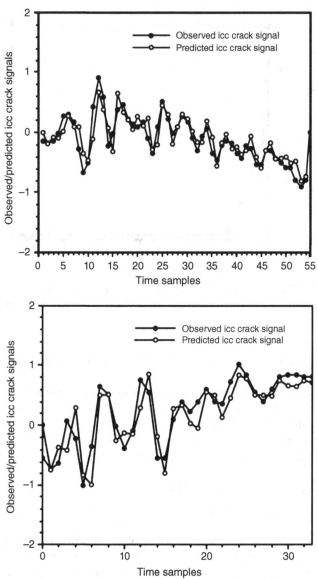

FIGURE 12.7   The predicted and observed amplitudes for the three samples of ice cracks. (From Fogel (1992b).)

attention was focused just on the same model structure as found by evolution (four lags in AR, one lag in MA) and RPEM returned the following coefficients:

$$
\begin{aligned}
A(q) &= [\,1 \quad -1.0113 \quad -0.6949 \quad -0.4579 \quad -0.1056\,] \\
C(q) &= [\,1 \quad 0.3525\,]
\end{aligned}
\tag{12.20}
$$

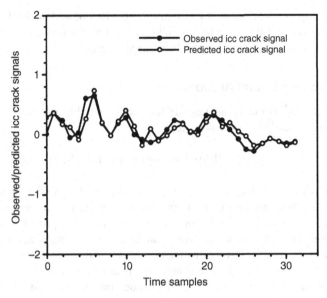

FIGURE 12.7 (*Continued*)

which translated to a mean squared error per sample of 0.0426, close but slightly higher than the 0.0402 that was attained by evolution.

The coefficients for the RPM appear quite different from those evolved, but one way to compare them is to examine what is called the *frequency response* of each model, which shows amplitude response as a function of frequency. Figure 12.8

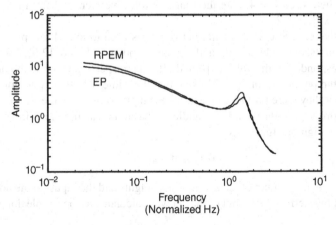

FIGURE 12.8 The frequency response of the best evolved model (EP) and a model of the same order optimized by recursive prediction error, a traditional method. In this case, the frequency response for both models is very similar despite having very different model coefficients. (From Fogel (1992b).)

shows the comparison of frequency response for the RPEM and evolved model, and the results are very similar. This makes an important broader point: models that look different in structure can have similar behavioral effects.[8]

12.2.2 Evolving Neural Networks

Although ARMA(X) models can generate enormously flexible behaviors, particularly when put in a so-called Box-Jenkins form:

$$A(q)y[t] = (B(q)/D(q))u[t] + (C(q)/F(q))e[t] \qquad (12.21)$$

where $A(q)$, $B(q)$, $C(q)$, $D(q)$, and $F(q)$ all are polynomials in the shift operator q^{-1}, it is often necessary to expand the possible form of model to a general nonlinear function. As described in the beginning of this book, neural networks can provide a convenient representation for such functions. Here's an example that combines evolutionary optimization with neural networks and also employs AIC to assess whether or not the evolved models explained the available data better than a noise-only model.

Consider the problem of modeling a chaotic time series such as the logistic difference equation [Fogel and Fogel, 1996]

$$x[t + 1] = \lambda \times x[t] \times (1 - x[t]) \qquad (12.22)$$

which is capable of diverse behavior [May, 1976] and depends on the value of λ. If $1 < \lambda < 3$, the fixed point for the equation is $x = 1 - \lambda^{-1}$. At $\lambda = 3$, the system bifurcates to give a cycle of period 2, which is stable for $3 < \lambda < 1 + 6^{0.5}$. As λ increases beyond this range, bifurcations give rise to a cascade of period doublings that lead to an apparently chaotic sequence for $3.57 < \lambda \leq 4$. The experiments presented here used $\lambda = 4$. One thousand samples were created from (12.22) using $\lambda = 4$ and $x[0] = 0.2$.

A multilayer feedforward neural network was used to model this process. It had two input units, two hidden units, and a single output unit (Figure 12.9). The two input units corresponded to the values $x[t]$ and $x[t - 1]$, respectively, while the output unit provided the prediction of $x[t + 1]$. Note that the values of $x[t - 1]$ do not enter into (12.22), but they were provided as extraneous input to examine the manner in which these uncorrelated data would be handled. The nonlinear functions in both hidden nodes were sigmoid functions

$$f(d) = (1 + e^{-d})^{-1} \qquad (12.23)$$

where d is the dot product of the appropriate weights and the inputs to the node, offset by a given bias term. The output of the net was calculated as the simple dot product of

---

[8] For example, within neural networks you could create a multilayer perceptron that performs very similar to a radial-basis function network, even though the structures of the two networks would be very different. Also, you may find in constructing evolutionary algorithms that the behavioral effects of mutation operators may be very similar to some other operator, such as crossover, in certain cases.

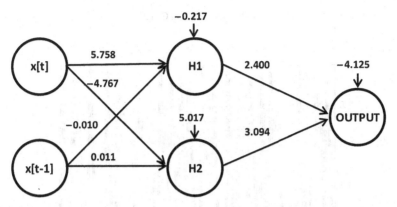

FIGURE 12.9   The neural network structure used to model the logistic system with $\lambda = 4$. There are two input nodes, two hidden nodes (H1 and H2), and one output node. The figure shows the best evolved neural network after 1000 generations. Note that the weights from the input node associated with $x[t - 1]$ are very close to zero, indicating that the evolutionary process learned that $x[t - 1]$ was not involved in the system. (From Fogel and Fogel (1996).)

the weights and inputs to the final node, offset by the associated bias term. No sigmoid function was applied to the output.

The evolutionary approach included 50 parent neural networks, each initialized with uniformly randomized weights and biases from $[-2, 2]$. Offspring were created from the parents by adding a Gaussian random variable with zero mean and variance equal to the mean squared error of the parent to every weight in the parent network.[9] The process generated one offspring from each parent and was iterated for 1000 generations.

Two experimental setups were employed to examine two different facets of this modeling problem. The first did not employ any noise in (12.22) and the experiment sought to determine if the evolutionary process would learn that the input associated with $x[t - 1]$ was not relevant for future predictions. The second employed an additional noise term

$$x[t + 1] = \lambda \times x[t] \times (1 - x[t]) + e[t] \tag{12.24}$$

where $e[t]$ was zero mean Gaussian noise with a standard deviation that was approximately the same as the standard deviation of the logistic process without noise in (12.22). This creates a signal-to-noise ratio problem of being able to identify whether or not there was a chaotic signal embedded in the noise.

For the first case, Figure 12.9 shows the best-evolved weights and bias terms. The weights from the input node for $x[t - 1]$ were set very close to zero, which was as expected because $x[t - 1]$ does not enter into the generation of $x[t + 1]$. The agreement between the actual values generated by the system and the neural network model are shown in Figure 12.10.

---

[9] This procedure allows a smaller step-sized mutation as performance improves; however, it requires an estimate of a lower bound on the error. In this case, the estimated lower bound was 0.0, but this is not the case generally, and thus self-adaptive variation methods are favored as a more general approach.

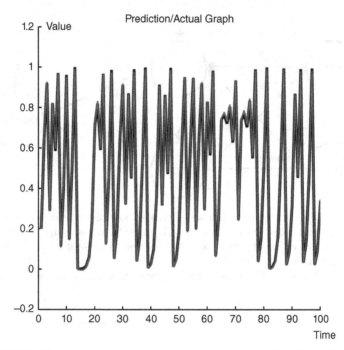

FIGURE 12.10 The agreement between the actual logistic system (dark line) with $\lambda = 4$ over 100 time steps and the output of the best evolved neural network (red line) shown in Figure 12.9. (From Fogel and Fogel (1996).)

For the second case, Table 12.1 shows a comparison of the AIC score for each of five trials using the noisy logistic model (12.24). In this case, the best-evolved neural network was compared against a Gaussian noise-only model. Recall that the AIC incorporates both the goodness-of-fit of a model and also the number of parameters used. The evolved neural networks could be expected to generate a better fit than a noise-only model, but it was not known ahead of time whether or not a signal could be detected in noise by comparing the evolved neural networks with a noise-only model. The expected AIC generated by assuming only a Gaussian noise process for (12.24) is -1368.1. In each of the five evolutionary trials, the AIC for the evolved neural

TABLE 12.1   The AIC Value for Each Trial with the Noisy Logistic Function

Trial	AIC for Evolved Neural Network	Expected AIC Noise-Only Hypothesis
1	-1443.3	-1368.1
2	-1478.2	-1368.1
3	-1398.1	-1368.1
4	-1452.8	-1368.1
5	-1502.1	-1368.1

Source: From Fogel and Fogel (1996).

network was lower, indicating that it was a better explanation for the data than was a noise-only model. Thus, this provided a confirmation that the evolutionary method was able to detect a chaotic signal in noise, in this case at a level where the standard deviation of the signal was on par with the standard deviation of the noise.[10]

12.2.3   Evolving Multiple Interacting Programs as Networks

There's an extension to the concept of evolving neural networks that hasn't received comparable attention, but possesses a flexibility that makes it interesting. Recall that neural networks can include feedback loops, rather than just be feedfoward, and also recall that when using evolution to optimize neural networks, the functions inside the nodes do not necessarily need to be differentiable because no calculus methods are being used to find points of optimality. With that in mind, consider the possibility for having arbitrarily connected networks, in which each node is itself an arbitrary function.

For example, suppose a tree-based representation [Koza, 1992; Chellapilla, 1998] were used to represent the transfer functions inside the nodes of a neural network (see Section 12.2.2). A tree structure might represent a function such as

$$T_0 = (x + T_1)T_0 \qquad (12.25)$$

in the following form:

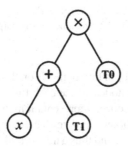

where T_1 is an input from another node, x is an observed variable, and T_0 is the prior output the last time the function was called. Suppose there were several functions like this, each acting as the transfer function in a recursive neural network. The result might appear as shown in Figure 12.11.

In this "neural network," the values of T_0, T_1, and T_2 are initialized at specified values, such as 1.0. The value x is an observed value, which could be a function of time (i.e., x[t]), and one node is designated as the output node for the network (in this case the node that defines the function for T_0). The single input x is passed to all three other nodes, but each of these nodes performs a mapping function (which is later subject to evolution by variation and selection). T_0 depends on previous values of

---

[10] This is also known as 0 decibels or 0 dB. The results are similar to those offered in Fogel and Fogel (1996).

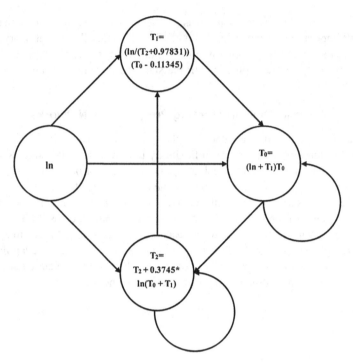

FIGURE 12.11   A multiple interactive programs (MIPs) network. The structure looks much like a neural network. The value of T_0 is presumed as the output of the network. The value "In" represents an input.

itself, as well as the value T_1. T_1 in turn depends on the results from T_0 and T_2, and T_2 depends on all three noninput nodes. Since each function inside of each node can be represented as a tree, and since tree representations were developed early in evolutionary computation within a branch known as *genetic programming*, these functions are often described as *programs*, and thus a network like this is described as a network of multiple interacting programs, or a *MIPs net*.

Evolution can operate on MIPs networks by adjusting both the topology of the network, including adding and deleting nodes, and the connectivity of the nodes, as well as the functions inside the nodes. This is constrained by having to include the input variables and output variables. (In the diagram above, the net would always include an input node and it would always include a function for T_0, which was designated as the output.) The functions inside the nodes can be modified via recombination and/or mutation (see Section 11.1).

One example [Angeline, 1998] used evolutionary MIPs nets to predict annual sunspot data.[11] Figure 12.12 shows the rate of optimization of the best MIPs net in terms of the mean squared error when using 100 parents to generate 150 offspring

---

[11] As of this writing, sunspot data are available at http://www.heatonresearch.com/wiki/Workbench_Time-Series_Example.

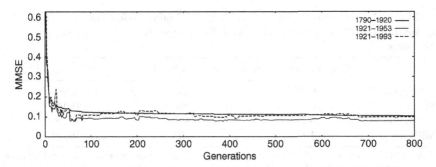

FIGURE 12.12 The normalized mean squared error of the best-evolved MIPs network applied to the problem of predicting sunspots. (From Angeline (1998).)

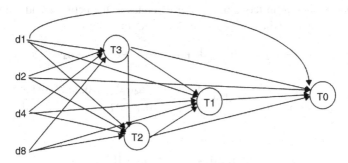

FIGURE 12.13 The structure and equations representing the best-evolved MIPs network after 800 generations. (From Angeline (1998).) The values d_1, d_2, d_4, and d_8 indicate previous observations from 1, 2, 4, and 8 time steps in the past.

each generation over 800 generations. Figure 12.13 shows the architecture that resulted from the evolution and Figure 12.14 shows the performance of the best model on future data in the sunspot series. Table 12.2 compares results from other literature

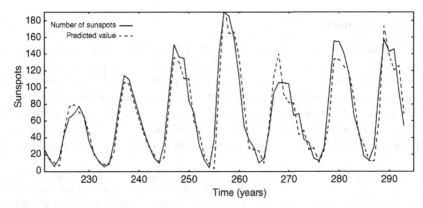

FIGURE 12.14 The goodness-of-fit to the number of sunspots per year that was obtained by the best MIPs network. (From Angeline (1998).)

TABLE 12.2   A Comparison of the Training and Testing Performance Obtained in Angeline (1998) Using MIPs Networks and Other Publications Using Neural Networks

Study	Total Number of Nodes	Training Set NMSE	1921–1950 Test Set NMSE
Weigend *et al.*	16(12-3-1)	0.082	0.086
Svarer *et al.*	16(12-3-1)	0.090	0.082
Aerrabotu *et al.*	16(12-3-1)	0.100	0.106
MIPs	8(4-1-1-1-1)	0.103	0.079

Although the MIP networks had a slightly higher training error, the test set performance in years 1921–1993 was better than published results. NMSE: normalized mean squared error.

on sunspot prediction in terms of number of nodes in the network and the training/testing performance, which favored the MIPs net on testing.

$$T_3 = \sigma(-0.8637 - (d_2 - 0.1021 + d_4 + d_4 d_4 + (d_2 + d_1 + d_4 + d_4 + d_4)d_8)) \quad (12.26)$$

$$T_2 = \sigma\left(\frac{d_2}{T_3/(d_1 + d_4 + T_3 d_1)} - d_4 + \frac{d_8 + d_2}{-0.9370} + \frac{d_4 d_4 d_4}{0.1120 T_3} - \frac{0.3027 - (0.0657 + d_2)}{d_2 + d_1}\right)$$
$$(12.27)$$

$$T_1 = \sigma\left(\frac{T_2 d_1/(d_4 + d_8) + (d_4 - 0.4183)\times}{(d_8 + d_8/(T_2 + 0.4601 + T_3 - (T_3 + 0.9195)) - T_2) + d_4 + (T_2/d_1) - d_8} - d_8\right)$$
$$(12.28)$$

$$T_0 = \frac{d_1 - T_1 T_3 d_2}{T_3 + T_2} \quad (12.29)$$

12.2.4 Summary

Input–output modeling is central to many scientific activities. Traditional methods of input–output modeling have required a person (the investigator) to use judgment to offer a model form for consideration and then use gradient-based methods to optimize parameters for the model. This results in an iterative approach, as it's rarely the case that someone can consider the "best" model on an initial attempt, and gradient search methods may stall at local optima even if the model is "correct."[12] The evolutionary approach described here is also iterative; however, it offers the possibility of

---

[12] The word *correct* is put in quotes because there is something known as "the myth of the mathematical model" in that nature presents us with physical systems, not mathematical systems. We employ mathematics to gain a better understanding of those systems; however, it is a mistake to believe that mathematical models are equivalent to nature. Thus, in most cases of modeling nature, there is no such thing as a "correct" model, only models that provide better or worse predictive understanding of the phenomena of interest.

automating much of the exploration for model form and associated parameters, particularly when coupled with information-theoretic criteria that can trade off the goodness-of-fit of the model for the number of parameters it employs.

Time series prediction models provide an opportunity to illustrate an important point in evolutionary computation. The effectiveness of a particular variation operator depends directly on the match between the representation and the evaluation function. Research in Chellapilla *et al.* (1997a, 1997b) showed that four completely equivalent forms of linear filters (i.e., mathematical models) generated completely different error landscapes in the neighborhood of the optimum instantiation for a given set of data (see Figure 12.15). Each of these representations involved the use of continuous

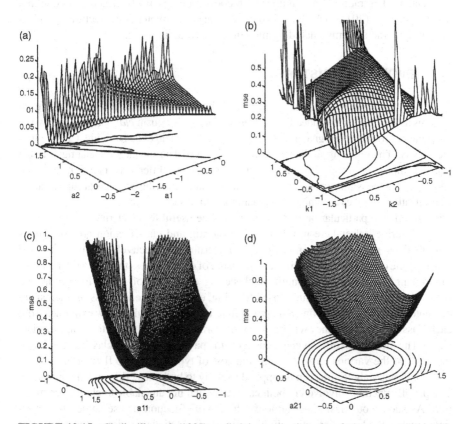

FIGURE 12.15 Chellapilla et al. (1997) studied the application of evolutionary system identification applied to a system governed by the transfer function $H(z) = [0.0154(1 + 3z^{-1} + 3z^{-2} + z^{-3})]/[1.0 - 1.99z^{-1} + 1.5720z^{-2} - 0.4583z^{-3}]$ driven by white noise. The letter z is employed to be consistent with system identification literature. There are four equivalent model forms that can be used to describe the system, known as *direct, lattice, cascade,* and *parallel.* Each uses continuous-valued coefficients. The graphs (a)–(d) show the mean squared error surface plotted in the neighborhood of the optimal solution for each of these model forms, respectively. Even though each model form is equivalent in function, the surface that each generates varies dramatically from a ridge function to a quadratic bowl. The proper variation operators to use differ in each case.

parameters, and yet the appropriate variation operators to use would certainly be much different in each case. Thus, there is plain evidence that neither the cardinality of the representation (coding in real numbers versus, say, binary numbers) nor the model form itself is sufficient to indicate which variation operator(s) will be best. It is the degree of fit between the operator, the representation, and the error surface that ultimately determines the optimization performance of an evolutionary algorithm.

The application of evolutionary algorithms to time series prediction is wide ranging and very engaging. Some recent publications of interest include Mirmomeni and Punch (2011), which focused on forecasting chaotic times series using coevolutionary models (coevolution is described later in this chapter), Huang et al. (2009), which analyzed financial market trading opportunities with coevolutionary fuzzy predictive modeling, and Braun et al. (2011), which studied evolutionary parameter and structure optimization in dynamic systems, including modeling a hydraulic valve.

12.3   EVOLVING CLUSTERS

Clustering has already been described within the areas of neural networks and fuzzy systems as a core activity. Indeed, *clustering* and its close cousin *classification* are two of the principal concerns in signal processing or the broader field of data mining. Clustering is the activity of recognizing similarities between observed data and then associating different data together as a group. Also important is recognizing differences between observed signals so that the within-group variance is much smaller than the between-group variance. Classification involves assigning particular labels to observed signals. Such labels may or may not refer to particular properties that would be useful for clustering.

As described elsewhere in this book, clustering and classification are sometimes described as *unsupervised* and *supervised* learning, respectively. In unsupervised learning, the clustering algorithm uses a measure of effectiveness that typically involves minimizing the description length of clustering or a trade-off of within-group and between-group variance. In supervised learning, examples of each class of interest are presented and the task is to find a mapping function such that when new examples from each class are observed, they will be classified correctly (i.e., assigned to the appropriate class). The measure of effectiveness concerns the payoffs for correct classifications, and the costs of the various errors (at least the cost of type I and type II errors).

Evolutionary algorithms can be applied usefully to both clustering and classification. The payoff functions involved in both cases are often not amenable to classic optimization. As such, you must either simplify the payoff function or use some other non-traditional method of optimization. Evolutionary optimization provides one possibility for consideration.

12.3.1   Evolutionary Information: Theoretic Clustering with Rotatable Hyperboxes

Suppose data are observed in two continuous dimensions, x_1 and x_2, and must be clustered into groups with similar properties. One approach to this clustering problem was offered in Simpson (1992, 1993), which suggested using boxes as clusters to

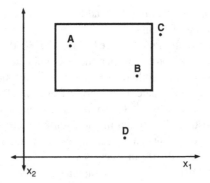

FIGURE 12.16 In the diagram, there are two measurable parameters x_1 and x_2. A box is drawn that defines a cluster. Points contained within the boundaries of the box (e.g., points A and B) are said to be complete members of that cluster, or possess membership 1.0. Points outside the box (e.g., C and D) could still be considered members of the cluster defined by the box, but they would possess membership less than 1.0. Typically, membership is based on a metric such that D is given less membership than C.

surround grouped data. (In higher dimensions, these would be "hyperboxes," but in two dimensions as we have here, they are simply rectangles.) Others, such as Buckles et al. (1994), suggested a similar idea of using ellipses (or hyperellipses), but for the sake of illustration, we'll focus on using boxes. Simpson viewed these boxes as fuzzy clusters, with elements inside a box having a membership of 1.0 and elements outside a box having a membership that depended on a rule for that box (see Figure 12.16).

In Simpson (1992, 1993), the boxes were always aligned with the coordinate axes, but data are not always aligned with coordinate axes, so let's consider an extension in which the boxes can be rotated and thus not aligned with coordinate axes. As reported in Ghozeil and Fogel (1996a), a population of 50 parents was created in which each solution encoded a complete clustering: a set of hyperboxes (in this case, rectangles), represented as a five-tuple:

$$(x, y, \theta, w, h)$$

where x and y defined the (x, y) position of the center of the hyperbox, θ was the rotation in radians anticlockwise of the hyperbox around its center, w was the length of the sides of the hyperbox in the direction of θ (i.e., the width), and h was the length of the sides of the hyperbox in the direction perpendicular to θ (i.e., the height). This evolutionary approach also employed self-adaptation (see Section 11.5), incorporating a five-tuple of self-adaptive parameters that were used to control the generation of an offspring:

$$(\sigma_x, \sigma_y, \sigma_\theta, \sigma_w, \sigma_h)$$

where each element was defined as the standard deviation of a Gaussian mutation for the corresponding component (explained in more detail below).

Each parent was also defined by a parameter, Nbox, indicating the number of hyperboxes in its solution, and two additional self-adaptive parameters, σ_{AddBox} and

σ_{DelBox}, for controlling the likelihood of adding boxes or deleting boxes from a solution when creating an offspring.

So, for example, one solution might have been (1, 20, 30, 0, 10, 40, 0.25, 0.25, 5, 5, 0.5, 5, 5). This would, in order of parameters, indicate one box, centered at $(x, y) = (20, 30)$, with $\theta = 0$, width of 10, height of 40, and with the remaining parameters governing how to create an offspring from the offspring of this solution $(\sigma_{AddBox}, \sigma_{DelBox}, \sigma_x, \sigma_y, \sigma_\theta, \sigma_w, \sigma_h)$.[13] Of course, a clustering solution with only one cluster isn't very interesting. The actual method started each parent with five boxes, and initialized the parameters at values that were considered reasonable given the range of the observed data (0–100 in x and y).

Mutation worked by creating a new solution via Gaussian random variation of each existing box. For example, the x location of a new offspring was set to

$$x' = x + N(0, \sigma_x) \tag{12.30}$$

and similarly for the other parameters of each box. The number of boxes was raised (up to the maximum of five) or lowered (down to the minimum of one) probabilistically. New boxes were initialized at random. The self-adaptive parameters were updated using a lognormal approach akin to that described in Chapter 11.5, such as

$$\sigma'_{AddBox} = \sigma_{AddBox} \times exp[N(0.3, \tau_1) + N(0.3, \tau_2)] \tag{12.31}$$

where $\tau_1 = [2(Nbox)p]^{0.5}$ and $\tau_2 = [2(Nbox)p^{0.5}]^{-0.5}$, and there were $p = 5$ dimensions per box. The product of Nbox and p yields the total dimensions involved in that particular solution. The other self-adaptive parameters were determined by

$$\sigma'_x = \sigma_x \times exp[N(0.3, \tau_1) + N(0.3, \tau_2)] \tag{12.32}$$

and similarly for $\sigma'_y, \sigma'_\theta, \sigma'_w$, and σ'_h. (*Note:* The value 0.3 that was included in the self-adaptive update was determined empirically and appeared to offer a bias toward increasing the mutation step sizes in balance to selective pressures that reduce step sizes.)

Unlike the prior time series modeling described in Section 12.2, the work in Ghozeil and Fogel (1996a) relied on the minimum description length principle. The details of the implementation are lengthy, but in summary,

1. MDL requires an assumption about the likelihood function for the data. For the experiments in Ghozeil and Fogel (1996a), the likelihood function was a uniform distribution, where the likelihood for any point in a hyperbox was the inverse of the volume of the hyperbox. It was required that hyperboxes have at least one more point in them than there were dimensions in the data.

2. Any hyperboxes that contained no data were pruned and not evaluated as part of any proposed clustering solution.

---

[13] In this case, the box parameters were generated first and then the self-adaptive parameters were varied. So the new self-adaptive parameters could only be tested after the offspring had made it through a round of selection. This is perhaps an atypical approach. The more typical approach varies the self-adaptive parameters first and then uses the new self-adaptive parameters to create the offspring.

3. An "outlier" hyperbox was always included in every clustering to take care of scoring any points not contained in other hyperboxes. This "outlier" hyperbox was defined to cover the entire range of the available data. (In Ghozeil and Fogel (1996a), data ranged from [0,100] in x and y.)

Figure 12.17a and b shows the result of evolving the clustering for data generated uniformly in a rotated hyperbox. Figure 12.17a shows the rate of optimization for a 50

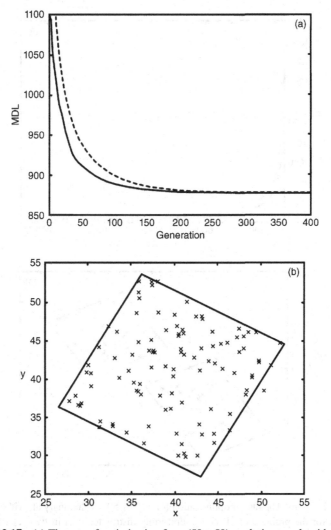

FIGURE 12.17   (a) The rate of optimization for a (50 + 50) evolutionary algorithm over 400 generations for the problem of clustering uniformly distributed data in a rotated box. The solid line is the best score at each generation, while the dashed line is the mean score of all surviving parent solutions. (b) The final best-evolved clustering for the data generated by placing data uniformly at random in a rotated box. (From Ghozeil and Fogel (1996a).)

parent/50 offspring population over 400 generations. Figure 12.17b shows that the optimization of the MDL score generated a box that contained almost all of the data. One point at approximately (52,38) is outside the main box, and is therefore considered part of the outlier box.

Figure 12.18a–c shows a similar result for data generated uniformly in two rotated boxes. Figure 12.18b shows a typical best-evolved clustering, which could be improved still by tightening the box in the lower left to fit the data in its cluster

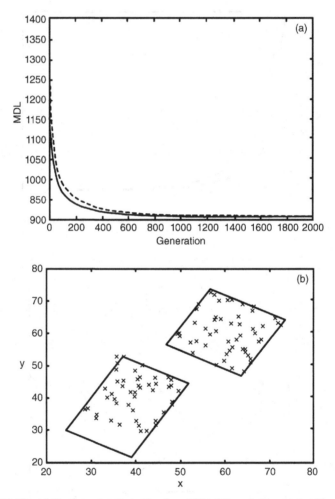

FIGURE 12.18 (a) The rate of optimization for a (50 + 50) evolutionary algorithm over 2000 generations for the problem of clustering uniformly distributed data in two separated rotated boxes. The solid line is the best score at each generation, while the dashed line is the mean score of all surviving parent solutions. (b) A typical best-evolved clustering for the case of data uniformly distributed from two rotated boxes. These results could be improved still by tightening the box in the lower left to fit the data in its cluster more closely. (c) An unusual (1 in 50) result from the evolution on the problem in part (a). Occasionally, the evolution would fail to separate the two boxes, opting for a single box instead. (From Ghozeil and Fogel (1996a).)

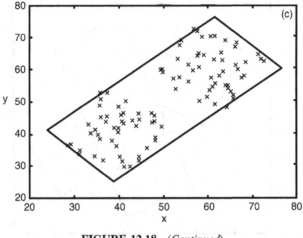

FIGURE 12.18   (*Continued*)

more closely. In 1 of 50 trials, the result in Figure 12.18c was obtained, which was suboptimal; however, it was within one standard deviation of the mean of all the best clusterings in 49 other trials.

Figure 12.19a and b shows the results of clustering two overlapping rectangles. In this case, because the MDL penalizes boxes that overlap—it isn't a minimum description of a data point to place it in two or more boxes—the best-evolved

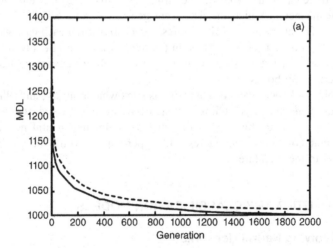

FIGURE 12.19   (a) The rate of optimization for a (50 + 50) evolutionary algorithm over 2000 generations for the problem of clustering uniformly distributed data in a rotated box. The solid line is the best score at each generation, while the dashed line is the mean score of all surviving parent solutions. (b) A typical best-evolved clustering for the case of data uniformly distributed from two rotated boxes that overlap in a cross. These results could be improved still by tightening the box on the left to fit the data in its cluster more closely while picking up the two points that remain outside the cluster. (From Ghozeil and Fogel (1996a).)

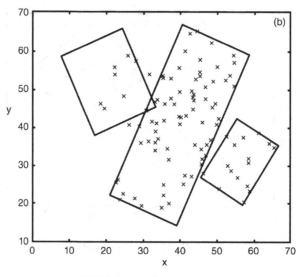

FIGURE 12.19   (*Continued*)

clustering after 2000 generations shows three boxes and the leftmost could be optimized further to produce a tighter cluster, which would ultimately remove the two outlier data points from the outlier box.

For the three cases, the MDL of the "true" data clustering (i.e., a clustering that reflects the manner in which the data were generated) were 887, 893, and 1025, respectively. The average best MDL scores across 50 trials in each case were 877.7, 909.1, and 1004.6, respectively. Thus, in the final case, the best-evolved MDL was actually lower (i.e., better) than "truth" because of the disadvantage truth has in having overlapping boxes.

The choice of hyperboxes for clustering is somewhat arbitrary, and other shapes could be used, notably hyperellipses, which have received more attention in the literature (for example, this is akin to radial basis function neural networks). An extension that combines hyperboxes and hyperellipses would be an interesting contribution to the literature.

12.4   EVOLUTIONARY CLASSIFICATION MODELS

12.4.1   Evolving Neural Networks

There are voluminous examples of applying evolutionary optimization to adjust the weights and/or topology of neural networks. One example described later in this chapter treats the problem of controlling a cart–pole system comprising multiple poles. That system uses a fully connected recursive neural network. For an example in classification, however, the case illustrated here involves a feed-forward neural network applied to a problem in breast cancer detection.

Carcinoma of the breast is second only to lung cancer as a tumor-related cause of death in women. The U.S. Centers for Disease Control and Prevention (CDC) indicate that over 220,000 women were diagnosed with breast cancer in 2011 in the United States (the most recent available year) and over 40,000 deaths occurred. Breast cancer begins as a focal curable disease, but it is usually not identifiable by palpation at this stage. Mammography remains a mainstay in effective screening.

The use of neural networks to assist mammographers in screening mammograms began in the mid-1990s. In one example, Fogel et al. (1998) employed evolutionary algorithms to optimize the weights of a fixed, small, neural network. In this case, data were acquired from 216 cases of suspicious mammograms. These were cases that were potentially suggestive of a malignancy; however, in each case, the woman had subsequently undergone a fine needle aspiration and assessment—a procedure in which cells are removed at a particular location and examined under a microscope. Thus, the "truth" of whether or not each woman had cancer could be known about each of these cases, to the extent possible: It's always possible that the fine needle aspiration may be taken and miss some cancer cells, or that a reading may be performed incorrectly. Of the 216 cases, 111 were associated with a biopsy-proven malignancy and 105 cases were indicated to be negative by biopsy.

A domain expert (Eugene Wasson, MD, Maui Memorial Hospital, Maui, HI) assessed these mammograms in terms of the following features:

1. Mass size: either zero or in millimeters
2. Mass margin: (each subparameter rated as none (0), low (1), medium (2), or high (3))
   a. Well circumscribed
   b. Microlobulated
   c. Obscured
   d. Indistinct
   e. Spiculated
3. Architectural distortion: none or distortion
4. Calcification number: none (0), <5 (1), 5–10 (2), or >10 (3)
5. Calcification morphology: none (0), not suspicious (1), moderately suspicious (2), or highly suspicious (3)
6. Calcification density: none (0), dense (1), mixed (2), faint (3)
7. Calcification distribution: none (0), scattered (1), intermediate (2), clustered (3)
8. Asymmetric density: either zero or in millimeters

These features, along with patient age, were recorded for each case as rated by the domain expert and used as inputs to a neural network with 13 inputs (12 radiographic features and patient age). The neural network had only two hidden nodes (this was as compared to other literature using backpropagation to train networks in similar cases that employed an order of magnitude more nodes, for example, Wu et al. (1993)). The hidden nodes were sigmoid functions:

$$f(y) = (1 + e^{-y})^{-1} \tag{12.33}$$

where y was the sum of the bias term and the dot product of the input feature vector and the associated weight vector.

Rather than use a typical train–test–validate approach, leave-one-out cross-validation was employed. This procedure removes the first data example from the available data and trains on the remainder. It then employs the best evolved neural network to predict the case for the held-out data. After that, the held-out example is returned and the next in sequence is held-out. The process is repeated for all data. In the end, for n examples to learn from, there are n held-out cases, each based on training all the remaining data.[14]

In the available data, there were 158 cases that involved suspicious masses; the other cases involved suspicious calcifications without suspicious masses. For these 158 cases, a population of 250 neural networks was evolved over 200 generations in a (250 + 250)-EA using standard Cauchy mutations on weights and biases and self-adaptation to adjust the scaling on the Cauchy mutation. The self-adaptation step for the ith scaling factor was

$$\sigma_i' = \sigma_i \exp\left(\tau N(0, 1) + \tau' N_i(0, 1)\right) \tag{12.34}$$

where n = 31 parameters, $\tau = (2n)^{-0.5}$, $\tau' = (2n^{0.5})^{-0.5}$, $N(0, 1)$ is a standard normal random variable sampled once for all n parameters of the vector σ, and $N_i(0, 1)$ is a standard normal random variable sampled anew for each parameter. (As described in Chapter 11, this is a standard approach in self-adaptation.) These updated self-adaptive parameters were then used to generate new weight values for the offspring according to the rule

$$x_i' = x_i + \sigma_i' C \tag{12.35}$$

where C is a standard Cauchy random variable. Selection eliminated half of the total parent and offspring weight sets based on their observed error performance using a form of tournament selection.

Figure 12.20 shows a typical rate of optimization in each training run, while the effectiveness of the overall process is shown in Figure 12.21 using a receiver operating characteristic (ROC) curve, which shows the trade-off between the probability of correctly identifying a malignancy and the probability of a false positive (type I error). Often in ROC analysis, attention is given to the area under the curve (AUC), for which bigger numbers are better. (The best would be a 1.0 probability of detection and a 0.0 probability of false alarm, which would yield an AUC = 1.0.) In this case, the AUC was approximately 0.9196. Another way to

---

[14] There are other versions of cross-validation that hold out more than 1 case at a time. These are often called k-fold cross-validation, where k is the number held-out at each "fold."

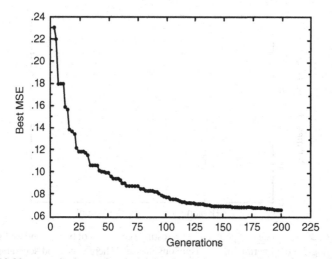

FIGURE 12.20   A typical rate of optimization for the best-evolved neural network in terms of minimizing mean squared error between the network's output and the target value (0 = benign, 1 = malignant) on features describing suspicious masses in mammograms. (From Fogel et al. (1998).)

examine the discriminatory capability of a classifier is a cell point chart (see Figure 12.22), which shows good separation between the two classes for this approach.

It's important to compare different-sized neural networks in classification because of the classic problem of overfitting versus undergeneralizing. (A function—like a neural network—with too many parameters can overfit training data and give poor

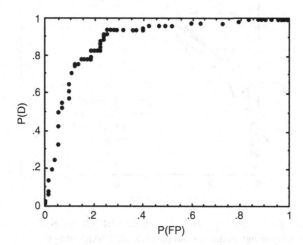

FIGURE 12.21   A typical receiver operating characteristic (ROC) curve for a best-evolved neural network with two hidden nodes on the mammogram data describing suspicious masses. (From Fogel et al. (1998).)

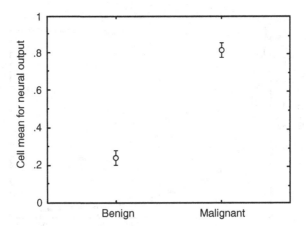

FIGURE 12.22 A typical cell point chart for testing separation of benign and malignant cases in the mammogram data pertaining to suspicious masses. There was good separation between the classes. (From Fogel et al. (1998).)

performance on new data. But a function with too few parameters may not be able to generalize sufficiently from the available training data, and therefore also give poor performance on new data.) Figure 12.23 shows a comparison of ROC curves for evolved neural networks using 1–5 hidden nodes when applied in leave-one-out cross-validation to the entire data set of masses and calcifications. The performance of the neural network with two hidden nodes dominates the others, and notably the performance of a network with only one hidden node was much worse.

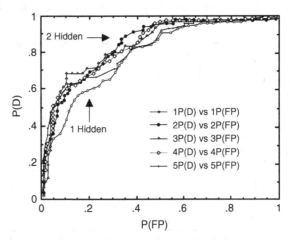

FIGURE 12.23 The receiver operating characteristic (ROC) curves obtained when classifying all available data (including masses and microcalcifications) using neural networks with one to five hidden nodes. The results show both the benefit of using multiple hidden nodes as compared to a single hidden node and that there was no benefit for having more than two hidden nodes. (From Fogel et al. (1998).)

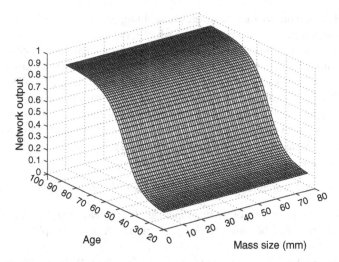

FIGURE 12.24   The output of the best-evolved neural network for the case of suspicious masses in which mass size is varied from 0 to 80 mm and the patient age varied from 20 to 100 years of age with all other inputs held at their mean values. The network's output did not tend to vary with mass size, but exhibited a sigmoid relationship to patient age. (From Fogel et al. (1998).)

Neural networks sometimes present a challenge in determining the relationships between input and output data that have been captured by the distributed network. In Fogel et al. (1998), efforts were made to better understand the relationship between the output of the best evolved neural networks and the inputs of patient age and mass size. To do this, over the 158 cases of suspicious masses, all variables other than patient age and mass size were held constant at their average values, while age and size were varied from 20 to 100 (years) and from 0 to 80 (mm), respectively. Figure 12.24 shows the result, in which the neural network's output increased as a sigmoid function of patient age, but appeared independent of mass size. The relationship to patient age was not surprising—the recommendation for annual screening for mammograms in the United States is presently controversial between the ages of 40 and 50, which is reflected in Figure 12.24 as the range of time when risk is likely to increase. The lack of relationship of mass size to malignancy was potentially surprising; however, this may reflect early detection of smaller cancers and the presence of large benign masses (fibro adenomas) in the available data.

12.4.2 Evolving Rules

Evolutionary algorithms can be used to optimize decision trees or sets of rules just as well as neural networks, fuzzy clusters, or other data structures. One example of this is found in Porto et al. (2005), in which evolution was used to optimize rule sets for classifying sonar returns from man-made or natural objects underwater. The data used in this example were from active sonar units that produced various signals in a shallow water field containing four metal spheres resembling sea mines positioned at different depths. Signals were taken from settings in which these man-made objects were present or absent, in which case the sonar would pick up natural sources of reflectance. There were four input features that were derived from the sonar returns at different times. These features measured how "peaked" a particular return might be. (The more the "peaked," the more likely the return to have come from a metal sphere.) Data consisted of 425 examples, which were divided into training, testing, and validation sets.

The process of evolving rules was in essence that of evolving a program. Rules were determined as sets of expressions, connecting logical operators, and consequents. For example, a rule might take the form of

$$\text{IF (Input Variable \#1} > 0.5) \text{ THEN (TempVar \#2} \mathrel{+}= 0.2) \qquad (12.36)$$

which means that if the value of the first input variable is greater than 0.5, the value in the second temporary variable must be increased by 0.2. A more complex rule might be

$$\text{IF ((Input Variable \#1} > 0.5) \text{ AND (TempVar \#3} < 0.5) \text{ OR (Input Variable \#3} >$$
$$\text{Input Variable \#4) THEN (Output} \mathrel{+}= \text{TempVar \#3)}$$

$$(12.37)$$

Porto et al. (2005) arbitrarily limited the number of expressions in any rule from one to twice the number of input nodes. Each rule had to contain at least one and up to a maximum (user-defined) number of expressions, where each expression contained the values from at least one of some number of descriptors and temporary variables (TV). For convenience, a procedure was adopted where if any expression or logic with an antecedent before or after an AND statement was false, then the entire antecedent was considered false. Thus, in the more complex rule above, if the first or second condition were false, then the entire rule would be considered false regardless of the relationship between the third and fourth input variables.

An expression had three parts: (i) a left-hand side (LHS), which was either a descriptor or a temporary variable (TV), (i) an operator, which was either less than, less than or equal to, greater than, or greater than or equal to, and (iii) a right-hand side (RHS), which was either a fixed real number or a TV or another descriptor. Two or more expressions in a rule were connected by a logical operator AND or OR. The full combination of expressions and logic operators was termed an antecedent.

For each rule there were two possible consequents: THEN or ELSE. If an antecedent was true, what follows the THEN was considered, otherwise what follows the ELSE (if it exists) wasn't considered. There were three possible actions following the consequent: (i) do nothing, (ii) update the output, and (iii) alter a TV. Every rule was required to have a THEN consequent that was used when the rule as evaluated was true. The rule could have an optional ELSE consequent. A consequent had three parts: (i) a LHS, which was an output or TV, (ii) an operator, which was an increment ($+=$), decrement ($-=$), or set equal to ($=$), and (iii) a RHS, which was either a TV or a real number.

The rule set generated an output value that was used for classifying data. The value of the output was used when all of the rules in a rule set were executed in order. There were numerous parameters governing the evolution of the rule sets, including the typical number of parents, offspring, generations, inputs used, and so forth. In addition, the following specific conditions were also applied:

- The initial number of temporary variables (set to 2)
- The maximum number of temporary variables (set to 20)
- The initial number of rules (set to 20)
- The maximum number of rules (set to 30)
- The possible initial values for temporary variables (set to 0.0 or 0.5, at random)
- The starting sigma values for evolving real values (set to 0.1)
- A collection of 32 self-adaptable probabilities for (i) applying operators, including operators to add, delete, swap, reverse, reinsert, or copy rules, or reinsert a block of rules, (ii) mutating rules by modifying the LHS, RHS, negation, add/delete an ELSE, changing logic, add/delete a temporary variable, or changing a temporary variable value, and (iii) combining rules through one-point, two-point, or uniform crossover, and also through each of the corresponding so-called "headless chicken" or "random" forms of these operators.

Figure 12.25 shows a comparison of training and testing while evolving rule sets for classifying the sonar data. Test set performance was computed at each generation of training. While training performance improved consistently over time, test set performance achieved a best result around generation 350, indicating the possibility for overfitting data in the training set. Ultimately, Figure 12.26 shows a representative separation between classifying the spheres versus natural objects; the separation is clear, and is reflected in the ROC curve shown in Figure 12.27.

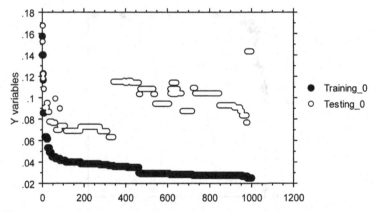

FIGURE 12.25 The training and testing performance of the evolved rules sets when classifying sonar data. Training set performance improved through 1000 generations (x-axis); however, test set performance started to degrade after about the 300th generation. This indicates the possibility of overfitting after that iteration. Y-variable refers to the training and testing results. (From Porto et al. (2005).)

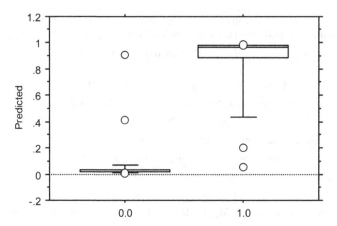

FIGURE 12.26   The test set performance (shown with box and whiskers plot, where circles represent outliers in the top or bottom 5% of the data) of the best-evolved rule set for classifying metal spheres (0) versus normal background (1). There was good separation between the two classes. (From Porto et al. (2005).)

The evolved rule-based classifiers demonstrated excellent ability to discriminate between simulated sea mines and background reflections. To offer one example of a rule-based classifier, the following set of eight rules were evolved for the second shuffle of the data. Both TV#1 and TV#2 were initialized at 0.5.

1: IF (attribute#2 < TV#1) ELSE (TV#2 = 0.5)

2: IF (attribute#1 < 0.724314) OR (attribute#1 > 0.197255) AND (attribute#1 ≤ 0.172157) OR (attribute#1 < TV#2) OR (attribute#4 ≥ TV#1) THEN (TV#1 = 0.5)

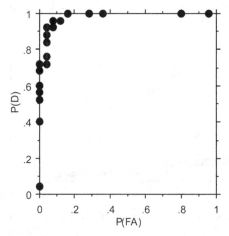

FIGURE 12.27   The receiver operating characteristic (ROC) curve for the best-evolved rule sets in classifying metal spheres versus normal background. The area under the curve was close to 1.0. (From Porto et al. (2005).)

3: IF (attribute#1 > attribute#3) OR (attribute#4 > TV#2) AND (attribute#4 > TV#2) AND (attribute#1 ≥ TV#2) THEN (TV#2 += 0.130713)

4: IF (attribute#3 > TV#2) THEN (TV#1 += TV#2)

5: IF (attribute#2 > 0.661569) AND (attribute#2 > TV#2) THEN (TV#2 += TV#2)

6: IF (attribute#3 ≤ 0.846667) AND (attribute#1 ≤ attribute#2) AND (attribute#3 < TV#2) OR (attribute#2 > attribute#4) OR (attribute#2 ≥ 0.539216) THEN (TV#1 −= 0.111959)

7: IF (attribute#4 > TV#2) THEN (TV#1 −= 0.1)

8: IF (attribute#4 ≥ 0.344706) AND (attribute#3 > 0.793333) AND (attribute#4 ≤ attribute#1) OR (attribute#4 > 0.517255) AND (attribute#2 < attribute#4) THEN (Output#1 = TV#1) ELSE (Output #1 = 0.1)

In this case, rule #1 is a functional no-op because TV#2 starts at 0.5 and there is no other consequence to this rule. It could be removed without affecting any outcome. Other rules are not as plain to examine.

While the example here used precise terms, the rules could just as well be fuzzy. For example, temporary variables could represent thresholds of membership in fuzzy membership functions. The comparisons of an attribute versus a threshold could be written in the form of an attribute versus a fuzzy membership term, such as "IF (attribute is HIGH)" or "IF (attribute is SUFFICIENTLY CLOSE)." The membership functions could also be optimized by evolutionary algorithms both in terms of the shape of the functions and the scaling of the functions.

12.5 EVOLUTIONARY CONTROL SYSTEMS

12.5.1 Cart–Pole Systems

Figure 12.28 shows a single pole atop a cart that is placed on a track. The goal in this problem is to push and pull on the cart so as to keep the pole from falling over (or falling beyond a threshold number of degrees) while simultaneously keeping the cart from hitting either end of the track. Wieland (1991a, 1991b) examined the application of evolutionary algorithms to optimize neural controllers of single, multiple, and jointed cart–pole systems (see Figures 12.29 and 12.30). The single-pole system is controllable with a linear system of the position and velocity of the cart $(x, \dot{x})$ and the angle and angular velocity of the pole $(\theta, \dot{\theta})$. The problem is more challenging when having to control multiple or jointed cart–pole systems.

Wieland (1991a, 1991b) described the multiple-pole system with the following equations:

$$\ddot{x} = \frac{F - \mu_c \operatorname{sgn}(\dot{x}) + \sum_{i=1}^{N} \tilde{F}_i}{M + \sum_{i=1}^{N} \tilde{m}_i} \qquad (12.38)$$

$$\dot{\theta}_i = -\frac{3}{4l_i}\left(\ddot{x}\cos\theta_i + g\sin\theta_i + \frac{\mu_{p_i}\dot{\theta}_i}{m_i l_i}\right) \qquad (12.39)$$

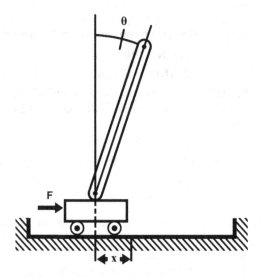

FIGURE 12.28 A typical cart–pole system. A single pole is mounted at the center of a cart. The cart is on a track. The control goal is to apply a series of forces F to the cart to keep the cart within limits on the track and keep the pole from exceeding a displacement angle θ. (From Wieland (1991a, 1991b).)

where x is the distance of the cart from the center of the track, θ_i is the angle of the ith pole from vertical, N is the number of poles on the cart, g is the acceleration due to gravity, m_i and l_i are the mass and the half-length of the ith pole, M is the mass of the cart, μ_c is the coefficient of friction of the cart on the track, μ_{p_i} is the coefficient of friction for the ith hinge, F is the force applied to the cart, and $\tilde{F}_i$ is the effective

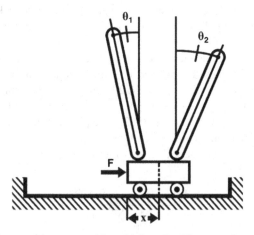

FIGURE 12.29 A cart–pole system with multiple poles. The control goal remains to apply a series of forces F to the cart to keep the cart within limits on the track and keep both poles from exceeding displacement angles. (From Wieland (1991a, 1991b).)

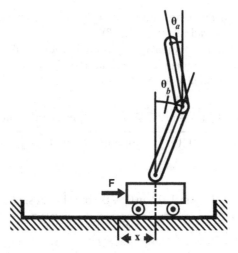

FIGURE 12.30   A cart–pole system with a hinged pole. The control goal remains to apply a series of forces F to the cart to keep the cart within limits on the track and keep both poles from exceeding displacement angles. (From Wieland (1991a, 1991b).)

force from the ith pole on the cart:

$$\tilde{F}_i = m_i l_i \dot{\theta}_i^2 \sin \theta_i + \frac{3}{4} m_i \cos \theta_i \left(\frac{\mu_{p_i} \dot{\theta}_i}{m_i l_i} + g \sin \theta_i \right) \tag{12.40}$$

and $\tilde{m}_i$ is the effective mass of the ith pole:

$$\tilde{m}_i = m_i \left(1 - \frac{3}{4} \cos^2 \theta_i \right) \tag{12.41}$$

Fully connected neural networks with 10 nodes were evolved via a process of mutation, crossover, inversion, and proportional selection. Each network had neurons defined for input and output and these values along with each weight and threshold were encoded in eight bits (binary). The weights and threshold values were concatenated to form each candidate neural controller. Selection was based on how long a neural controller could keep the cart–pole system in balance.

The challenge of a multiple-pole system is greatest when the poles are about the same length. When one is much smaller than the other, the "region of controllability" is much larger. Wieland (1991a, 1991b) invented a learning process in which a system with one pole of 1 m and another pole of 0.9 m was eventually kept in balance by starting with poles of 1 and 0.1 m and incrementally elongating the shorter pole by 1% after the neural networks demonstrated an ability to control the system for a long period of time. Specific details regarding the mass of the cart, length of the track, and maximum angle of deflection for each pole were omitted; however, standard values of these parameters in other literature include a 1.0 kg cart, a 4.8 m track, and a maximum angle of $\pm 12°$ or $\pm \pi/16$ rad (about $\pm 11.25°$).

Fogel (1996) used an evolutionary approach to combine the modeling methods described in Section 12.2 in order to control a cart–pole system with one pole. In this case, the system was modeled as

$$\dot{\theta}_t = \frac{g \sin \theta_t + \cos \theta_t \left[(F_t - m_p l \dot{\theta}_t^2 \sin \theta_t + \mu_c \, \mathrm{sgn}(\dot{x}_t))/m_c + m_p \right] - (\mu_p \dot{\theta}_t / m_p l)}{1 \left[(4/3) + (m_p \cos \theta_t)/(m_c + m_p) \right]}$$

(12.42)

$$\ddot{x}_t = \frac{F_t + m_p l \left[\dot{\theta}_t^2 \sin \theta_t - \dot{\theta}_t \cos \theta_t \right] - \mu_c \, \mathrm{sgn}(\dot{x}_t)}{m_c + m_p}$$

(12.43)

where $m_c = 1.0$ kg (mass of the cart), $m_p = 0.1$ kg (mass of the pole), $1 = 0.5$ m (half-pole length), $\mu_c = 0.0005$ (coefficient of friction of the cart on the track), $\mu_p = 0.000002$ (coefficient of friction of the pole on the cart), $F_t =$ the force applied to the cart's center of mass at time t, and $g = 9.8$ m/s^2. The maximum allowable limit for the cart was ± 1 m and for the pole was $\pm \pi/16$ rad, with a control force limited to ± 10 N.

The goal was to estimate a linear model of the cart–pole system given the observed sequential stimulus–response pairs in closed loop and generate a control input to the system to keep it balanced as long as possible. The linear equations were of the following form:

$$A_1(q)x(t) = B_1(q)u(t) + C_1(q)e_1(t) + D_1(q)\theta(t)$$

(12.44)

$$A_2(q)\theta(t) = B_2(q)u(t) + C_2(q)e_2(t) + D_2(q)x(t)$$

(12.45)

where the first equation described the position of the cart (which is coupled with the dynamics of the pole) and the second equation described the angle of the pole (which is coupled with the dynamics of the cart). The process u(t) represented the control input to the system at time t and $e_1(t)$ and $e_2(t)$ represented the residual error terms for each equation. As in Section 12.2, the polynomials $A_1, A_2; B_1, B_2; C_1, C_2;$ and D_1, D_2 were functions of their respective terms, going back in time to some number of lag values. The size of each polynomial and the corresponding parameters were evolved while controlling the cart–pole system, up until the point of system failure.

For the sake of space here, if you're interested in replicating the procedure, you should refer to the details in Fogel (1996). The results for a case of a pole with half-length of 0.5 m and a cart of mass 1.0 kg are shown in Figure 12.31, which shows the failures of the system as the cart hit the end of the track or the pole exceeded the maximum allowable threshold; however, after about 60 s the system was stabilized.

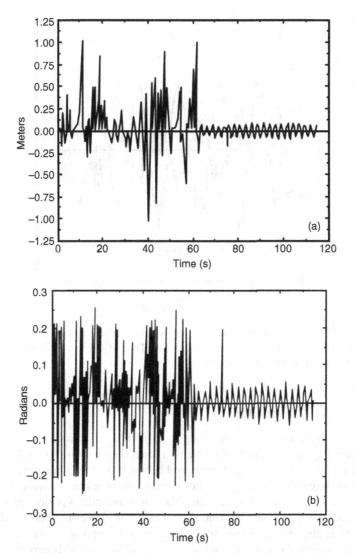

FIGURE 12.31   The behavior of the cart–pole system during evolutionary learning. The cart (a) and pole (b) exceed their control limits, while the evolutionary process learns the closed-loop system. The last 40 s of the depicted time show the cart and pole oscillating within the control limits. (From Fogel (1996).)

Figure 12.32 shows how the cart and the pole oscillated during the period of stability, with the cart and pole moving in opposition. In this case, the evolution generated not only a control input to move the cart appropriately but also a set of equations that modeled the dynamics of the unstable system, even though it was being observed in closed loop.

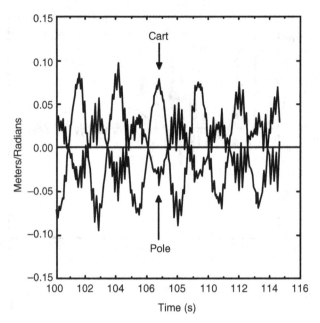

FIGURE 12.32 The behavior of the cart–pole system after the evolutionary controller had learned to control the system within the control limits. The cart and pole oscillated in opposition to maintain control. (From Fogel (1996).)

12.5.2 Truck Backer–Upper

Here's another system that's interesting to explore as an introduction to applying evolutionary algorithms to control problems. It involves a semi-trailer truck. The semi-trailer is hitched to the tractor and can swivel behind it, as shown in Figure 12.33. The objective of the control problem is to take the semi-trailer truck from an arbitrary starting location and move it so that the midpoint of the rear of the trailer is positioned next to the loading dock at the designated position and such that the trailer is perpendicular to the loading dock. Since the trailer can only be affected by the application of control through the hinged system, this can make for some complicated system dynamics.

The typical problem in published papers uses four dimensions to describe the state of the system: the (x, y)-coordinates of the midpoint of the rear of the trailer and θ_t and θ_d, the angle of the trailer with respect to the loading dock, and the angle of the tractor relative to the angle of the trailer, respectively [Chellapilla, 1998; Koza, 1992]. An additional variable θ_c is used to describe the angle of the tractor (also known as a "cab") to the line perpendicular to the loading dock.

To make the problem more manageable, the truck is assumed to back up at a constant speed, and steering is accomplished by changing the angle $u(t)$, which represents the angle of the front tires with respect to the tractor at time t. (The standard nomenclature here uses t both as an index for time and for the subscript on the angle of the trailer with respect to the line perpendicular to the loading dock.)

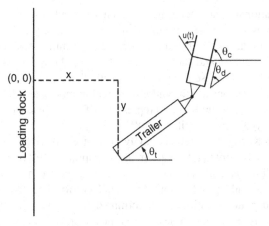

FIGURE 12.33 The truck backer–upper problem. The trailer is attached to the cab via a hitch. The midpoint of the back of the trailer is described as a point (x,y). The angles describe the configuration of the trailer and the cab, and u(t) describes the input control.

The control goal is to modify u(t) at each time step (typically with a $t = 0.02$ s interval) such that x is within 0.1 m of the loading dock, y is within 0.42 m of the loading dock, and θ_t is within about 0.12 rad (7°) of horizontal. These values follow earlier work that used neural networks for performing this task [Nguyen and Widrow, 1989].

The system dynamics for the semi-trailer involve the following seven equations:

$$A = r \cos(u(t)) \tag{12.46}$$

$$B = A \cos (\theta_c(t) - \theta_t(t)) \tag{12.47}$$

$$C = A \sin (\theta_c(t) - \theta_t(t)) \tag{12.48}$$

$$x(t + 1) = x(t) - B \cos (\theta_t(t)) \tag{12.49}$$

$$y(t + 1) = y(t) - B \sin (\theta_t(t)) \tag{12.50}$$

$$\theta_c(t + 1) = \tan^{-1} \left(\frac{d_c \sin \theta_c(t) - r \cos \theta_c(t)\sin u(t)}{d_c \cos \theta_c(t) + r \sin \theta_c(t) \sin u(t)} \right) \tag{12.51}$$

$$\theta_t(t + 1) = \tan^{-1} \left(\frac{d_s \sin \theta_t(t) - C \cos \theta_t(t)}{d_s \cos \theta_t(t) - C \sin \theta_t(t)} \right) \tag{12.52}$$

where d_c is the length of the tractor (set at $d_c = 6$ m) and d_s is the length of the trailer (set at $d_s = 14$ m). A controller is given a certain amount of time (e.g., 60 s) to move the tractor-trailer from an arbitrary starting point to within the target area for the loading dock.

Various evolutionary approaches have been made to address this problem that rely on symbolic expressions in the form of tree-based data structures, for example, Koza (1992) and Chellapilla (1998). The trees operated on terminal states of {X, Y, TANG, DIFF, NumericConstant} and the function set of {ADD, SUB, MUL, DIV, ATG, IFLTZ}. Each of these is self-explanatory with the exception possibly of ATG (arctangent), NumericConstant (a randomized real number), and IFLTZ (if less than zero, which takes three arguments and represents "if arg1 is less than zero then return arg2 else return arg3") [Chellapilla, 1998].

The system in Koza (1992) relied heavily on subtree crossover to create new trees from existing trees. In contrast, the system in Chellapilla (1998) relied on six different mutation operations, and no subtree crossover, to generate new solutions. Importantly, both methods addressed the problem satisfactorily. The method of Chellapilla (1998) generated better solutions as determined by an overall objective function (assessed as "an order of magnitude" better); however, no statistical significance of the difference to the solutions in Koza (1992) was presented in Chellapilla (1998).

12.6 EVOLUTIONARY GAMES

Games are characterized by rules that govern the behavior of one or more players. Most often we think of games between two or more players, but many games can be played single-handed, such as solitaire. Behavior in games is a matter of stimulus–response, that is, for a given state—the stimulus—what move should be made?—the response. Generally, there is an overall goal to a game, which may be defined in terms of winning, losing, or playing to a draw; however, games are very general and life itself can be viewed as a game, with each person having a complex time-varying objective function that describes something fuzzy, like overall happiness.

Rational players seek to maximize the payoffs that they receive when making moves in a game, hopefully placing themselves in more favorable states that lead to even more successful moves. Some games are called zero-sum, meaning that what one player wins, another necessarily loses (such as with poker). But not all games are necessarily competitive. The players on a soccer team, for example, must work together to collectively achieve a goal. Even then, there is winning and losing or playing to a draw. But if one again considers life, there are many opportunities for successful teamwork that do not depend on winning or losing, such as marriage and raising children.

Sometimes players do not have the same objective function. Unlike checkers, chess, backgammon, or other competitive games, sometimes one player may be ambivalent to another—such as the game of courtship, in which one person tries to win the attention of another, and sometimes the other is entirely disinterested. The concept of gaming is very flexible and can be used to treat problems in economics, evolution, social dilemmas, board games, video games, and so forth. Any situation in which purpose-driven entities can recognize a state of existence and determine how to allocate resources to move to a different more desirable state can be viewed as a game.

Intelligence itself has been described in terms of games: the ability to make appropriate decisions in order to adapt behavior to meet specific goals in a range of environments [Fogel et al., 1966]. Whereas the rules dictate what allocations of resources are available to players, strategies determine how those plays in the game will be undertaken. Game theory [von Neumann and Morgenstern, 1944] can treat certain classes of mathematical games; however, many real-world circumstances go beyond the mathematical limits and assumptions of game theory. In these cases, evolutionary algorithms can be used to determine suitable or even optimal strategies to play as a function of the behavior of the opponent(s).

Evolution in these cases is often conducted in a format called coevolution. Instead of evolving solutions against a fixed objective function, solutions are judged in terms of how well they compete against other solutions in the same or a different population. Thus, the objectives may change as the players themselves change. This is more akin to what occurs in nature, as the environment is a function not only of physical processes but also of the other organisms that it comprises.

This section highlights one particular game that has received considerable attention: the iterated prisoner's dilemma. Other games are mentioned as well, but are beyond the scope of replication in an introductory class. Still, it is important for students in evolutionary computing to have a broad appreciation of the types of problems that can be addressed by evolutionary algorithms, and also the comparably low level of domain knowledge that may be necessary, even in games that would be challenging to many adults.

12.6.1 Iterated Prisoner's Dilemma

A classic game that has received many decades of research interest is the iterated prisoner's dilemma. The game is structured with two players, each having two options: cooperate with each other or defect against each other. These options are described typically with the symbols C and D. The rationale for the game, and why it's called a prisoner's dilemma, comes from imaging two prisoners who have been caught for a crime. During separate interrogation, the prosecutor offers each a lesser sentence for "ratting out" his cohort in crime. If neither criminal rats out his partner, then the sentence will be pretty low for each one (as the evidence the prosecutor has isn't that strong). If one criminal defects against the other, while the other refuses to defect, then the rat goes free and the other criminal gets a long sentence (based on the rat's testimony). But if both criminals rat each other out, then they each get somewhat long sentences.

It's often easier to view games from the perspective of maximizing payoffs (rather than minimizing how long you go to jail). So, the prisoner's dilemma is often recast as follows:

Player 1/Player 2	C	D
C	R/R	S/T
D	T/S	U/U

Player 1's play of C or D is in column 1. Player 2's play is in column 2. If both players choose to cooperate (C), then they each get a reward of R points. If they both defect (D), they each get U points for being uncooperative. If one chooses to cooperate (C) and the other chooses to defect (D), then the one cooperating gets the sucker's choice of S points and the one defecting gets the temptation choice of T points. One specific example of a well-studied prisoner's dilemma is

Player 1/Player 2	C	D
C	3/3	0/5
D	5/0	1/1

Thinking about the game logically, if you are a player in the game, you may consider it this way. If the other player cooperates, you will do better if you defect. If the other player defects, you again will do better if you defect. Therefore, you should defect. Of course, the other player would have the same thoughts and thus both of you would choose to defect and get only 1 point each, when you each could have cooperated and received 3 points each. Research has shown that when this game is iterated over many plays between two players, the propensity for mutual defection is often reduced. Thus, the *iterated* prisoner's dilemma (IPD) is fundamentally different from the *one-shot* prisoner's dilemma.

There is a vast literature on evolutionary approaches to exploring the IPD. A thorough treatment would comprise a complete chapter in a textbook. For space, the focus here will be on two early approaches that contrast different representations of the iterated prisoner's dilemma game. The first is from Axelrod (1987), which is a fundamental contribution in the history of IPD research. Robert Axelrod, a game theoretician, conducted tournaments with other academics and practitioners in 1979 that showed a robust strategy in a long iterated game that was called *tit-for-tat*, which cooperates on the first move and then mirrors whatever the other player does after that. Tit-for-tat will meet cooperation with more cooperation but will defect against a player who defects until that other player changes to cooperate. Axelrod was interested to see if tit-for-tat, or something like it, would evolve spontaneously in a population of competing strategies.

The approach in Axelrod (1987) used a binary encoding of all moves with a history limited to the prior three moves. The encoding described what to do on the first move (C or D), what to do on the second move based on what was done on the first move (CC/C or D, CD/C or D, DC/C or D, DD/C or D, where the pair of symbols is the first and second player's moves, respectively, from the first move, and the C or D after the slash represents the choice of whether or not to cooperate on the next move), and so forth up to move histories of length 3. Axelrod (1987) used a 70-bit string to encode each of these possibilities, and used a formulation much like a simple genetic algorithm to let the strings evolve against each other.

One experiment involved a population of these strings competing against each other. This is a form of *coevolution*, because there is no fixed objective function to maximize or minimize. The right strategy for one population may not be the right strategy for all populations. For example, if all but one strategy would always cooperate no matter what, then the best choice for the last strategy is to always defect.

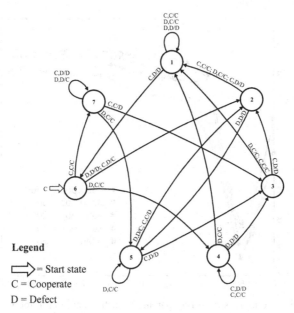

FIGURE 12.34 A finite-state machine for the iterated prisoner's dilemma. The machine starts in state 6 and cooperates on the first move. If the opponent also cooperates (C,C), then the machine cooperates and transitions to state 7. (From Fogel (1993).)

But if all but one strategy is tit-for-tat, then the worst choice for the last strategy is to always defect. Axelrod (1987) had strings multiply in proportion (up to the limit size of the population) to their earned scores, and used mutation and crossover to vary the strings as they multiplied. The experiment generated binary strings that would diverge initially away from cooperation but then over time would tend toward mutual cooperation. This was a fundamentally important result because it showed that no intervention is required in an iterated prisoner's dilemma to have cooperation emerge (at least in the particular IPD that Axelrod explored).

One limitation with this approach is that it is defined to act only on histories of moves that go back three plays in time. In order to explore the possibility of having longer time histories, Fogel (1991, 1993) created a similar coevolutionary algorithm but used finite-state machines to represent strategies. For example, Figure 12.34 shows a finite-state strategy. This machine starts in state 6 and cooperates. Then, if the other player cooperates, this machine will again cooperate and go to state 7. If the other player defects, this machine will cooperate again but go to state 2. You can follow the different possibilities through the trajectory of states.

In Fogel (1993), coevolutionary trials with the IPD and finite-state machines using populations ranged from 50 to 1000 parents, and the average results were very similar. Figure 12.35 shows that the mean score of the surviving parents fell initially (indicating that most survivors were doing a lot of defecting) and then rose up to about 3.0 (indicating mutual cooperation). The results were robust across population sizes, and similar to what was seen in Axelrod (1987).

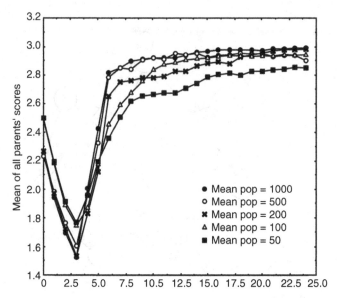

FIGURE 12.35   The mean score of all parents at each generation in the iterated prisoner's dilemma as a function of the population size. The pattern was same for all tested population sizes: initial evolution toward mutual defection followed by a rise to mutual cooperation. (From Fogel (1993).)

The iterated prisoner's dilemma has received considerable attention with evolutionary methods. The basic prisoner's dilemma model is a simplification of real-world circumstances. It leaves out many potentially important facets. For example, if you are playing a game with someone, it might be that you have the option to leave the game, so you have the problem of determining how long you should stay engaged [Fogel, 1995]. You may find that your options aren't as simple as a dichotomy of cooperate or defect. You may have a range of behaviors that allow you to "cooperate a little bit." This can be modeled as a continuum from −1 (complete defection) to +1 (complete cooperation) using evolutionary neural networks rather than finite-state machines [Harrald and Fogel, 1996] as well as by fuzzy descriptions of behaviors [Borges et al., 1997] or with intermediate precise levels [Darwen and Yao, 2002]. Results have shown that more options in behavior or a continuum of behavioral choices can cause less mutual cooperation [Darwen and Yao, 2001] (see also studies in Ashlock et al. (2010) and the reviews and contributions in Kendall et al. (2007)).

12.6.2   Board Games and Video Games

Traditional board games present challenging environments for machine learning, and there have been many interesting applications of evolutionary computing on a wide range of these games. In contrast to knowledge-based methods, which generally seek to craft the best possible playing program by capturing existing human expertise, research using evolutionary algorithms to evolve and adapt strategies is often aimed at determining what concepts can be learned autonomously and what level of play can be achieved without human intervention (or with as little as feasible).

For example, the world's best checkers program is called Chinook [Schaeffer et al., 1996] and it is perfect. Checkers has been proven to be draw from the standard starting position [Schaeffer et al., 2007], and Chinook contains all the information required to play perfectly. However, suppose that there were no human experts in checkers, no way of looking up the right moves from an endgame database, and no suggestions for how to judge the quality of different positions of checkers on the board. Could a machine still learn how to play the game at some level of competence?

This was the question posed and answered in the "Blondie24" research offered in Fogel (2002). By combining evolutionary algorithms with neural networks to evaluate alternative checkerboard positions, the computer was able to compete different ideas about which moves to favor. Random mutation of neural network weights served to generate new ideas.[15] Over hundreds of generations, based on the information contained in the number of pieces, the location of the pieces, the types of pieces, and the rules of checkers, along with a minimax decision strategy,[16] the evolutionary program created a neural network, called Blondie24, that was evaluated as being in the top 500 of 120,000 checkers players on www.zone.com. For specifics on the implementation of Blondie24, see Chellapilla and Fogel (2001).

The Blondie24 line of research was extended to chess and dubbed "Blondie25." Starting from a publicly available chess program rated below the master level and with additional neural networks added to evaluate positions, the evolutionary approach adapted chess playing strategies that were eventually able to defeat Fritz8 (which was in the top five computer programs in the world at the time) as well as a human master, James Quon (see Fogel et al. (2005, 2006)).[17]

Checkers and chess are deterministic games: Random chance does not play a part in the outcome. Many games, however, rely on an element of random chance, such as blackjack, Monopoly®, Risk®, or backgammon. Each of these games (and more) has been addressed with evolutionary algorithms; see Darwen (2001), Fogel (2004), Azaria and Sipper (2005), Frayn (2005), and Mukhar (2013)). This reference list is merely a small sample of all the work that has been done in these areas, and these areas represent a small sample of the space of board games that has been considered. Readers who are interested in learning more about this line of research should review papers published in the *IEEE Symposium on Computational Intelligence and Games* and the *IEEE Transactions on Computational Intelligence and AI in Games*.

In addition to board games, evolutionary algorithms can be used to control the behavior of players or even the content in video games. For many years, there have been competitions held at the major annual conferences in evolutionary computation that encourage researchers to explore evolving faster and smoother racing cars using

---

[15] The neural networks started with random weight and the piece count was entered in the output node. Thus, the initial generation was governed by the piece differential and random noise.

[16] Minimax chooses a move that minimizes the maximum damage that an opponent can be expected to do in the future.

[17] Sadly, James Quon passed away in 2010 at the age of 42 from a brain aneurysm. His chess legacy is remembered today by the James Quon Chess Foundation.

The Open Racing Car Simulator (TORCS), more deadly bots in the first-person shooter game Unreal (or more lifelike bots, or bots that are harder to differentiate from a human-controlled bot), team play in games such as Pac-man® (controlling the ghost team), and many others. Some of the relevant references include, but are not limited to, Wittkamp et al. (2008), Hastings et al. (2009), Perez et al. (2009), Galanopoulos et al. (2012), Pena et al. (2012), and Cotta et al. (2013).

It's natural to think about extending the evolution of strategies or designs in simulations to evolving physical designs of real-world devices. Some examples of that come in an area called *evolvable hardware*, which is reviewed in the next chapter; however, the Web site www.boxcar2d.com has an interesting and simple look at evolving moving vehicles using simple physical models, which follows seminal work by Karl Sims that can be seen at https://www.youtube.com/watch?v=JBgG_VSP7f8. Lipson and Pollack (2000) evolved mobile robots in a simulator and then used a three-dimensional printer to actually create the evolved designs and proved their viability. You can see a video of this at https://www.youtube.com/watch?v=qSI0HSkzG1E.

12.7   SUMMARY

Evolutionary algorithms can be applied to a very wide variety of optimization problems. Here, we started with the basics of regression analysis and identified that evolutionary optimization can be applied not only to adjust parameters of a regression model but also to do so in light of criteria that are not related to the mean squared error. This served as a foundation for extending evolutionary modeling to time series prediction in which the model coefficients and structure can be varied simultaneously and evaluated in light of information criteria. Similarly, evolutionary algorithms can adjust the weights and topology of neural networks, or the membership of functions of fuzzy systems.

The evolutionary approach isn't restricted to the time domain. It can also be applied to the spatial domain and used for clustering and classification. Again, information theoretic criteria can be used to find an optimal number of clusters and the objective functions in classification applications do not need to be based on squared error or any differentiable function.

System modeling can then be extended to control systems. If a model can be created to anticipate what a system will do next, then it becomes feasible to ask what must be done to put that system in a more desired state. In the case here, the cart–pole system and truck backer–upper were offered as canonical examples.

The extension from control is gaming, which involves two or more intelligent adversaries trying to control each other. There have been many applications in gaming and the focus in this chapter was on games of cooperating versus defecting, as well as more traditional games such as checkers, chess, and video games. It is now possible to also evolve physical devices in simulation and then construct them using three-dimensional printing or other technologies. This broad range of applications is testament to the flexibility of evolutionary computing.

EXERCISES

12.1. Generate 100 (x, y) points from a uniform distribution over $(-10, 10)$. Write an evolutionary rule-based classifier to determine which quadrant each point is in.

12.2. Create an evolutionary algorithm operating on neural networks to replicate the chaos detection experiment offered in Section 12.2.2; however, instead of using sigmoid transfer functions, use radial basis functions $(f(x) = (1/\sqrt{2\pi})\exp(-x^2))$ and see if this transfer function provides as good an ability to differentiate between signal and noise-only models.

12.3. Generate 100 samples from a sine wave $y = \sin(2x + 3)$ as $x = 1, \ldots, 100$. Add Cauchy-distributed random noise to the samples. (Recall that a standard Cauchy random variable is obtained by taking the ratio of two independent standard normal random variables.) Use an evolutionary algorithm to estimate an equation of $y = \sin(ax + b)$ based on minimizing the squared error between the predicted and actual values of y, and then also using the absolute error. Which objective function gives an answer that you think is better?

12.4. Create two distributions of 50 points in (x, y). The first is distributed as a two-dimensional Gaussian random variable centered at $(2, 2)$ with standard deviation of $(1, 1)$, respectively, for each dimension. The second is also Gaussian, centered at $(5, 5)$ with standard deviation of $(1, 1)$ for each dimension. Create an evolutionary algorithm that assigns each of the 100 points to one of two clusters. Devise a metric for assessing the quality of the clustering and then use an evolutionary algorithm to optimize for that metric. What does the final assignment look like? Is it reasonable? Now repeat the process assigning points to one of three clusters. What adjustment do you think would be appropriate to compare the clustering with two assigned groups versus three assigned groups?

12.5. Search on the Internet for data that are suitable for a classification problem. For example, you might search for "Wisconsin breast cancer data machine learning." For the set of data that you choose, create a classification algorithm that uses fuzzy membership functions to interpret the inputs. For example, with the Wisconsin breast cancer data, the first feature is the radius of a cell nucleus. This could be described as "small," "medium," or "large," with various membership functions. Use an evolutionary algorithm to adjust the membership functions based on training and test set performance. Describe the results in terms of class separation and/or receiver operating characteristic curves.

12.6. Using the same data set, write an evolutionary algorithm that performs feature selection. That is, have the evolutionary algorithm determine which features to include in the analysis and which to omit. One way to do this is to evolve a bit string in which a 1 indicates that a feature will be used and a 0 indicates that it will not be used. Choose one of the information-theoretic statistics to help trade off the goodness-of-fit that you can obtain for the number of features that you include.

12.7. Write an evolutionary algorithm to learn to play tic-tac-toe against a fixed rule-based opponent that makes perfect moves, but has a 0.1 probability of playing at random to any open square. The evolutionary algorithm could use a neural network, fuzzy logic system, rule-based finite-state automata, or symbolic expression to determine the next move. Any data structure is fine as long as it is subject to variation and selection. Determine a metric for measuring improvement against the fixed opponent and identify whether or not your evolutionary algorithm is making progress toward being a perfect player.

12.8. Write an evolutionary algorithm for the iterated prisoner's dilemma taking into account that players reside in physical locations. Using a 3×3 grid, let each of the nine cells in the grid be represented by an evolving population of prisoner's dilemma players. Each player is evaluated against another randomly selected player from a neighboring population. Selection then eliminates the lower scoring players in each population and variation operators (recombination and/or mutation, as you determine) are used to create new players. Does mutual cooperation emerge from this arrangement of players? If so, decrease the payoff for mutual cooperation until it does not; if not, increase the payoff for mutual cooperation until it does.

CHAPTER 13

Collective Intelligence and Other Extensions of Evolutionary Computation

Chapters 10–12 have introduced the basic concepts of evolutionary algorithms, particularly as they are applied for optimization. In this chapter, we'll cover other aspects of evolutionary algorithms and methods that are related to simulating evolution on computers.

It's important to understand upfront that there are very many extensions of evolutionary algorithms. This text is intended to provide an introduction to the field of evolutionary computation that would be suitable as part of a broader introduction to computational intelligence within the framework of a one semester college course. By consequence, we cannot cover everything here and do it in any level of depth that would be appropriate.

This chapter begins with a population-based optimization approach called particle swarm optimization (PSO), which models the flocking behavior we see in certain animals. We then focus on another population-based approach called differential evolution, which searches a landscape for optima by using different vectors between existing solutions. Another approach follows that is based on modeling how ants search and find food sources. It turns out that the strategy ants use can be applied for finding short paths through graphs and other engineering problems.

After that, we'll focus on specific applications of evolution that are found in hardware, as opposed to only in software. We'll also address problems for which having a person act as the fitness function provides possibilities that are beyond what evolutionary algorithms can do solely in software. Finally, we'll conclude with the application of evolutionary algorithms to multiple criteria optimization problems.

13.1 PARTICLE SWARM OPTIMIZATION

When considering how a flock of birds or a school of fish moves, it's easy to think that the actions of each individual are somehow coordinated with the actions of the others, and that's exactly the case. Research shows, for example, that starlings coordinate

Fundamentals of Computational Intelligence: Neural Networks, Fuzzy Systems, and Evolutionary Computation, First Edition. James M. Keller, Derong Liu, and David B. Fogel.

their movements based on seven neighboring birds [Young *et al.*, 2013]. Thousands of starlings may form a flock, but the movements of the flock are based on the local interactions of overlapping groups of just seven birds.

It's interesting that modeling the way birds flock, or fish school, or things in general "swarm" can lead to an optimization algorithm. The organisms that we might model in these cases are likely optimizing something, such as group cohesiveness versus individual effort [Young *et al.*, 2013]. That might not correlate to what we'd like to do in terms of using swarming as a method to find points of interest on an objective function, but we can tailor a swarming method toward that end.

Particle swarm optimization was introduced as such an approach [Kennedy and Eberhart, 1995]. Suppose you have a collection of possible solutions to a problem, which we'll call *particles*. The particles reside in $\Re^n$ and the goal is to find the minimum, maximum, or some other point of an objective function $f(\mathbf{x})$, where $\mathbf{x} \in \Re^n$. Much like the solutions in the population of an evolutionary algorithm, these particles are located at various positions and can be evaluated in light of some objective criteria. Each particle is denoted by $\mathbf{x}_i(t)$, where $\mathbf{x}_i$ is the ith particle located at $\mathbf{x}$ (a vector) at a particular time t.

Next comes the swarming part: How do the particles move? They each have a velocity vector, denoted by $\mathbf{v}_i(t)$, where $\mathbf{v}_i$ is the velocity of the ith particle at a particular time t. If there were no swarming, each of the particles would move according to their unchanging velocity forever. The fact that the particles will swarm means that their velocities will change as a function of what is known about other particles in the collective, and also what a particle remembers about where it has been.

Each particle is given a memory. It remembers the location that has yielded the best result from the objective function. Each particle also has knowledge about the results of other particles in a neighborhood (akin to starlings being aware of other starlings), and each particle knows the location of the particle in its neighborhood that has the best result from the objective function. That information is used to change the velocities of the particles and thereby having them move to different locations, searching for a better location.

Each particle's new velocity is a function of (i) its current velocity, (ii) the vector that points to the particle's own best location, and (iii) the vector that points to the best location of the particles in its neighborhood. The vectors that point to known best locations are weighted by random variables. With starlings, the neighborhood is believed to be seven birds, but in PSO the neighborhood can be small, like a particle and its two closest neighbors, or it can expand to be the entire collection of particles. Each particle then moves according to its new velocity.

A little pseudocode can help clarify the update procedure for each particle:

$$v_i(t + 1) = a \times v_i(t) + b \times U_1 \times (\text{PersonalBest}_i - x_i(t)) + c \times U_2$$
$$\times (\text{NeighborhoodBest}_i - x_i(t))$$

$$x_i(t + 1) = x_i(t) + v_i(t + 1)$$

where U_1 and U_2 are distributed $U(0, 1)$, PersonalBest_i is the location where the ith particle found the best score in its memory, $\text{NeighborhoodBest}_i$ is the location where the best score was found in the ith particle's neighborhood of particles, and a, b, and c are scaling terms. Some basic settings for the scalars are $a = 1$, and $b + c = 4$. The trade-off of b and c essentially weights the relative importance of a particle's own experience for the importance of its neighborhood's experience.

One issue that comes with these update equations is that the particles can see their velocities increase in magnitude without bound. That's not particularly realistic for modeling real flocking and empirically it's not helpful when searching for optima in our typical objective functions. So, a limit is placed on the velocity in each dimension, v_{max}. Some experimentation may be required to have a good setting for this maximum value.

Another issue that arises is the choice of the number of particles. Many publications have used collections of 10–50 particles, but the appropriate size is problem dependent, as is the choice of how to construct a neighborhood for each particle.

Here's an example that uses 50 particles with a neighborhood of 3 particles and $|v| = 1$ applied to the function $f(x, y) = x^2 + y^2 - 20[\cos(\pi x) + \cos(\pi y) - 2]$, which is depicted in Figure 13.1. The global minimum value of the function is $f(x, y) = 0$ and the second-best minimum value is $f(x, y) = 4$.

In this example, all the particles were initialized uniformly at random between -10 and 10 in each dimension. After 500 iterations, the best score found was 0.0355. Figure 13.2 shows the best score of all the particles at each generation and the average of each particle's historical best score. The graph shows continual improvement toward the global optimum.

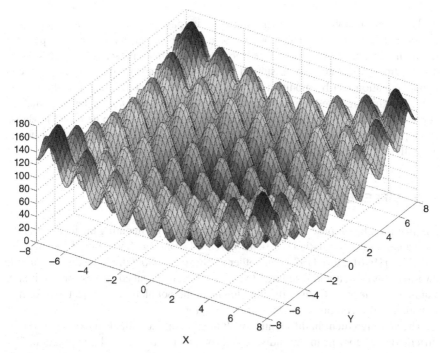

FIGURE 13.1 A surface with multiple local optima used for testing the particle swarm optimization algorithm.

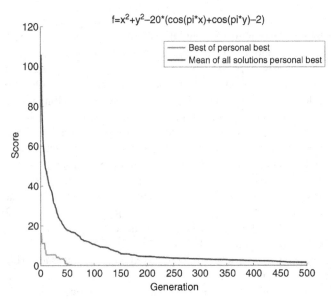

FIGURE 13.2   The rate of optimization for the best score ever found and the mean of all particles' best scores for the problem shown in Figure 13.1.

There are many areas for investigation in particle swarm optimization. For example, it may be that neighborhood size would benefit by varying by particle, or varying by particle over time. The learning factors that represent the weights on the acceleration terms can also be subject to online learning per particle. The velocity update equation can be modified by applying a "constriction" factor to the overall velocity, which relieves the need for setting a value of v_{max}. Other extensions include applications in discrete space and dynamic settings. Reviews can be found in Banks *et al.* [2007], Kiranyaz *et al.* [2013], and Kaveh [2014] with current research found annually in the IEEE Swarm Intelligence Symposium.

13.2   DIFFERENTIAL EVOLUTION

Another population-based search algorithm that relies on updating the location of the individuals in the population is called *differential evolution*. It was introduced in Storn and Price [1996], Storn [1996], and other publications about the same time as particle swarm optimization and is tailored to searching in real-valued spaces for maxima or minima of functions. Each of the individuals in the population is subject to a form of mutation and recombination, as well as selection.

The key ingredient in differential evolution is that individuals move based on the differential vectors from the individual to other individuals in the population. The population size needs to be at least four individuals and as with all evolutionary or related methods, the population is initialized at random or based on some prior knowledge of where to search for a solution to the problem posed.

Each of the individuals in the population is subjected first to mutation, which is based on the individual's relationship to three other distinct individuals in the population, chosen at random each time. Let's call these three individuals x_1, x_2, and x_3, and let's call the individual that we are mutating x_0.

First we pick a random dimension of the problem, d, uniformly from 1 to n, where the problem has n dimensions. We'll remember that dimension. Then, for each dimension $i = 1, \ldots, n$, we create a uniform random number $u_i \sim U(0, 1)$. If $u_i < p_c$ or $i = d$, then the new value of the solution in the ith dimension is given by

$$x_{0i} = x_{1i} + a(x_{2i} - x_{3i})$$

where a is a scalar value between $[0, 2]$ called a *differential weight*; otherwise, the new value of the solution in the ith dimension is retained from x_{0i}. The value p_c is a "crossover probability," which really means that it is transplanting a value for that dimension from another random solution added with a weighted combination of a difference between two other solutions.

Finally, for this solution, if the new version of x_0 is better than the original, it replaces the original in the next generation; otherwise the original x_0 is retained for the next generation. This process continues until a solution of sufficient quality is found, or a maximum number of generations is met.

Let's see how this approach works on two simple functions. The first is the usual quadratic bowl defined by $f(x, y) = x^2 + y^2$ (Figure 13.3), and the second is the same one that we just saw in Figure 13.1, which has multiple local optima.

We'll use 50 solutions in the population in each case, with our scalar value $a = 0.8$ and $p_c = 0.02$, with all solutions initialized uniformly at random between -20 and 20.

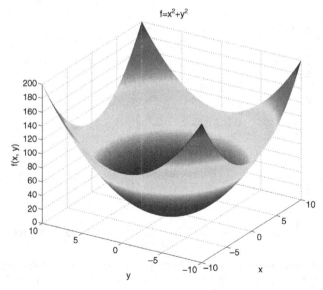

FIGURE 13.3 A simple quadratic function $f(x, y) = x^2 + y^2$ to be used in the differential evolution experiments.

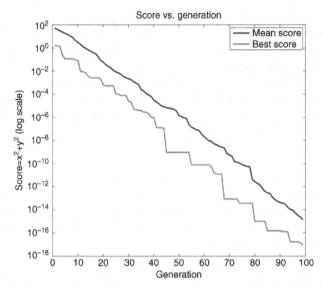

FIGURE 13.4   The average of all scores in the population (n = 50) and the best score in the population when applying differential evolution to the quadratic bowl. Note that the y-axis is on a log scale.

Figure 13.4 shows the average of all the scores in the population and the best score as a function of the number of generations for the quadratic bowl and Figure 13.5 shows these data when applying differential evolution on the multimodal surface.

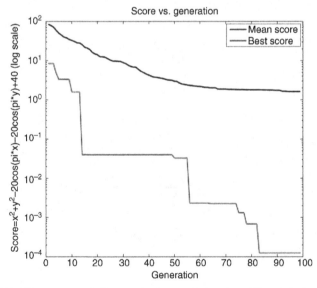

FIGURE 13.5   The average of all scores in the population (n = 50) and the best score in the population when applying differential evolution to the multlimodal function $f(x, y) = x^2 + y^2 - 20[\cos(\pi x) + \cos(\pi y) - 2]$. Note that the y-axis is on a log scale.

For the quadratic bowl, the best solution has less than 10^{-6} error in about 40 generations. For the multimodal function, the method is still able to find a solution that has less than 10^{-4} error in about 80 generations. A quick comparison with the results using PSO on the same multimodal function favors the use of differential evolution; however, it's always important to remember that this may not generalize to other functions, and may be sensitive even simply to the initialization.

13.3   ANT COLONY OPTIMIZATION

Another biologically inspired method for solving problems is called *ant colony optimization* (ACO), and it simulates how ants discover food sources and communicate their discoveries with other ants. ACO predates PSO and differential evolution [Dorigo, 1992] and is used generally for combinatorial optimization problems.

Here's the thinking underlying the approach. Suppose you're an ant and you have to get from your colony's nest to a food source. The problem is that you don't know where the food source is. So you have to search for it. Once you find it, you have to get back home, so you have to search for that too.

It would be great if you could share what you learn as you search with other ants, so that they could search more efficiently. Nature provides a way to do that. It's called a pheromone trail, and ants leave pheromone trails as they move. The strength of the trail is dependent on the number of ants that traverse the trail and how recently the ants have traversed the trail.

More ants on a trail means more pheromone, which entices other ants to follow that trail rather than search somewhere else. The pheromone evaporates over time, so once food sources are exhausted, the ants don't continue to travel to an "empty refrigerator" forever.

Let's see how we can model these principles to address a classic 30-city traveling salesman problem. The ants will start at a particular city and they have to find a complete path that visits every other city once and only once and then returns home. The objective is to complete the path with the shortest possible distance. Let's say we have a set of 50 ants in our population.

It's time for the first ant to start its trek. The probability of it visiting any of the other available city is given by

$$p(i,j) = \frac{ph(i,j) \times cost(i,j)^{-1}}{\sum_{k \in All} ph(i,k) \times cost(i,k)^{-1}}$$

where $p(i,j)$ is the probability of going from city i to city j, $ph(i,j)$ is pheromone factor between city i and city j, and $cost(i,j)$ is the distance between city i to city j. When starting, we set the value of $ph(i,j)$ to 1.0 for all pairs of cities.

From the formula, the probability that the ant will go to city j is dependent in part on the pheromone factor. The higher the value of $ph(i,j)$, the higher the likelihood of traveling to city j. Also, the probability is inversely related to the distance between the current city i and the new city j. Each ant wanders through the cities in traveling salesman problem according to the probability equation, which is computed for all available (unvisited) cities at each step through the path for each ant.

The pheromone factor can be updated in different ways. For our example, let's say we update the pheromone factor after each of the 50 ants has traversed a tour of the cities. The new pheromone strength is computed as

$$\text{new ph}(i, j) = \alpha \text{ph}(i, j) + \Delta \text{ph}(i, j)$$

where α is the evaporation rate and $\Delta \text{ph}(i, j)$ is the delta increment that comes from ants going from city i to city j. Here, let's say that the evaporation rate $\alpha = 0.9$ and

$$\Delta \text{ph}(i, j) = \sum_{k=1}^{50} \begin{cases} Q, & \text{if kth ant traveled between city i and city j} \\ 0, & \text{else} \end{cases}$$

and $Q = 1/50$ (i.e., the inverse of the number of ants).

After each set of 50 ants traverses the graph, the pheromone factors are updated, encouraging the ants to favor shorter paths that have been well traveled.

Here are some results from this approach on a 30-city traveling salesman problem. Figure 13.6 shows the best path found after 800 iterations of the 50 ants traversing the graph. The problem was created by distributing the cities uniformly at random in a 100×100 square. The final best path is 411.41 units long. Statistical mechanics can be used to estimate the optimum path length for n randomly distributed cities in an

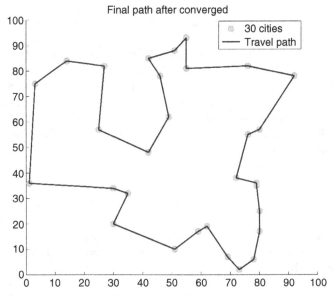

FIGURE 13.6   The best path found by a group of 50 ants iterated 800 times using the ant colony optimization routine on a randomly generated 30-city traveling salesman problem.

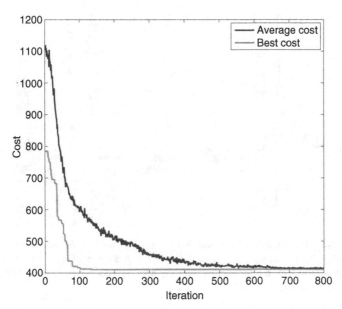

FIGURE 13.7   The average cost over each of the 50 ants at each iteration and cost of the best path found when iterating 800 times using the ant colony optimization routine on the traveling salesman problem shown in Figure 13.6.

area A by the formula $0.749 \times (nA)^{0.5}$ [Bonomi and Lutton, 1984]. In this example, the expected best solution is 410.24. So, the solution is close to the expected best tour length.

Figure 13.7 shows the rate of optimization of the best and average cost of tours found during the search. By about the 100th iteration, the best solution already has nearly the expected best score. By the 800th iteration, the average tour length across all 50 ants is also close to this score, indicating that the search has converged.

Ant colony optimization has been applied to a diverse set of engineering problems. For example, the method has been employed to construct neural network topologies [Salama and Abdelbar, 2014], design type-2 fuzzy controllers [Hsu and Juang, 2013], induce decision trees [Otero *et al.*, 2012], perform fingerprint analysis [Cao *et al.*, 2012], and many other application areas. For reviews, see Dorigo and Gambardella [1997] and Dorigo and Stutzle [2004].

13.4   EVOLVABLE HARDWARE

We've spent a lot of time addressing specific implementations of evolutionary algorithms or other variations of evolutionary computing, all in software. But it's interesting to consider implementing evolutionary principles in the design of hardware too.

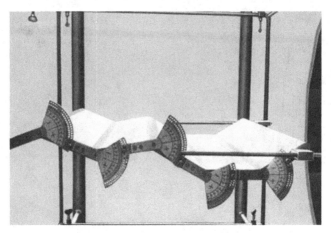

FIGURE 13.8 The hinged plates assembled as a physical device. Angles between the plates were described by x_1 through x_5. A ball falling through a series of nails was used to create random numbers to vary the protractors. From geneticargonaut.blogspot.com/2006/03/evolutionary-computation-classics-vol.html.

Some of the earliest experiments in evolutionary computation accomplished by a group in Berlin, Germany, involved manipulating physical devices [Rechenberg, 1973]. This was performed by constructing a device, such as series of planes attached at hinges with measured protractor settings (Figure 13.8), and then throwing dice or using a pachinko-like device to generate a random number, and making random changes to the physical structures.[1]

Scoring the set of hinged plates was accomplished in a *de facto* wind tunnel, where the drag of the plates was measured using a pitot tube. The objective was to adjust the angles between the plates such that the overall set of plates would have minimum drag. This method of throwing dice and making adjustments based on the random rolls, while retaining the best-found configuration as the next starting point, was able to find the series of flat plates that provides minimum drag.

More complicated experiments involved evolving the mechanical design of a bent pipe and a nozzle that would offer maximum energy efficiency of fluids flowing through the devices. These experiments are reviewed in Fogel [1988].[2]

More recently, experiments in Lohn *et al.* [2015] describe the evolution of S-band omnidirectional and medium gain antennas, which were evolved in simulation and

---

[1] Ingo Rechenberg and Hans-Paul Schwefel worked in tandem, with a third contributor Peter Bienert, at Technical University of Berlin starting in 1963–1964. The first computer available to the group was a Zuse Z23 that arrived in 1965. The experiments in hardware evolution were conducted in real hardware primarily because the researchers conducting the experiments had no electronic computers available to model the dynamics of the systems and execute the algorithms in software. In addition, whether or not all of the relevant physics in more complicated structures were known was uncertain. Thus, performing experiments *in situ* provided real validation of the physics of the system (I. Rechenberg (1996) personal communication to D. Fogel at Technical University of Berlin, Germany).

[2] Also see http://geneticargonaut.blogspot.com/2006/03/evolutionary-computation-classics-vol.html for a video of the evolution of the flashing nozzle created by H.-P. Schwefel.

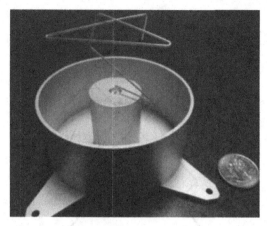

FIGURE 13.9   The final prototype of the S-band antenna from Lohn *et al.* [2015].

then verified in real-world settings. Figure 13.9 shows a sample design of the antenna, which is unconventional. The final evolved designs were launched into space as part of NASA's Lunar Atmosphere and Dust Environment Explorer (LADEE), which launched on September 6, 2013 and orbited the moon from October 2013 through April 2014. The antennas provided "65% increased downlink coverage and 44% cost savings for the mission" [Lohn *et al.*, 2015]. For other related work, see Homby *et al.* [2011].

Other interests in evolving hardware have focused on adapting field programmable gate arrays (FPGAs) [Thompson, 1996, 1998; Stoica *et al.*, 2003] and recovering from faults that may occur [Greenwood, 2005]. Additional research has been performed in evolving electronic circuits in simulation [Koza *et al.*, 1996; Vasicek and Sekanina, 2014]. While a more in-depth treatment of evolving hardware is beyond the scope of this text, interested readers should see Greenwood and Tyrrell [2006] for more information.

13.5   INTERACTIVE EVOLUTIONARY COMPUTATION

Sometimes it's not very easy or even impossible to know how to write a fitness function in a mathematical equation. For example, suppose you wanted to use an evolutionary algorithm to create a piece of artwork or music. What fitness function would you use?

There have been some thoughts on this. For example, symmetry in artwork can be aesthetically pleasing [Fogel, 1992]. Certain known chord progressions resonate with our expectations when listening to music [Fukumoto, 2014]. These and other aspects might be measureable and therefore could be put in a heuristic for scoring a particular drawing or musical composition.

But, as the cliché goes, sometimes beauty really is in the eye of the beholder. In such cases, it's possible to use a human's judgment as the fitness function and guide

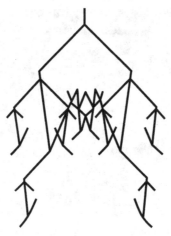

FIGURE 13.10   A starting biomorph described as a "frog" at www.emergentmind.com/
biomorphs

evolution to solutions that fit the individual's judgment. The result is something called
interactive evolutionary computation, because it involves the interaction of the human
operator for the scoring function.

One early example is found in the *Biomorphs* program introduced in Dawkins
[1986]. The program is replicated presently at www.emergentmind.com/biomorphs
so that you can try it for yourself. The program provides you nine different stick figure
objects. You can guide an evolutionary search to see if you can create various shapes.

One of the authors (D. Fogel) tried this and started with the frog-like figure shown in
Figure 13.10. Over a succession of 25 generations, he was able to create the rocketship-
like drawing shown in Figure 13.11. His goal was the prototype spaceship shown in

FIGURE 13.11   The biomorph D. Fogel evolved over 25 generations with the idea of making
the image look like a rocketship.

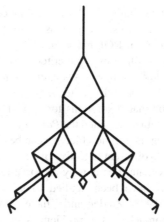

FIGURE 13.12 The image that D. Fogel had in mind while adapting the original biomorph from Figure 13.10 over 25 generations. This prototype image was provided at the Web site as an example of a possible outcome.

Figure 13.12. Although the fit between the two images is not exact, the image in Figure 13.11 is qualitatively closer to that of Figure 13.12 than is Figure 13.10.

In this same vein, one of the first practical applications of interactive evolutionary computation was in area of police sketch artistry [Caldwell and Johnson, 1991], where someone can say whether or not a given face looks more or less like the suspect he/she has in mind and the algorithm can adjust that face until it comes close to the person's mental image.

Interactive evolutionary algorithms have been used in designing ergonomic systems [Brintrup *et al.* 2008], portions of video games [Walsh and Gade, 2010; Cardamone *et al.* 2011], personal hearing aids [Takagi and Ohsaki, 2007; also see Fogel and Fogel [2012]], and even finding a nice blend of coffee [Herdy, 1996]. For a review of many other examples, see Takagi [2001].

13.6 MULTICRITERIA EVOLUTIONARY OPTIMIZATION

The final topic to cover involves finding solutions that satisfy multiple criteria. This is often the case in real-world problem solving, where a solution may be measured in multiple ways. For example, a financial asset management algorithm might be measured in terms of the return on investment, but also in terms of its volatility (such as the annualized standard deviation of its monthly returns). In this case, we'd like the ROI to be high, but we also want the volatility (which is a way of measuring risk) to be low.

One approach to handling multiple criteria is to combine them in a single utility function that returns a real value. For example, we could say that the value of the asset management algorithm was determined by

$$f(\mathbf{x}) = ax_1/x_2$$

where x_1 is the ROI, x_2 is the volatility, and a is a scaling constant. As ROI increases, so does $f(\mathbf{x})$. As volatility decreases, $f(\mathbf{x})$ increases. So, the best values of $f(\mathbf{x})$ will represent some trade-off between ROI and volatility.[3]

A different approach to handling multiple criteria involves finding solutions that are not "dominated" by other solutions. A solution dominates another when it is equally good or better in all measurable criteria, but better in at least one criterion. There may be many nondominated solutions for a given multicriteria problem.

The entire set of these solutions is called the *Pareto set*.[4] This is the set of solutions for which, for any solution in the set, no objective can be improved without reducing performance with respect to a different objective. It's sometimes of interest to find the entire Pareto set, or approximate it as closely as possible.

Evolutionary algorithms have been applied to this problem for many years [Fonseca and Flemming, 1993; Zitzler and Thiele, 1998; Knowles and Corne, 1999]. One interesting approach to the problem is called NSGA-II [Deb *et al.*, 2002], which stands for nondominated sorting genetic algorithm.

The procedure starts by looping over all individuals x_i, $i = 1, \ldots, k$ in the population. Each solution begins with a set S_i initialized to the empty set and a number n_i set equal to 0. The set S_i ultimately contains all the individuals that are dominated by x_i. The number n_i is the number of individuals in the solution that dominates x_i. When the loop is complete, every individual that has $n_i = 0$ belongs to the "first front" (another set) and its rank $r_i = 1$.

After determining the members in the first front (presuming it's not empty), a procedure is implemented to determine the individuals to be stored in the second front. For each individual x_i in the first front, a loop is performed over all individuals in the corresponding S_i (which is the set of individuals dominated by x_i). For each individual j in S_i, its value n_j is decremented. If that value is zero, then that individual is assigned to the second front. Once that loop is completed, the second front is also complete. The procedure is repeated until the subsequent front remains empty.

Next, the NSGA-II algorithm works to determine the crowding of individuals within each front. Initially, all crowding scores are set to zero. Then, for each objective function m, individuals are sorted based on the objective. Individuals at the extremes of each front are assigned a value of positive infinity so that they will always be propagated into the next generation. In between the extremes, the crowding scores are incremented for each individual based on the Euclidean distance between each pair of neighboring individuals (within in front) across all of the objectives. The crowding procedure sets up a tournament selection function, which is based on nondomination and crowding criteria.

Selection can be implemented to fill the population for each next generation from each front in turn (that is, first from the first front, then from the second front, and so on) until the number of solutions in the remaining front would exceed the population

---

[3] A similar form of measure is called the Sharpe Ratio, which is computed as (ROI–RFR)/STD, where ROI is the return on investment, RFR is the risk-free rate of return, and STD is the standard deviation of the return on investment. In essence, the ratio measures how many standard deviations above the risk-free rate of return the asset is producing.

[4] Named after Vilfredo Pareto (1848–1923).

size. Solutions from this last front can be selected at random or based on crowding criteria.

New offspring solutions can be created from the new parents by any appropriate evolutionary algorithm. For example, here is an implementation of a 50 parent–50 offspring NSGA-II procedure on the problem:

Find the Pareto front for x_1, x_2, x_3 for the two criteria:

$$f_1 = 1 - \exp\left(-\sum_{i=1}^{3}\left(x_i - \frac{1}{\sqrt{3}}\right)^2\right)$$

$$f_2 = 1 - \exp\left(-\sum_{i=1}^{3}\left(x_i + \frac{1}{\sqrt{3}}\right)^2\right)$$

The solutions were initialized at random in the range $[-5, 5]$. Mutation was based on a fixed Gaussian random variation of a normal random variable with zero mean and standard deviation of 1 added to each dimension of a parent.[5]

For this problem, we can compute the actual Pareto front before we start the evolution, so we can determine how well the NSGA-II algorithm can fit the Pareto front. The equation of the Pareto front is

$$A^2 - 2AB + B^2 + 8A + 8B + 16 = 0$$

where

$$A = \log(1 - f_1)$$

$$B = \log(1 - f_2)$$

with f_1 and $f_2 \in [0, 1 - e^{-4}]$.

The population after generation 10 is shown in Figure 13.13. By generation 30, you can see the population starting to converge toward the Pareto-optimal front (Figure 13.14). At generation 70, the front is well estimated (Figure 13.15) and there are only four solutions remaining in the second front. By generation 100, all solutions are in the first front and the process has evidently converged close to the correct answer (Figure 13.16).

Remember that this problem and the approach are provided only for illustration. The variation operator here was not constructed to converge quickly. It used a fixed step size, with a standard deviation that was clearly larger than optimal given the distribution of solutions shown in Figure 13.16. Still, the NSGA-II process was able to guide the simple evolutionary algorithm to come close to the optimal Pareto set in only 100 generations.

---

[5] This was done to keep the illustration simple. Faster convergence could be expected by a more appropriate choice of mutation, as discussed in earlier chapters.

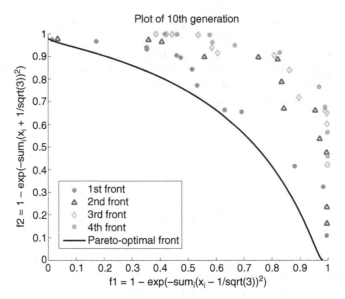

FIGURE 13.13 The population at the 10th generation on the multicriteria problem described in the text. The legend indicates which front each of the solutions is in. The solid black line represents the mathematical optimum solution for this problem.

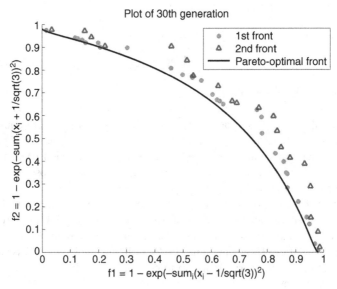

FIGURE 13.14 The population at the 30th generation on the multicriteria problem described in the text. Solutions in the population are now contained in one of two fronts. The solid black line represents the mathematical optimum solution for this problem.

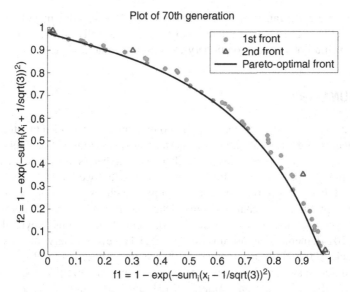

FIGURE 13.15   The population at the 70th generation on the multicriteria problem described in the text. Only four solutions remain in the second front. The first front is close to the solid black line that represents the mathematical optimum solution for this problem.

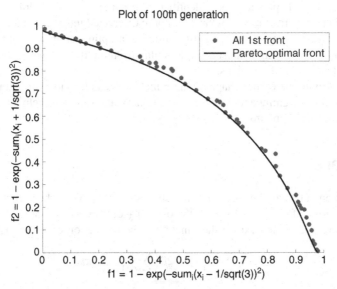

FIGURE 13.16   The population at the 100th generation on the multicriteria problem described in the text. All surviving solutions are in the first front, which has converged close to the solid black line that represents the mathematical optimum solution for this problem.

Readers are encouraged to review Deb *et al.* [2002] for implementing NSGA-II and Deb and Jain [2014] for an extension of NSGA-II, as well as Zhou *et al.* [2011] for a survey of multiobjective evolutionary algorithm techniques.

13.7 SUMMARY

This and the prior chapters on evolutionary computation are intended to introduce the main concepts of the field as part of a semester course on computational intelligence. We've covered many topics in these four chapters. But there's more to learn about, as there are many aspects of evolutionary computation that we haven't covered.

If you'd like to discover more about topics such as evolutionary search on constrained problems [Michalewicz and Schoenauer, 1996; Coello Coello, 2012], classifier systems [Wilson, 1995; Bull, 2015], cultural algorithms [Reynolds, 1994; Ali *et al.* 2014], memetic algorithms [Mei *et al.* 2011], evolutionary robotics [Lipson and Pollack, 2000; Nolfi and Floreano, 2000; Nouyan *et al.*, 2009; Lipson, 2014], artificial immune systems [Dasgupta *et al.*, 2011], artificial life [Conrad and Pattee, 1970; Ray, 1992; Ostman and Adami, 2013], and coevolution [Fan *et al.*, 2014; Omidvar *et al.*, 2014], please see the references included for you. Even when including those topics, that wouldn't cover everything in evolutionary computation. But as they say in statistical regression analysis: "You have to draw the line somewhere." And in evolutionary computation, that line is always evolving.

This chapter has presented several extensions of evolutionary computation that are based on modeling natural search mechanisms and other population-based search procedures. The chapter also has identified that sometimes it's helpful to evolve solutions in hardware or use a human operator to offer a judgment about the quality of evolved solutions. Finally, the topic of multicriteria optimization was introduced in terms of both optimizing against a single multiattribute utility (fitness) function and finding a Pareto set of solutions.

Together with the earlier chapters, these materials can provide you with a strong foundation for designing and applying evolutionary algorithms to help address your own challenging problems.

EXERCISES

13.1. Implement a particle swarm optimization algorithm to find the minimum of the function $f(x, y) = x^2 + y^2 - 20[\cos(\pi x) + \cos(\pi y) - 2]$ as shown in Figure 13.1. Then extend the function to be a function of n dimensions:

$$f(\mathbf{x}) = \sum_{i=1}^{n} x_i^2 - 20\left(\sum_{i=1}^{n}(\cos(\pi x_i) - 1)\right)$$

Estimate the convergence rate in terms of number of generations to reach a target level of the function for an initialization that you choose as a function of n for $n = 2, 3, 5,$ and 10.

13.2. Refer back to software you have written or use an available feed forward neural network simulator that employs sigmoid functions for the nodes and instead of using backpropagation, substitute a differential evolution algorithm to do the training. Test the results on data from http://archive.ics.uci.edu/ml/datasets/ Wine, which provide 13 attributes about wine from three different wineries. The object is to have the neural network identify which winery made the wine based on the attributes. Compare your performance with differential evolution with that of backpropagation. Based on your results, what is your intuition about the relative convergence rate and reliability of the two approaches on these data?

13.3. Implement an ant colony optimization algorithm to demonstrate the double-bridge problem. The problem has ants that start on one side of a double bridge and must go across one of the two bridges to get to a food source. Start with one of the bridges at half the length of the other. How quickly does the population of ants converge to traversing the shorter bridge? Now increase the length of the shorter bridge and estimate convergence time as a function of the ratio of the two bridge lengths. For more details on the double-bridge experiment, see http://www.scholarpedia.org/article/File:SameLengthDoubleBridge.png.

13.4. Imagine that you wanted to design a possible improvement to the winglets that are now ubiquitous on commercial airplane wingtips. Winglets are a relatively recent invention, having been developed in the 1970s at NASA. Think about a series of experiments you could conduct using evolvable hardware that might assist in finding a new design to improve the lift and drag performance of a wing that has winglets.

13.5. Consider the problem of finding a personalized shoe for elite professional athletes. How could you use interactive evolutionary computation to help design such a shoe? Detail the steps and experiments you would conduct.

13.6. For the Pareto problem in Section 13.6, code the NSGA-II algorithm and see if you can replicate the results presented in the chapter to your own satisfaction. Next, increase the number of variables to four and five and rerun the algorithm. What is your expectation about the algorithm's run time until convergence as a function of the number of variables on this problem?

■ REFERENCES

Akaike, H. (1974) A new look at the statistical model identification. *IEEE Transactions on Automatic Control*, *19*(6), 716–723.

Ali, M., Awad, N., and Reynolds, R. (2014) Balancing search direction in cultural algorithm for enhanced global numerical optimization. *Proceedings of the 2014 IEEE Symposium on Swarm Intelligence*, 1–7.

Altenberg, L. (1995) The schema theorem and price's theorem. In L. D. Whitley and M. D. Vose (Eds.), *Foundations of Genetic Algorithms 3*. San Mateo, CA: Morgan Kaufmann, 23–49.

Angeline, P. (1997) Comparing subtree crossover with macromutation. In P. J. Angeline, R. G. Reynolds, McDonnell, and R. Eberhart (Eds.), *Evolutionary Programming VI*. Berlin: Springer, 101–111.

Angeline, P. (1998) Evolving predictors for chaotic time series. *Proceedings of the Application on Science and Computational Intelligence*, *3390*, 170–180.

Angeline, P., Fogel, D., and Fogel, L. (1996) A comparison of self-adaptation methods for finite state machines in a dynamic environment. In L. Fogel, P. Angeline, and T. Baeck (Eds.), *Evolutionary Programming V*. Cambridge, MA: MIT Press, 441–449.

Ashlock, D., Eun-Youn, K., and Ashlock, W. (2010) A fingerprint comparison of different prisoner's dilemma payoff matrices. *Proceedings of the 2010 IEEE Symposium on Computational Intelligence and Games*, 219–226.

Atmar, J. (1979) *The Inevitability of Evolutionary Invention*, unpublished manuscript.

Atmar, W. (1992) On the rules and nature of simulated evolutionary programming. *Proceedings of the 1st Annual Conference on Evolutionary Programming*, La Jolla, CA.

Atmar, W. (1994) Notes on the simulation of evolution. *IEEE Transactions on Neural Networks*, *5*(1), 130–147.

Autumn, K., Sitti, M., Liang, Y., Peattie, A., Hansen, W., Sponberg, S., *et al.* (2002) Evidence for van der Waals adhesion in gecko setae. *Proceedings of the National Academy of Sciences of the United States of America*, *99*(19), 12252–12256.

Axelrod, R. (1987) The evolution of strategies in the iterated prisoner's dilemma. In L. Davis (Ed.), *Genetic Algorithms and Simulated Annealing*. Pitman, London, 32–41.

Azaria, Y. and Sipper, M. (2005) GP-gammon: genetically programming backgammon players. *Genetic Programming and Evolvable Machines*, *6*(3), 283–300.

Bäck, T. (1994) Selective pressure in evolutionary algorithms: a characterization of selection mechanisms. *Proceedings of the 1st IEEE Conference on Evolutionary Computation*, 57–62.

Fundamentals of Computational Intelligence: Neural Networks, Fuzzy Systems, and Evolutionary Computation, First Edition. James M. Keller, Derong Liu, and David B. Fogel.
© 2016 by The Institute of Electrical and Electronics Engineers, Inc. Published 2016 by John Wiley & Sons, Inc.

Bäck, T. (1996) *Evolutionary Algorithms in Theory and Practice*. New York: Oxford University Press.

Bäck, T., Hoffmeister, F., and Schwefel, H. (1991) A survey of evolution strategies. *Proceedings of the 4th International Conference on Genetic Algorithms*, 2–9.

Baker, J. (1985) Adaptive selection methods for genetic algorithms. *Proceedings of the 1st International Conference on Genetic Algorithms and Their Applications*, 101–111.

Banerjee, T., Keller, J., Skubic, M., and Stone, E. (2013) Day or night activity recognition from video using fuzzy clustering techniques. *IEEE Transactions on Fuzzy Systems*, 22(3), 483–493.

Banks, A., Vincent, J., and Anyakoha, C. (2007) A review of particle swarm optimization. Part I: background and development. *Natural Computing*, 6, 467–484.

Banon, G. (1981) Distinction between several subsets of fuzzy measures. *Fuzzy Sets and Systems*, 5, 291–305.

Banzhaf, W., Nordin, P., Keller, R., and Francone, F. (1998) *Genetic Programming: An Introduction*. San Francisco, CA: Morgan Kaufmann.

Barni, M., Cappellini, V., and Mecocci, A. (1996) Comments on 'A possibilistic approach to clustering'. *IEEE Transactions on Fuzzy Systems*, 4(3), 393–396.

Barron, A. (1984) Predicted squared error: a criterion for automatic model selection. In S. J. Farlow (Ed.), *Self-Organizing Methods in Modeling*. New York: Marcel Dekker.

Battle, D. and Vose, M. (1993) Isomorphisms of genetic algorithms. *Artificial Intelligence*, 60, 155–165.

Bellman, R. and Zadeh, L. (1970) Decision-making in a fuzzy environment. *Management Science*, 17(4), 141–164.

Beyer, H. (2010) *The Theory of Evolution Strategies*. Berlin: Springer.

Bezdek, J. (1981) *Pattern Recognition with Fuzzy Objective Function Algorithms*. New York: Plenum.

Bezdek, J. (1993) Fuzzy models: what are they, and why. *IEEE Transactions on Fuzzy Systems*, 1(1), 1–16.

Bezdek, J. and Hathaway, R. (2002) VAT: a tool for visual assessment of (cluster) tendency. *Proceedings of the International Joint Conference on Neural Networks (IJCNN)*, 2225–2230.

Bezdek, J., Keller, J., Krishnapuram, R., and Pal, N. (1999) *Fuzzy Models and Algorithms for Pattern Recognition and Image Processing*. Norwell, MA: Kluwer Academic Publishers.

Binns, C. (2006) *How flies walk on ceilings*. Retrieved from http://www.livescience.com/10536-flies-walk-ceilings.html.

Black, M. (1937) Vagueness: an exercise in logical analysis. *Philosophy of Science*, 4(4), 427–455.

Bondagula, R. and Xu, D. (2007) MUPRED: a tool for bridging the gap between template based methods and sequence profile based methods for protein secondary structure prediction. *Proteins: Structure, Function, and Bioinformatics*, 66(3), 664–670.

Bonomi, E. and Lutton, J. (1984) The N-city travelling salesman problem: statistical mechanics and the metropolis algorithm. *SIAM Review*, 26(4), 551–568.

Borges, P., Pacheco, R., Barcia, R., and Khator, S. (1997) A fuzzy approach to the prisoner's dilemma. *BioSystems*, 41, 127–137.

Borkowski, L. (1970) *Jan lukasiewicz: selected works*. Amesterdam: North Holland Pub. Co.

Braun, J., Krettek, J., Hoffmann, F., and Bertram, T. (2011) Structure and parameter identification of nonlinear systems with an evolution strategy. *Proceedings of the 2011 IEEE Congress on Evolutionary Computation*, 2444–2451.

Bremermann, H., Rogson, M., and Salaff, S. (1966) Global properties of evolution processes. In H. H. Pattee, E. A. Edlsack, L. Fein, and A. B. Callahan (Eds.), *Natural Automata and Useful Simulations*. Washington DC: Spartan Books, 3–41.

Briceño, R. and Eberhard, W. (2012) Spiders avoid sticking to their webs: clever leg movements, branched drip-tip setae, and anti-adhesive surfaces. *Naturwissenshaften, 99* (4), 337–341.

Brintrup, A., Ramsden, J., and Takagi, H. (2008) Ergonomic chair design by fusing qualitative and quantitative criteria using interactive genetic algorithms. *IEEE Transactions on Evolutionary Computation, 12*(3), 343–354.

Brunk, C. (1991) Darwin in an age of molecular revolution. *Contention, 1*(1), 131–150.

Buckles, B., Petry, F., Prabhu, D., George, R., and Srikanth, R. (1994) Fuzzy clustering with genetic search. *Proceedings of the 1st IEEE Conference on Evolutionary Computation*, 46–50.

Buckley, J. (1992) Universal fuzzy controllers. *Automatica, 28*, 1245–1248.

Bull, L. (2015) A brief history of learning classifier systems: from CS-1 to XCS and its variants. *Evolutionary Intelligence, 8*, 55–70.

Buzas, M. and Culver, S. (1984) Species duration and evolution: benthic Foraminifera on the Atlantic continental margin of North America. *Science, 225*(4664), 829–830.

Caldwell, C. and Johnson, V. (1991) Tracking a criminal suspect through 'Face-Space' with a genetic algorithm. *Proceeding of the 1991 International Conference on Genetic Algorithms*, 416–421.

Cao, Z. (1990) Mathematical principle of fuzzy reasoning. *Proceedings of the North America Fuzzy Information Processing Society (NAFIPS'90)*, 362–365.

Cao, K., Yang, X., Chen, X., Zang, Y., Liang, J., and Tian, J. (2012) A novel ant colony optimization algorithm for large-distorted fingerprint matching. *Pattern Recognition, 45*(1), 151–161.

Cardamone, L., Loiacono, D., and Lanzi, P. (2011) Interactive evolution for the procedural generation of tracks in a high-end racing game. *Proceedings of the 13th Annual Conference on Genetic and Evolutionary Computation*, 395–402.

Chellapilla, K. (1997) Evolving computer programs without subtree crossover. *IEEE Transactions on Evolutionary Computation, 1*(3), 209–216.

Chellapilla, K. (1998) Evolving nonlinear controllers for backing up a truck-and-trailer using evolutionary programming. In V. W. Porto, N. Saravanam, D. Waagen, and A. E. Eiben (Eds.), *Evolutionary Programming VII*. Berlin: Springer, 417–426.

Chellapilla, K. and Fogel, D. (1997a) Exploring self-adaptive methods to improve the efficiency of generating approximate solutions to the traveling salesman problems. *Evolutionary Programming VI*, 361–371.

Chellapilla, K. and Fogel, D. (1997b) Two new mutation operators for enhanced search and optimization in evolutionary programming. *Applications of Soft Computing, 3165*, 260–269.

Chellapilla, K. and Fogel, D. (1999) Fitness distributions in evolutionary computation: motivation and examples in the continuous domain. *BioSystems, 54*(1–2), 15–29.

Chellapilla, K. and Fogel, D. (2001) Evolving an expert checkers playing program without using human expertise. *IEEE Transactions on Evolutionary Computation*, *5*(4), 422–428.

Chellapilla, K., Fogel, D., and Rao, S. (1997) Gaining insight into evolutionary programming through landscape visualization: an investigation into IIR filtering. In P. J. Angeline, R. G. Reynolds, J. R. McDonnell, and R. Eberhart (Eds.), *Evolutionary Programming VI*. Springer, 407–415.

Chiang, J. and Gader, P. (1997) Hybrid fuzzy-neural systems in handwritten word recognition. *IEEE Transactions on Fuzzy Systems*, *5*(4), 497–510.

Chua, L. and Roska, T. (2004) *Cellular Neural Networks and Visual Computing*. Cambridge University Press.

Clerc, M. (2004) Discrete particle swarm optimization. *New Optimization Techniques in Engineering*. Springer.

Coello Coello, C. (2012) Constraint-handling techniques used with evolutionary algorithms. *Proceedings of the 14th Annual Conference Companion on Genetic and Evolutionary Computation*, 849–872.

Conrad, M. and Pattee, H. (1970) Evolution experiments with an artificial ecosystem. *Journal of Theoretical Biology*, *28*, 393–409.

Corne, D. and Knowles, J. (2003) No free lunch and free leftovers theorems for multiobjective optimisation problems. *Evolutionary Multi-Criterion Optimization*. Berlin: Springer, Vol. *2632*, 327–341.

Cotta, C., Fernandez-Leiva, A., Sánchez, A., and Lara-Cabrera, R. (2013) Car setup optimization via evolutionary algorithms. *Advances in Computational Intelligence*, Springer, 346–354.

Cover, T. and Hart, P. (1967) Nearest neighbor pattern classification. *IEEE Transactions on Information Theory*, *13*, 21–27.

Darwen, P. (2001) Why co-evolution beats temporal difference learning at backgammon for a linear architecture, but not a non-linear architecture. *Proceedings of the 2001 Congress on Evolutionary Computation*, 1003–1010.

Darwen, P. and Yao, X. (2001) Why more choices cause less cooperation in iterated prisoner's dilemma. *Proceedings of the 2001 Congress on Evolutionary Computation*, 987–994.

Darwen, P. and Yao, X. (2002) Co-evolution in iterated prisoner's dilemma with intermediate levels of cooperation: application to missile defense. *International Journal of Computational Intelligence and Applications*, *2*, 83–107.

Dasarathy, B. (1991) *Nearest Neighbor (NN) Norms: NN Pattern Classification Techniques*. Los Alamatos, CA: IEEE Computer Society Press.

Dasgupta, D., Yu, S., and Nino, F. (2011) Recent advances in artificial immune systems: models and applications. *Applied Soft Computing*, *11*(2), 1574–1587.

Davis, T. and Principe, J. (1993) A Markov chain framework for the simple genetic algorithm. *Evolutionary Computation*, *1*(3), 269–288.

Dawkins, R. (1976) *The Selfish Gene*. Oxford: Oxford University Press.

Dawkins, R. (1986) *The Blind Watchmaker*. London: Penguin.

Deb, K. and Jain, H. (2014) An evolutionary many-objective optimization algorithm using reference-point-based nondominated sorting approach. Part I: solving problems with box constraints. *IEEE Transactions on Evolutionary Computation*, *18*(4), 577–601.

Deb, K., Pratap, A., Agarwal, S., and Meyarivan, T. (2002) A fast and elitist multiobjective genetic algorithm: NSGA-II. *IEEE Transactions on Evolutionary Computation*, 6(2), 182–197.

Dorigo, M. (1992) *Optimization, learning and natural algorithms*. Doctoral dissertation, Politecnico di Milano, Milan, Italy.

Dorigo, M. and Gambardella, M. (1997) Ant colony system: a cooperative learning approach to the traveling salesman problem. *IEEE Transactions on Evolutionary Computation*, 1(1), 53–66.

Dorigo, M. and Stützle, T. (2004) *Ant Colony Optimization*. Cambridge, MA: MIT Press.

Dubois, D. (2005) Special Issue on "40th anniversary of fuzzy sets". *Fuzzy Sets and Systems*, 156(3), 331–333.

Dubois, D. and Prade, H. (1985) A review of fuzzy set aggregation connectives. *Information Sciences*, 36, 85–121.

Dubois, D. and Prade, H. (2001) Possibility theory, probability theory and multiple-valued logics: a clarification. *Annals of Mathematics and Artificial Intelligence*, 32, 35–66.

Duenez-Guzman, E. and Vose, M. (2013) No free lunch and benchmarks. *Evolutionary Computation*, 21(2), 293–312.

Eberhart, R. and Kennedy, J. (1995) A new optimizer using particle swarm theory. *Proceedings of the 6th International Symposium on Micro Machine and Human Science*, 39–43.

Eiben, A., Aarts, F., and Van Hee, K. (1991) Global convergence of genetic algorithms: a Markov chain analysis. In H.-P. Schwefel and R. Manner (Eds.), *Parallel Problem Solving from Nature*. Berlin: Springer, 4–12.

English, T. (1996) Evaluation of evolutionary and genetic optimizers: no free lunch. In L. J. Fogel, P. J. Angeline, and T. Bäck (Eds.), *Evolutionary Programming V: Proceedings of the 5th Annual Conference on Evolutionary Programming*. Cambridge, MA: MIT Press, 163–169.

Fan, J., Wang, J., and Han, M. (2014) Cooperative coevolution for large-scale optimization based on kernel fuzzy clustering and variable trust region methods. *IEEE Transactions on Fuzzy Systems*, 22(4), 829–839.

Fisher, R. (1930) *The Genetical Theory of Natural Selection*. Oxford: Clarendon Press.

Fisher, R. (1936) The use of multiple measurements in taxonomic problems. *Annals of Eugenics*, 7(2), 179–188.

Fogel, L. (1964) *On the Organization of Intellect*. Ph.D. dissertation, UCLA, Los Angeles.

Fogel, D. (1988) *Evolutionary Computation: The Fossil Record*. Piscataway, NJ: IEEE Press.

Fogel, D. (1990) *System Identification through Simulated Evolution*. Master's thesis, UCSD, La Jolla, CA.

Fogel, D. (1991) The evolution of intelligent decision making in gaming. *Cybernetics and Systems*, 22(2), 223–236.

Fogel, L. (1992) Evolving art. *Proceedings of the 1st Annual Conference on Evolutionary Programming*, 183–184.

Fogel, D. (1992a) *Evolving artificial intelligence*. Doctoral dissertation, UCSD, La Jolla, CA.

Fogel, D. (1992b) Using evolutionary programming for modeling: an ocean acoustic example. *IEEE Journal of Oceanic Engineering*, 17(4), 333–340.

Fogel, D. (1993) Evolving behaviors in the iterated prisoner's dilemma. *Evolutionary Computation*, 1(1), 77–97.

Fogel, D. (1995) On the relationship between the duration of an encounter and the evolution of cooperation in the iterated prisoner's dilemma. *Evolutionary Computation, 3*, 349–363.

Fogel, D. (1996) A 'correction' to some cart-pole experiments. *Evolutionary Programming V: Proceedings of the 5th Annual Conference on Evolutionary Programming*, 67–71.

Fogel, D. (1998) *Evolutionary Computation: The Fossil Record*. New York: IEEE Press.

Fogel, D. (2000) *Evolutionary Computation: Principles and Practice in Signal Processing*. Bellingham, WA: SPIE.

Fogel, D. (2002) *Blondie24: Playing at the Edge of AI*. San Francisco, CA: Morgan Kaufmann.

Fogel, D. (2004) Evolving strategies in blackjack. *Proceedings of the 2004 Congress on Evolutionary Computation*, IEEE Press, 1427–1434.

Fogel, D. (2006) *Evolutionary Computation: Toward a New Philosophy of Machine Intelligence*, 3rd ed. New York: IEEE Press.

Fogel, D. (2008) *Evolving tinnitus masks*.

Fogel, D. and Atmar, J. (1990) Comparing genetic operators with Gaussian mutations in simulated evolutionary processes using linear systems. *Biological Cybernetics, 63*, 111–114.

Fogel, L. and Burgin, G. (1969) *Competitive goal-seeking through evolutionary programming*. Final report, Contract AF 19(628)-5927, Air Force Cambridge Research Laboratories.

Fogel, D. and Fogel, L. (1996) Preliminary experiments on discriminating between chaotic signals and noise using evolutionary programming. *Proceedings of the 1996 Conference on Genetic Programming*, 512–520.

Fogel, G. and Fogel, D. (2011) Simulation natural selection as a culling mechanism on finite populations with the hawk–dove game. *BioSystems, 104*(1), 57–62.

Fogel, D. and Fogel, G. (2012) Method and device for tinnitus masking. U.S. Patent 8,273,034.

Fogel, D. and Ghozeil, A. (1996) Using fitness distributions to design more efficient evolutionary computations. *Proceedings of the IEEE International Conference on Evolutionary Computation*, 11–19.

Fogel, D. and Ghozeil, A. (1997) A note on representations and variation operators. *IEEE Transactions on Evolutionary Computation, 1*(2), 159–161.

Fogel, D., Hays, T., Hahn, S., and Quon, J. (2005) Further evolution of a self-learning chess program. *Proceedings of the 2005 IEEE Symposium on Computational Intelligence and Games*.

Fogel, D., Hays, T., Hahn, S., and Quon, J. (2006) The Blondie25 chess program competes against Fritz 8.0 and a human chess master. *Proceedings of the 2006 IEEE Symposium on Computational Intelligence and Games*, 230–235.

Fogel, D. and Jain, A. (2000) Case studies in applying fitness distributions in evolutionary algorithms. *Proceedings of the IEEE 1st Symposium on Combinations of Evolutionary Computation and Neural Networks*, 91–97.

Fogel, L., Owens, A., and Walsh, M. (1966) *Artificial Intelligence through Simulated Evolution*. New York: John Wiley & Sons, Inc.

Fogel, D. and Robinson, C. (2003) *Computational Intelligence: The Experts Speak*. Piscataway, NJ: IEEE Press.

Fogel, D. and Stayton, L. (1994) On the effectiveness of crossover in simulated evolutionary optimization. *BioSystems, 32*(3), 171–182.

Fogel, D., Wasson, E., Boughton, E., and Porto, V. (1998) Evolving artificial neural networks for screening features from mammograms. *Artificial Intelligence in Medicine, 14*, 317–326.

Fonseca, C. and Fleming, P. (1993) Genetic algorithms for multi-objective optimization: formulation, discussion and generalization. *Proceedings of the 5th International Conference on Genetic Algorithms*, 416–423.

Fraser, A. (1968) The evolution of purposive behavior. In H. von Foerster, D. White, L. Peterson, and J. Russell (Eds.), *Purposive Systems*, Washington, DC: Spartan Books, 15–23.

Fraser, A. and Burnell, D. (1970) *Computer Models in Evolution*. McGraw Hill.

Frayn, C. (2005) An evolutionary approach to strategies for the game of monopoly. *Proceedings of the 2005 IEEE Symposium on Computational Intelligence and Games*.

Friedberg, R. (1958) A learning machine: Part I. *Research and Development*, 2(1), 2–13.

Friedman, G. (1956) *Selective feedback computers for engineering synthesis and nervous system analogy*. Master's thesis, UCLA.

Frigui, H. and Gader, P. (2009) Detection and discrimination of land mines in ground-penetrating radar based on edge histogram descriptors and a possibilistic-nearest neighbor classifier. *IEEE Transactions on Fuzzy Systems*, 17(1), 185–199.

Fukumoto, M. (2014) Creation of music chord progression suited for user's feelings based on interactive genetic algorithm. *Proceedings of the 3rd International Conference on Advanced Applied Informatics*, 757–762.

Gader, P., Keller, J., Krishnapuram, R., Chiang, J., and Mohamed, M. (1997) Neural and fuzzy methods in handwriting recognition. *IEEE Computer*, 30(2), 79–86.

Gader, P., Keller, J., and Nelson, B. (2001) Recognition technology for the detection of buried land mines. *IEEE Transactions on Fuzzy Systems*, 9(1), 31–43.

Gader, P., Mohamed, M., and Chiang, J. (1995) Comparison of crisp and fuzzy character neural networks in handwritten word recognition. *IEEE Transactions on Fuzzy Systems*, 3(3), 357–363.

Gader, P., Mohamed, M., and Keller, J. (1996) Dynamic-programming-based handwritten word recognition using the Choquet fuzzy integral as the match function. *Journal of Electronic Imaging*, 5(1), 15–24.

Galanopoulos, D., Athanasiadis, C., and Tefas, A. (2012) Evolutionary optimization of a neural network controller for car racing simulation. *Artificial Intelligence: Theories and Applications*. Springer, Vol. 7297, 149–156.

Gath, I. and Geva, A. (1989) Unsupervised optimal fuzzy clustering. *IEEE Transactions on Pattern Analysis and Machine Intelligence*, 11(7), 773–780.

Ghozeil, A. and Fogel, D. (1996a) Discovering patterns in spatial data using evolutionary programming. In J. R. Koza, D. E. Goldberg, D. B. Fogel, and R. L. Riolo (Eds.), *Genetic Programming 1996*. Cambridge, MA: MIT Press, 521–527.

Ghozeil, A. and Fogel, D. (1996b) A preliminary investigation into directed mutations in evolutionary algorithms. In H.-M. Voigt, W. Ebeling, I. Rechenberg, and H.-P. Schwefel (Eds.), *Parallel Problem Solving from Nature IV*. Berlin: Springer, 329–335.

Giarratano, J. and Riley, G. (2004) *Expert Systems: Principles and Programming*, 4th ed. Cambridge, MA: Course Technology.

Goldberg, D. (1989) *Genetic Algorithms in Search, Optimization and Machine Learning*. Reading, MA: Addison Wesley.

Goldberg, D. and Lingle, R. (1985) Alleles, loci, and the traveling salesman problem. *Proceedings of the 1st International Conference on Genetic Algorithms and Their Applications*, 154–159.

Grabisch, M. and Nicolas, J. (1994) Classification by fuzzy integral: performance and tests. *Fuzzy Sets and Systems*, *65*(2–3), 255–271.

Grabisch, M., Sugeno, M., and Murofushi, T. (2000) *Fuzzy Measures and Integrals: Theory and Applications*. New York: Springer.

Greenwood, G. (2005) On the practicality of using intrinsic reconfiguration for fault recovery. *IEEE Transactions on Evolutionary Computation*, *9*(4), 398–405.

Greenwood, G. and Tyrrell, A. (2006) *Introduction to Evolvable Hardware: A Practical Guide for Designing Self-Adaptive Systems*. New York: IEEE Press.

Grefenstette, J. (1995) Predictive models using fitness distributions of genetic operators. In L. D. Whitley and M. D. Vose (Eds.), *Foundations of Genetic Algorithms 3*. San Mateo, CA: Morgan Kaufmann, 139–161.

Grossberg, S. (1988) *Neural Networks and Natural Intelligence*. Cambridge, MA: MIT Press.

Grossberg, S. (1998) *Neural Networks and Natural Intelligence*. Cambridge, MA: MIT Press.

Guralnik, J., Simonsick, E., Ferrucci, L., Glynn, R., Berkman, L., Blazer, D., and Wallace, R. (1994) A short physical performance battery assessing lower extremity function: association with self-reported disability and prediction of mortality and nursing home admission. *Gerontology*, *49*(2), 85–94.

Gustafson, E. and Kessel, W. (1979) Fuzzy clustering with a fuzzy covariance matrix. *Proceedings of the IEEE Conference on Decision and Control*, 761–766.

Hagan, M., Demuth, H., and Beale, M. (1996) *Neural Network Design*. PWS Publishing Company.

Harrald, P. and Fogel, D. (1996) Evolving continuous behaviors in the iterated prisoner's dilemma. *BioSystems*, *37*(1), 135–145.

Hartl, D. and Clark, A. (1989) *Principles of Population Genetics*, 2nd ed. Sunderland, MA: Sinauer.

Hartl, D. and Jones, E. (2005) *Genetics: Analysis of Genes and Genomes*, 6th ed. Boston, MA: Jones & Barlett Learning.

Harvey, N., Luke, R., Keller, J., and Anderson, D. (2008) Speedup of fuzzy logic through stream processing on graphics processing units. *Proceedings of the World Congress on Computational Intelligence*, CEC, 3810–3816.

Hastings, E., Guha, R., and Stanley, K. (2009) Evolving content in the galactic Arms Race video game. *Proceedings of the 2009 IEEE Symposium Computational Intelligence and Games*, 241–248.

Havens, T. and Bezdek, J. (2012) An efficient formulation of the improved visual assessment of cluster tendency (iVAT) algorithm. *IEEE Transactions on Knowledge and Data Engineering*, *24*(5), 813–822.

Haykin, S. (1999) *Neural Networks: A Comprehensive Foundation*, 2nd ed. Prentice Hall.

Haykin, S. (2009) *Neural Networks and Learning Machines*, 3rd ed. Prentice Hall.

Herdy, M. (1996) Evolution strategies with subjective selection. In H.-M. Voigt, W. Ebeling, I. Rechenberg, and H.-P. Schwefel (Eds.), *Parallel Problem Solving from Nature IV*. Berlin: Springer, 22–31.

Herdy, M. (1997) Evolutionary optimisation based on subjective selection: evolving blends of coffee. *Proceedings of the 5th European Congress on Intelligent Techniques and Soft Computing (EUFIT'97)*, 2010–2644.

Hicks, C., von Baeyer, C., Spafford, P., van Korlaar, I., and Goodenough, B. (2001) The Faces Pain Scale-Revised: toward a common metric in pediatric pain measurement. *Pain, 93*(2), 173–183.

Hoffman, A. (1989) *Arguments on Evolution: A Paleontologist's Perspective.* New York: Oxford University Press.

Holland, J. (1975) *Adaptation in Natural and Artificial Systems.* Ann Arbor, MI: University of Michigan Press.

Homby, G., Lohn, J., and Linden, D. (2011) Computer-automated evolution of an X-band antenna for NASA's space technology 5 mission. *Evolutionary Computation, 19*(1), 1–23.

Hsu, P., Ge, L., Li, X., Stark, A., Wesdemiotis, C., Niewiarowski, P., and Dhinojwala, A. (2011) Direct evidence of phospholipids in gecko footprints and spatula–substrate contact interface detected using surface-sensitive spectroscopy. *Journal of Royal Society, Interface, 9*(69), 657–664.

Hsu, C., and Juang, C. (2013) Evolutionary robot wall-following control using type-2 fuzzy controller with species-DE-activated continuous ACO. *IEEE Transactions on Fuzzy Systems, 21*(1), 100–112.

Hsu, P. and Robbins, H. (1947) Complete convergence and the law of large numbers. *Proceedings of the National Academy of Sciences of the United States of America, 33*(2), 25.

Huang, H., Pasquier, M., and Chai, Q. (2009) Financial market trading system with a hierarchical coevolutionary fuzzy predictive model. *IEEE Transactions on Evolutionary Computation, 13*(1), 56–70.

Ignizio, J. (1991) *Introduction to Expert Systems: The Development and Implementation of Rule-Based Expert Systems.* NY: McGraw-Hill.

Jacob, F. (1977) Evolution and tinkering. *Science, 196*(4295), 1161–1166.

Jang, J. (1993) ANFIS: adaptive-network-based fuzzy inference system. *IEEE Transactions on Systems, Man and Cybernetics, 23*(3), 665–685.

Jones, T. (1995a) Crossover, macromutation and population-based search. *Genetic Algorithms: Proceedings of the Sixth International Conference,* 73–80.

Jones, T. (1995b) *Evolutionary algorithms, fitness landscapes, and search.* Doctoral dissertation, University of New Mexico, Albuquerque, NM.

Kashyap, R. (1980) Inconsistency of the AIC rule for estimating the order of autoregressive models. *IEEE Transactions on Automatic Control, 25*(5), 996–998.

Kaveh, A. (2014) Particle swarm optimization. *Advances in Metaheuristic Algorithms for Optimal Design of Structures.* Berlin: Springer, 9–40.

Keller, J. and Carpenter, C. (1988) Image segmentation in the presence of uncertainty. *Proceedings of the NAFIPS'88,* 136–140.

Keller, J. and Downey, T. (1988) Fuzzy segmentation using fractal features. *Proceedings of the SPIE Symposium on Intelligent Robots and Computer Vision,* 369–376.

Keller, J., Gader, P., and Hocaoglu, A. (2000) Fuzzy integrals in image processing and recognition. In M. Grabisch, T. Murofushi, and M. Sugeno (Eds.), *Fuzzy Measures and Integrals: Theory and Applications,* Studies in Fuzziness and Soft Computing. Springer, 435–466.

Keller, J., Gader, P., Tahani, H., Chiang, J., and Mohamed, M. (1994) Advances in fuzzy integration for pattern recognition. *Fuzzy Sets and Systems, 65*(2–3), 273–283.

Keller, J., Gray, M., and Givens, J. (1985) A fuzzy k-nearest neighbor algorithm. *IEEE Transactions on Systems, Man and Cybernetics*, *15*(4), 580–585.

Keller, J. and Hunt, D. (1985) Incorporating fuzzy membership functions into the perceptron algorithm. *IEEE Transactions on Pattern Analysis Machine Intelligence*, *7*(6), 693–699.

Keller, J., Krishnapuram, R., Chen, Z., and Nasraoui, O. (1994) Fuzzy additive hybrid operators for network-based decision making. *International Journal of Intelligent Systems*, *9*(11), 1001–1023.

Keller, J., Krishnapuram, R., Gader, P., and Choi, Y.-S. (1996) Fuzzy rule-based models in computer vision. In W. Pedrycz (Ed.), *Fuzzy Modelling: Paradigms and Practice*. Kluwer Academic Publishers, 353–371.

Keller, J. and Osborn, J. (1996) Training the fuzzy integral. *International Journal of Approximate Reasoning*, *15*(1), 1–24.

Keller, J. and Qiu, H. (1988) Fuzzy set methods in pattern recognition. In J. Kittler (Ed.), *Pattern Recognition*. Berlin: Springer, Vol. *301*, 173–182.

Keller, J., Qiu, H., and Tahani, H. (1986) Fuzzy integral and image segmentation. *Proceedings of the NAFIPS'86*, 324–338.

Kendall, G., Yao, X., and Chong, S. (2007) *The Iterated Prisoners' Dilemma: 20 Years On*. Singapore: World Scientific Publishing.

Kennedy, J. and Eberhart, R. (1995) Particle swarm optimization. *Proceedings of the IEEE International Conference on Neural Networks*, 1942–1948.

Khalil, H. (1992) *Nonlinear Systems*, Englewood Cliffs, NJ: Prentice Hall.

Kiranyaz, S., Ince, T., and Gabbouj, M. (2013) Multidimensional particle swarm optimization for machine learning and pattern recognition. *Adaptation, Learning, and Optimization*. Berlin: Springer, Vol *15*, 45–83.

Kirkpatrick, S., Gelatt, C., and Vecchi, M. (1983) Optimization by simulated annealing. *Science*, *220*(4598), 671–680.

Klir, G. and Yuan, B. (1995) *Fuzzy Sets and Fuzzy Logic: Theory and Applications*. New Jersey: Prentice Hall.

Knowles, J. and Corne, D. (1999) The pareto archived evolution strategy: a new baseline algorithm for pareto multiobjective optimisation. *Proceedings of the 1999 Congress on Evolutionary Computation*, *1*, 98–105.

Kohonen, T. (1982) Self-organized formation of topologically correct feature maps. *Biological Cybernetics*, *43*(1), 59–69.

Kohonen, T., Kangas, J., and Laaksonen, J. (1992) *SOMPAK: The Self-Organizing Map Program Package Version 1.2*. Helsinki, Finland: Helsinki University of Technology.

Kosko, B. (1992) Fuzzy systems as universal approximators. *IEEE International Conference on Fuzzy Systems*, 1143–1162.

Koza, J. (1992) *Genetic Programming*. Cambridge, MA: MIT Press.

Koza, J., Andre, D., Bennett, F., III, and Keane, M. (1996) Use of automatically defined functions and architecture: altering operations in automated circuit synthesis with genetic programming. In J. R. Koza, D. E. Goldberg, D. B. Fogel, and R. Riolo (Eds.), *Genetic Programming 1996: Proceedings of the First Annual Conference on Genetic Programming*. Cambridge, MA: MIT Press, 132–140.

Kreinovich, V., Mouzouris, G., and Nguyen, H. (1998) Fuzzy rule based modeling as a universal approximation tool. In H. T. Nguyen and M. Sugeno (Eds.), *Fuzzy Systems: Modeling and Control*. Kluwer: Academic Publishers, 135–196.

Krishnapuram, R. and Keller, J. (1993) A possibilistic approach to clustering. *IEEE Transactions on Fuzzy Systems*, *1*(2), 98–110.

Krishnapuram, R. and Keller, J. (1996) The possibilistic c-means algorithm: insights and recommendations. *IEEE Transactions on Fuzzy Systems*, *4*(3), 385–393.

Krishnapuram, R. and Lee, J. (1992a) Fuzzy-set-based hierarchical networks for information fusion for decision making. *Fuzzy Sets and Systems*, *46*(1), 11–27.

Krishnapuram, R. and Lee, J. (1992b) Fuzzy-set-based hierarchical networks for information fusion in computer vision. *Neural Networks*, *5*(2), 335–350.

LeCun, Y., Bottou, L., Bengio, Y., and Haffner, P. (1998) Gradient-based learning applied to document recognition. *Proceedings of the IEEE*, *86*(11), 2278–2324.

Lewontin, R. (1974) *The Genetic Basis of Evolutionary Change*. New York: Columbia University Press.

Liow, L., Fortelius, M., Bingham, E., Lintulaakso, K., Mannila, H., Flynn, L., and Stenseth, N. (2008) Higher origination and extinction rates in larger mammals. *Proceedings of the National Academy of Sciences of the United States of America*, *105*(16), 6097–6102.

Lipson, H. (2014) Challenges and opportunities for design, simulation, and fabrication of soft robots. *Soft Robotics*, *1*(1), 21–27.

Lipson, H. and Pollack, J. (2000) Automatic design and manufacture of robotic lifeforms. *Nature*, *406*, 974–978.

Lohn, J., Linden, D., Blevins, B., Greenling, T., and Allard, M. (2015) Automated synthesis of a lunar satellite antenna system. *IEEE Transactions on Antennas and Propagation*, *63*(4), 1436–1444.

Lord, P., Stevens, R., Brass, A., and Goble, C. (2003) Semantic similarity measure as a tool for exploring the gene ontology. *Pacific Symposium on Biocomputing*, 601–612.

Luke, S. and Spector, L. (1998) A revised comparison of crossover and mutation in genetic programming. In J. R. Koza, W. Banzhaf, K. Chellapilla, K. Deb, M. Dorigo, D. B. Fogel, M. H. Garzon, D. E. Goldberg, H. Iba, and R. L. Riolo (Eds.), *Genetic Programming 98*. San Francisco, CA: Morgan Kaufmann, 208–213.

Lyytinen, A., Brakefield, P., Lindstrom, L., and Mappes, J. (2004) Does predation maintain eyespot plasticity in *Bicyclus anynnana*. *Proceedings of the Royal Society of London B: Biological Sciences*, *271*(1536), 279–283.

Macready, W. and Wolpert, D. (1998) Bandit problems and the exploration/exploitation tradeoff. *IEEE Transactions on Evolutionary Computation*, *2*(1), 2–22.

Mallows, C. (1973) Some comments on C_P. *Technometrics*, *15*, 661–675.

Mamdani, E. (1977) Application of fuzzy logic to approximate reasoning using linguistic synthesis. *IEEE Transactions on Systems, Man and Cybernetics*, *26*(12), 1182–1191.

Mamdani, E. and Assilian, S. (1975) An experiment in linguistic synthesis with a fuzzy logic controller. *International Journal of Man-Machine Studies*, *7*(1), 1–13.

Marquie, C., Duchemin, A., Klug, D., Lamblin, N., Mizon, F., Cordova, H., Boulo, M., Lacroix, D., Pol, A., and Kacet, S. (2007) Can we implant cardioverter defibrillator under minimal sedation? *Europace*, *9*, 545–550.

Marr, D. (1982) *Vision*. San Francisco, CA: W. H. Freeman and Company.

May, R. (1976) Simple mathematical models with very complicated dynamics. *Nature, 261* (5560), 459–467.

Mayr, E. (1960) The evolution of life. In S. Tax and C. Callender (Eds.), *Evolution after Darwin: Issues in Evolution.* Chicago: University of Chicago Press.

Mayr, E. (1963) *Animal Species and Evolution.* Cambridge, MA: Belknap Press.

Mayr, E. (1988) *Toward a New Philosophy of Biology: Observations of an Evolutionist.* Cambridge, MA: Belknap Press.

McGrath, P., Seifert, C., Speechley, K., Booth, J., Stitt, L., and Gibson, M. (1996) A new analogue scale for assessing children's pain: an initial validation study. *Pain, 64*(2), 435–443.

Mei, Y., Tang, K., and Yao, X. (2011) Decomposition-based memetic algorithm for multi-objective capacitated arc routing problem. *IEEE Transactions on Evolutionary Computation, 15*(2), 151–165.

Mendez-Vazquez, A., Gader, P., Keller, J., and Chamberlin, K. (2008) Minimum classification error training for Choquet integrals with applications to landmine detection. *IEEE Transactions on Fuzzy Systems, 16*(1), 225–238.

Micchelli, C. A. (1986) Interpolation of scattered data: distance matrices and conditionally positive definite functions. *Constructive Approximation, 2,* 11–22.

Michalewicz, Z. (1992) *Genetic Algorithms + Data Structures = Evolution Programs.* Berlin: Springer.

Michalewicz, Z. and Fogel, D. (2006) *How to Solve It: Modern Heuristics,* 2nd ed. Berlin: Springer.

Michalewicz, Z., Nazhiyath, G., and Michalewicz, M. (1996) A note on the usefulness of geometrical crossover. *Proceedings of the 5th Annual Conference on Evolutionary Programming,* 305–312.

Michalewicz, Z. and Schoenauer, M. (1996) Evolutionary algorithms for constrained parameter optimization problems. *Evolutionary Computation, 4,* 1–32.

Mirmomeni, M. and Punch, W. (2011) Co-evolving data driven models and test data sets with the application to forecast chaotic time series. *Proceedings of the 2011 IEEE Congress on Evolutionary Computation,* 14–20.

Mukhar, K. (2013) *Using evolutionary computing to develop game strategy for a non-deterministic game.* M.S. thesis, University of Colorado at Colorado Springs, Colorado Springs, CO.

Murofushi, T. and Sugeno, M. (1991) A theory of fuzzy measures: representations, the Choquet integral, and null sets. *Journal of Mathematical Analysis and Applications, 159,* 532–549.

Nguyen, D. and Widrow, B. (1989) Truck backer-upper: an example of self-learning in neural networks. *Proceedings of the International Joint Conference on Neural Networks, 2,* 357–362.

Niewiarowski, P., Lopez, S., Ge, L., Hagan, E., and Dhinojwala, A. (2008) Sticky gecko feet: the role of temperature and humidity. *PLoS One, 3*(5), e2192.2190.

Nolfi, S. and Floreano, D. (2000) *The Biology, Intelligence, and Technology of Self-Organizing Machines.* Cambridge, MA: MIT Press.

Nordin, P. and Banzhaf, W. (1995) Complexity compression and evolution. *Proceedings of the 6th International Conference on Genetic Algorithms,* 310–317.

Nordin, P., Francone, F., and Banzhaf, W. (1996) Explicitly defined introns and destructive crossover in genetic programming. In P. J. Angeline and K. E. Kinnear (Eds.), *Advances in Genetic Programming 2*. Cambridge, MA: MIT Press, 111–134.

Nouyan, S., Gross, R., Bonami, M., Mondada, F., and Dorigo, M. (2009) Teamwork in self-organized robot colonies. *IEEE Transactions on Evolutionary Computation, 13*(4), 695–711.

Omidvar, M., Li, X., Mei, Y., and Yao, X. (2014) Cooperative co-evolution with differential grouping for large scale optimization. *IEEE Transactions on Evolutionary Computation, 18* (3), 378–393.

Ostman, B. and Adami, C. (2013) Predicting evolution and visualizing high-dimensional fitness landscapes. In A. Engelbrecht and H. Richter (Eds.), *Recent Advances in the Theory and Application of Fitness Landscapes*. Berlin: Springer, 493–510.

Otero, F., Freitas, A., and Johnson, C. (2012) Inducing decision trees with an ant colony optimization algorithm. *Applied Soft Computing, 12*(11), 3615–3626.

Pal, N., Keller, J., Popescu, M., Bezdek, J., Mitchell, J., and Huband, J. (2005) Gene ontology-based knowledge discovery through fuzzy cluster analysis. *Neural, Parallel & Scientific Computations, 13*(3–4), 337–361.

Parekh, G. and Keller, J. (2007) Learning the fuzzy connectives of a multilayer network using particle swarm optimization. *Proceedings of the 2007 IEEE Symposium on Foundations of Computational Intelligence (FOCI 2007)*, 591–596.

Park, J. and Sandberg, I. (1991) Universal approximation using radial-basis-function networks. *Neural Computation, 3*, 246–257.

Pedrycz, W. and Gomide, F. (1998) *An Introduction to Fuzzy Sets: Analysis and Design*. Cambridge, MA: MIT Press.

Pena, L., Ossowski, S., Pena, J., and Lucas, S. (2012) Learning and evolving combat game controllers. *Proceedings of the 2012 IEEE Conference on Computational Intelligence and Games*, 195–202.

Perez, D., Recio, G., Saez, Y., and Isasi, P. (2009) Evolving a fuzzy controller for a car racing competition. Proceedings of the 2009 IEEE Symposium Computational Intelligence and Games, 263–270.

Plodpradista, P. (2012) A comprehensive comparison between auto-merging possibilistic C-means and possibilistic C-means. University of Missouri MS Project Report.

Poggio, T. and Girosi, F. (1990a) Networks for approximation and learning. *Proceedings of the IEEE, 78*(9), 1481–1497.

Poggio, T. and Girosi, F. (1990b) Regularization algorithms for learning that are equivalent to multilayer networks. *Science, 247*, 978–982.

Popescu, M., Keller, J., and Mitchell, J. (2005) Gene ontology automatic annotation using a domain based gene product similarity measure. *Proceedings of the 14th IEEE International Conference on Fuzzy Systems*, 108–111.

Popescu, M., Keller, J., and Mitchell, J. (2006) Fuzzy measures on the gene ontology for gene product similarity. *IEEE Transaction on Computational Biology and Bioinformatics, 3*(3), 1–11.

Porto, V., Fogel, D., Fogel, L., Fogel, G., Johnson, N., and Cheung, M. (2005) Classifying sonar returns for the presence of mines: evolving neural networks and evolving rules. *2005 IEEE Symposium on Computational Intelligence for Homeland Security and Personal Safety*, 123–130.

Qiu, H. and Keller, J. (1987) Multispectral segmentation using fuzzy techniques. *Proceedings of the NAFiPS'87*, 374–387.

Radcliffe, N. (1992) Non-linear genetic representations. In R. Manner and B. Manderick (Eds.), *Parallel Problem Solving from Nature 2*. Amsterdam: North Holland, 259–268.

Raven, P. and Johnson, G. (1986) *Biology*. St. Louis: Time Mirror/Moseby College.

Ray, T. (1992) An approach to the synthesis of life. In C. G. Langton, C. Taylor, J. D. Farmer, and S. Rasmussen (Eds.), *Artificial Life II*. Reading, MA: Addison Wesley, 371–408.

Rechenberg, I. (1965) *Cybernetic solution path of an experimental problem*. Royal Aircraft Establishment, Library Translation 1122.

Rechenberg, I. (1973) *Evolutionsstrategie: Optimierung technischer systeme nach Prinzipien der biologischen evolution*. Stuttgart, Germany: Frommann Holzboog.

Rechenberg, I. (1996) Personal communication with David Fogel, Berlin, Germany.

Reed, D. and Bryant, E. (2004) Phenotypic correlations among fitness and its components in a population of the housefly. *Journal of Evolutionary Biology*, *17*(4), 919–923.

Reed, J., Toombs, R., and Barricelli, N. (1967) Simulation of biological evolution and machine learning. *Journal of Theoretical Biology*, *17*, 319–342.

Reynolds, R. (1994) An introduction to cultural algorithms. *Proceedings of the 3rd Annual Conference on Evolutionary Programming*, 131–139.

Rissanen, J. (1978) Modeling by shortest data description. *Automatica*, *14*, 465–471.

Rissanen, J. (1984) Universal coding, information, prediction, and estimation. *IEEE Transactions on Information Theory*, *30*, 629–636.

Rosenberg, R. (1967) *Simulation of genetic populations with biochemical properties*. Doctoral dissertation, University of Michigan, Ann Arbor, MI.

Rosenblatt, F. (1958) The perceptron: a probabilistic model for information storage and organization in the brain. *Psychological Review*, *65*, 386–408.

Rosenblatt, F. (1962) *Principles of Neurodynamics*. Washington, DC: Spartan Books.

Rudolph, G. (1994) Convergence analysis of canonical genetic algorithms. *IEEE Transactions on Neural Networks*, *5*(1), 96–101.

Rudolph, G. (1997) Reflections on bandit problems and selection methods in uncertain environments. *Proceedings of the 7th International Conference on Genetic Algorithms*, 166–173.

Rumelhart, D., Hinton, G., and Williams, R. (1986a) Learning internal representations by error propagation. In D. Remelhart and J. McCleland (Eds.), *Parallel Distributed Processing: Explorations in the Microstructure of Cognition*. Cambridge, MA: MIT Press, Vol *1*, 318–362.

Rumelhart, D., Hinton, G., and Williams, R. (1986b) Learning representations by back-propagation errors. *Nature*, *323*, 533–536.

Rundle, H., Vamosi, S., and Schluter, D. (2003) Experimental test of predation's effect of divergent selection during character displacement in Sticklebacks. *Proceedings of the National Academy of Sciences of the United States of America*, *100*(25), 14943–14948.

Russell, B. (1923) Vagueness. *The Australasian Journal of Philosophy*, *88*, 84–92.

Salama, K. and Abdelbar, A. (2014) A novel ant colony algorithm for building neural network topologies. In M. Dorigo, M. Birattari, S. Garnier, H. Hamann, M. Montes de Oca, C. Solnon, and T. Stützle (Eds.), *ANTS 2014*, LNCS 8667, Springer, 1–12.

Saravanan, N. and Fogel, D. (1997) Multi-operator evolutionary programming: a preliminary study on function optimization. In P. J. Angeline, R. G. Reynolds, McDonnell, and R. Eberhart (Eds.), *Evolutionary Programming VI*. Berlin: Springer, 215–221.

Schaeffer, J., Burch, N., Björnsson, Y., Kishimoto, A., Müller, M., Lake, R., *et al.* (2007) Checkers is solved. *Science, 317*(5844), 1518–1522.

Schaeffer, J., Lake, R., Lu, P., and Bryant, M. (1996) Chinook: the world man–machine checkers champion. *AI Magazine, 17*(1), 21.

Schaffer, J. and Eshelman, L. (1991) On crossover as an evolutionarily viable strategy. *Proceedings of the 4th International Conference on Genetic Algorithm,* 61–68.

Schopf, J. (2006) Fossil evidence of Archaean life. *Philosophical Transactions of the Royal Society B: Biological Science, 361*(1470), 869–885.

Schwefel, H. (1981) *Numerical Optimization of Computer Models*. New York: John Wiley & Sons, Inc.

Schwefel, H. (1995) *Evolution and Optimum Seeking*. New York: John Wiley & Sons, Inc.

Seising, R. (2005) 'Fuzzy sets' appear: a contribution to the 40th anniversary. *Proceedings of the FUZZ-IEEE,* 5–10.

Shafer, G. (1976) *A Mathematical Theory of Evidence*. Princeton, NJ: Princeton University Press.

Simpson, P. (1992) Fuzzy min–max neural networks. I. Classification. *IEEE Transactions on Neural Networks, 3*(5), 776–786.

Simpson, P. (1993) Fuzzy min–max neural networks. II. Clustering. *IEEE Transactions on Fuzzy Systems, 1*(1), 32–45.

Smith, T. and Smith, R. (2006) *Elements of Ecology*. Benjamin Cummings.

Sober, E. (2000) *Philosophy of Biology*, 2nd ed. Boulder, CO: Westview Press.

Spears, W. (1998) *The role of mutation and recombination in evolutionary algorithms*. Ph.D. dissertation, George Mason University, Fairfax, VA.

Spears, W. and De Jong, S. (1997) Analyzing GAs using Markov models with semantically ordered and lumped states. In R. Belew and M. Vose (Eds.), *Foundations of Genetic Algorithms*. San Mateo, CA: Morgan Kaufman, 85–100.

Specht, D. (1991) A general regression neural network. *IEEE Transactions on Neural Networks, 2,* 568–576.

Steinger, T., Roy, B., and Stanton, M. (2003) Evolution in stressful environments II: adaptive value and costs of plasticity in response to low light in *Sinapsis arvensis*. *Journal of Evolutionary Biology, 16*(2), 313–323.

Stoica, A., Zebulum, R., Ferguson, M., Duong, V., and Guo, X. (2003) Evolvable hardware techniques for on-chip automated reconfiguration of programmable devices. *Soft Computing, 8,* 354–365.

Storn, R. (1996) On the usage of differential evolution for function optimization. *NAFIPS 1996,* 519–523.

Storn, R. and Price, K. (1996) Minimizing the real functions of the ICEC'96 contest by differential evolution. *Proceedings of the 1996 International Conference on Evolutionary Computation,* 842–844.

Sugeno, M. (1977) Fuzzy measures and fuzzy integrals: a survey. In M. Gupta, G. N. Saridis, and B. R. Gaines (Eds.), *Fuzzy Automata and Decision Processes*. Amsterdam: North Holland, 89–102.

Sugeno, M. (1985) *Industrial Applications of Fuzzy Control*. New York: Elsevier Science Inc.

Sugeno, M. and Kang, G. (1988) Structure identification of fuzzy model. *Fuzzy Sets and Systems*, 28(1), 15–33.

Syswerda, G. (1991) Uniform crossover in genetic algorithms. *Proceedings of the 3rd International Conference on Genetic Algorithms*, 2–9.

Tahani, H. and Keller, J. (1990) Information fusion in computer vision using the fuzzy integral. *IEEE Transactions on Systems, Man and Cybernetics*, 20(3), 733–741.

Takagi, H. (2001) Interactive evolutionary computation: fusion of the capabilities of EC optimization and human evaluation. *Proceedings of the IEEE*, 89(9), 1275–1296.

Takagi, H. and Ohsaki, M. (2007) Interactive evolutionary computation-based hearing aid fitting. *IEEE Transactions on Evolutionary Computation*, 11(3), 414–427.

Takagi, T. and Sugeno, M. (1985) Fuzzy identification of systems and its applications to modeling and control. *IEEE Transactions on Systems, Man and Cybernetics*, 15(1), 116–132.

Teller, A. (1996) *Evolving Programmers: The Coevolution of Intelligence Recombination Operators*. Cambridge, MA: MIT Press.

Templeton, A. (1982) Adaptation and the integration of evolutionary forces. In R. Milkman (Ed.), *Perspectives in Evolution*. Sunderland, MA: Sinauer, 15–31.

Theodoridis, S. and Koutroumba, K. (2009) *Pattern Recognition*, 4th ed. San Diego, CA: Academic Press.

Thompson, A. (1996) Silicon evolution. *Genetic Programming 1996: Proceedings of the First Annual Conference*, 444–452.

Thompson, A. (1998) On the automatic design of robust electronics through artificial evolution. In M. Sipper, D. Mange, and A. Perez-Urie (Eds.), *Evolvable Systems: From Biology to Hardware*. Berlin: Springer, 13–24.

Vasicek, Z. and Sekanina, L. (2014) How to evolve complex combinational circuits from scratch? *Proceedings of the 2014 IEEE International Conference on Evolvable Systems*, 133–140.

Verbruggen, H. and Babuska, R. (1999) *Fuzzy Logic Control: Advances in Applications*. Singapore: World Scientific.

von Neumann, J. and Morgenstern, O. (1944) *Theory of Games and Economic Behavior*. Princeton, NJ: Princeton University Press.

Vose, M. (1999) *The Simple Genetic Algorithm*. Cambridge, MA: MIT Press.

Walsh, P. and Gade, P. (2010) Terrain generation using an interactive genetic algorithm. *Proceedings of the 2010 IEEE Congress on Evolutionary Computation*, 1–7.

Wang, L. (1992) Fuzzy systems are universal approximators. *Proceedings of the IEEE International Conference on Fuzzy Systems*, 1163–1170.

Wang, S., Keller, J., Burks, K., Skubic, M., and Tyrer, H. (2006) Assessing physical performance of elders using fuzzy logic. *Proceedings of the 15th IEEE Internationl Conference on Fuzzy Systems*, 2998–3003.

Wang, Z. and Klir, G. (1993) *Fuzzy Measure Theory*. Norwell, MA: Kluwer Academic Publishers.

Werbos, P. (1974) *Beyond regression: new tools for prediction and analysis in the behavioral sciences*. Ph.D. thesis, Harvard University.

Werbos, P. (1994) *The Roots of Backpropagations: From Ordered Derivatives to Neural Networks and Political Forecasting.* New York: John Wiley & Sons, Inc.

White, D. and Poulding, S. (2009) A rigorous evaluation of crossover and mutation in genetic programming. In L. Vanneschi, S. Gustafson, A. Moraglio, I. de Falco, and M. Ebner (Eds.), *Genetic Programming: EuroGP 2009.* Berlin: Springer, 220–231.

Wieland, A. (1991a) Evolving controls for unstable systems. *Connectionist Models: Proceedings of the 1990 Summer School,* 91–102.

Wieland, A. (1991b) Evolving neural network controllers for unstable systems. *Connectionist Models: Proceedings of the 1990 Summer School,* 667–673.

Wilbik, A. and Keller, J. (2012) A distance metric for a space of linguistic summaries. *Fuzzy Sets and Systems, 208,* 79–94.

Wilbik, A. and Keller, J. (2013) A fuzzy measure similarity between sets of linguistic summaries. *IEEE Transactions on Fuzzy Systems, 21*(1), 183–189.

Wilbik, A., Keller, J., and Bezdek, J. (2012) Generation of prototypes from sets of linguistic summaries. *Proceedings of the World Congress on Computational Intelligence,* FUZZ-IEEE, 472–479.

Wilson, E. (1992) *The Diversity of Life.* New York: W.W. Norton.

Wilson, S. (1995) Classifier fitness based on accuracy. *Evolutionary Computation, 3*(2), 149–175.

Wittkamp, M., Barone, L., and Hingston, P. (2008) Using NEAT for continuous adaptation and teamwork formation in Pacman. *Proceedings of the IEEE Symposium on Computational Intelligence and Games,* 234–242.

Wolpert, D. and Macready, W. (1997) No free lunch theorems for optimization. *IEEE Transactions on Evolutionary Computation, 1*(1), 67–82.

Wong, D., Hockenberry-Eaton, M., Winkelstein, M., and Schwartz, P. (2001) *Wong's Essentials of Pediatric Nursing,* 6 ed. St. Louis: Mosby Inc.

Wright, S. (1932) The roles of mutation, inbreeding, crossbreeding, and selection in evolution. *Proceedings of the 6th International Congress of Genetics, 1,* 356–366.

Wu, Y., Giger, M., Doi, K., Vyborny, C., Schmidt, R., and Metz, C. (1993) Artificial neural networks in mammography: application to decision making in the diagnosis of breast cancer. *Radiology, 187,* 81–87.

Xu, D., Keller, J., Popescu, M., and Bondugula, R. (2008) *Applications of Fuzzy Logic in Bioinformatics.* London, UK: Imperial College Press.

Xu, L., Krzyżak, A., and Yuille, A. (1994) On radial basis function nets and kernel regression: statistical consistency, convergence rates, and receptive field size. *Neural Networks, 7,* 609–628.

Yager, R. (1980) On a general class of fuzzy connectives. *Fuzzy Sets and Systems, 4*(3), 235–242.

Yager, R. (1988) On ordered weighted averaging aggregation operators in multicriteria decision making. *IEEE Transactions on Systems, Man and Cybernetics, 18*(1), 183–190.

Yager, R. (1993) Families of OWA operators. *Fuzzy Sets and Systems, 59,* 125–148.

Yager, R. (1996) Quantifier guided aggregation using OWA operators. *International Journal of Intelligent Systems, 11,* 49–73.

Yager, R. (2004) Modeling prioritized multicriteria decision making. *IEEE Transactions on Systems, Man and Cybernetics, 34*(6), 2396–2404.

Yan, B. (1993) *Optimization of the fuzzy integral in computer vision and pattern recognition.* Ph.D. dissertation, University of Missouri, Columbia, MO.

Yan, B. and Keller, J. (1991) Conditional fuzzy measures and image segmentation. *Proceedings of the NAFIPS'91*, 32–36.

Yang, M. and Lai, C. (2011) A robust automatic merging possibilistic clustering method. *IEEE Transactions on Fuzzy Systems*, 19(1), 26–41.

Young, G., Scardovi, L., Cavagna, A., Giardina, I., and Leonard, N. (2013) Starling flock networks manage uncertainty in consensus at low cost. *PLoS Computational Biology.* doi: 10.1371/journal.pcbi.100289

Yurkovich, S. and Passino, K. (1999) A laboratory course on fuzzy control. *IEEE Transactions on Education*, 42(1), 15–21.

Zadeh, L. (1965) Fuzzy sets. *Information and Control*, 8(3), 338–353.

Zadeh, L. (1973) Outline of a new approach to the analysis of complex systems and decision processes. *IEEE Transactions on Systems, Man and Cybernetics*, 3(1), 28–44.

Zadeh, L. (1975a) The concept of a linguistic variable and its application to approximate reasoning: Part 1. *Information Sciences*, 8, 199–249.

Zadeh, L. (1975b) The concept of a linguistic variable and its application to approximate reasoning: Part 2. *Information Sciences*, 8, 301–357.

Zadeh, L. (1976) The concept of a linguistic variable and its application to approximate reasoning: Part 3. *Information Sciences*, 9, 43–80.

Zadeh, L. (1978) Fuzzy sets as a basis for a theory of possibility. *Fuzzy Sets and Systems*, 1, 3–28.

Zhou, A., Qu, B., Li, H., Zhao, S., Suganthan, P., and Zhang, Q. (2011) Multiobjective evolutionary algorithms: a survey of the state of the art. *Swarm and Evolutionary Computation*, 1(1), 32–49.

Zimmermann, H. and Zysno, P. (1980) Latent connectives in human decision making. *Fuzzy Sets and Systems*, 4(1), 37–51.

Zitzler, E. and Thiele, L. (1998) Multiobjective optimization using evolutionary algorithms: a comparative case study. In A. E. Eiben, T. Back, M. Schoenauer, and H.-P. Schwefel (Eds.), *Parallel Problem Solving from Nature V.* Berlin: Springer, 292–301.

Zurada, J., Marks, L., and Robinson, C. (1994) *Computational Intelligence: Imitating Life.* New York: IEEE Press.

Fundamentals of Computational Intelligence: Neural Networks, Fuzzy Systems, and Evolutionary Computation, First Edition. James M. Keller, Derong Liu, and David B. Fogel.
© 2016 by The Institute of Electrical and Electronics Engineers, Inc. Published 2016 by John Wiley & Sons, Inc.

IEEE Press Series on
COMPUTATIONAL INTELLIGENCE

Series Editor, **David B. Fogel**

The IEEE Press Series on Computational Intelligence includes books on neural, fuzzy, and evolutionary computation, and related technologies, of interest to the engineering and scientific communities. Computational intelligence focuses on emulating aspects of biological systems to construct software and/or hardware that learns and adapts. Such systems include neural networks, our use of language to convey complex ideas, and the evolutionary process of variation and selection. The series highlights the most-recent and groundbreaking research and development in these areas, as well as the important hybridization of concepts and applications across these areas. The audiences for books in the series include undergraduate and graduate students, practitioners, and researchers in computational intelligence.

Reinforcement Learning and Approximate Dynamic Programming for Feedback Control. Edited by Frank L. Lewis and Derong Liu. 2012. 978-1118-10420-0

Complex-Valued Neural Networks: Advances and Applications. Edited by Akira Hirose. 2013. 978-1118-34460-6

Unsupervised Learning: A Dynamic Approach. Matthew Kyan, Paisarn Muneesawang, Kambiz Jarrah, Ling Guan. 2014. 978-0470-27833-8

Introduction to Type-2 Fuzzy Logic Control: Theory and Applications. Jerry M. Mendel, Hani Hagras, Woei-Wan Tan, William W. Melek, Hao Ying. 2014. 978-1118-278291

Fundamentals of Computational Intelligence: Neural Networks, Fuzzy Systems, and Evolutionary Computation. James M. Keller, Derong Liu, David B. Fogel. 2016. 978-1119-214342

Printed in the United States
By Bookmasters